Surfactants from Renewable Raw Materials

Surfactants from Renewable Raw Materials

Edited by

Divya Bajpai Tripathy, Anjali Gupta,
Arvind Kumar Jain, Anuradha Mishra

CRC Press
Taylor & Francis Group
Boca Raton London New York

CRC Press is an imprint of the
Taylor & Francis Group, an **informa** business

First edition published 2022
by CRC Press
6000 Broken Sound Parkway NW, Suite 300, Boca Raton, FL 33487-2742

and by CRC Press
2 Park Square, Milton Park, Abingdon, Oxon, OX14 4RN

© 2022 selection and editorial matter, Divya Bajpai Tripathy, Anjali Gupta, Arvind Kumar Jain, Anuradha Mishra; individual chapters, the contributors

CRC Press is an imprint of Taylor & Francis Group, LLC

ISBN: 978-0-367-70165-9 (hbk)
ISBN: 978-0-367-70205-2 (pbk)
ISBN: 978-1-003-14487-8 (ebk)

DOI: 10.1201/9781003144878

Typeset in Times
by KnowledgeWorks Global Ltd.

Contents

Preface

A class of versatile molecules which find their application in various industries are known as 'Surfactants'. Surfactants, which may be completely invisible to us most of the time, are present in almost all the chemicals that we use in our daily life. They show their presence in washrooms from toothpastes to toiletries to bath and body recipes; in cosmetics from moisturizing and lubricating the skin to creating an attractive protective layer for nails, lips and hair to sunscreens; in perfumes, deodorants and room fresheners; in kitchen from food products such as mayonnaise, salad creams, dressings, deserts to baked food items and to washing dishes and cleaning the greased surfaces; from detergent liquids and powders to fabric softeners in laundry; and in an array of home/work place cleaning products.

Alkali metal soap-like substances (predecessors of surfactants) were in use for several thousand years. Vedas (1500–500 BCE), the earliest records of ancient India, notably Rig Veda and Atharva Veda, referred to applying soft earth, saponaceous fragrant soft plant materials on the body before a water bath to remove dirt and other impurities from human skin. Soaps in ancient India were made at home by gently heating a mixture of sweet almonds, bitter almonds, ghee, almond oil, sesame oil, aromatic oils, lather-producing fruits like reetha (*Sapindus mukorossi*), etc. Sanskrit texts, Sushruta Samhita and Charaka Samhita, also mentioned the preparation of lime and caustic soda from aqueous extracts of ashes.

In 2800 BC, the presence of soap-like substance was mentioned in ancient Babylonian inscriptions. Egyptians, way back in 1550 BC, used to create a soap-like substance for bathing purpose. During the time of Neo-Babylonian Empire (556–539 BC), a combination of ashes and oils was used for cleaning purposes. The word sapo first appeared in Pliny the Elder's Historia Naturalis, which discussed the manufacture of soap by the Phoenicians in 600 BC using tallow and ashes. Early Romans made soaps in the first century AD. The first authentic mention of soaps as cleansing agents appeared in the 200 AD in the script of Galen, a Greek physician. Zosimos of Panopolis, *circa* 300 AD, also described soap-making. In the eighth century, soap-making became common in Spain and Italy, and by the twelfth century, the Kingdom of England and Bedouins also started making it. The industrial production has started in Islamic world by the thirteenth century. Around the fifteenth century onwards, Europe took the center stage in soap-making. In the nineteenth century, liquid soap was invented. In the year 1886, Lever brothers came into soap business, and their company exists even now as one of the biggest players in the field. And gradually, the surfactants found their place beyond the realm of soap industry and entered into new domains of paints, adhesives, plastics, petroleum production, oilfield chemicals, dispersants, corrosion inhibitors, cosmetics and personal care products, food, agrochemicals, textile, drugs industries etc. It is expected that a healthy growth in the market for surfactants will be maintained due to developments in the end-user industries.

Traditional surfactants are known to have adverse environmental impact. Researchers are therefore looking at ways to produce eco-friendly surfactants from renewable resources. They are also looking at cost-effective production, considering widespread excessive demand of surfactants. Substitute of petrochemical-based surfactants with those based on renewable resources has become the major need of the day. Increased interest of researchers to synthesize eco-friendly and cost-effective surfactants attracted them to the world of plants. Plant product-derived surfactants have not only been proven as an eco-friendly and economical but their lesser toxicity also enables them to be exploited in various sectors. In addition, biodegradability of plant-based surfactants has reduced the environmental concerns by reducing water and soil pollution. However, complete scientific data for this group of surfactants is scarce and has not been compiled properly.

Idea behind the proposed book emerged from the clear demand of a detailed reference book summarizing the surfactants based on almost all available plant products, their properties and applications. This book will precisely cover the class of surfactants synthesized using plants-based raw materials. A total of 13 chapters are included in the book.

Chapter 1 is introductory that covers classification, history, mechanism, properties and applications of surfactants.

Chapters 2–8 cover the synthesis, properties and applications of surfactants based on fats/oils, proteins, steroids, terpenes, alkaloids, sugars, polysaccharides, flavonoids and microbes.

Chapters 9–10 describe nanosurfactants, an advanced class of surfactants.

Chapter 11 discusses the characteristics and medicinal applications of pulmonary surfactants in preterm babies as well as their probable contribution in COVID-19.

Chapter 12, the concluding chapter, discusses the environmental acceptability of these surfactants. Biodegradability of surfactants is also discussed in this chapter, which would help the researchers to label them as green or non-green surfactants.

Chapter 13 discusses the compilation of patents based on properties, synthesis and applications of surfactants from last decade's database.

It is expected that this book will stimulate the interest of many more investigators in the academic universe towards surfactant research. It will also stimulate the surfactant market to strengthen work in the area of surfactants based on renewable raw material. It will facilitate the speedy progress in generating novel eco-friendly and cost-effective surfactants from lab synthesis to industrial production level.

June, 2021
Anjali Gupta
Divya B. Tripathy
Anuradha Mishra
Arvind K Jain

Editors

Dr Divya Bajpai Tripathy is currently a full time Associate Professor in the Department of Chemistry, School of Basic and Applied Sciences, Galgotias University, Greater Noida, India. She has research and teaching experience of more than 11 years. Dr Tripathy has more than 30 research publications in reputed journals/book chapters/conference proceedings to her credit and two Indian and one published Australian patent. She has received scientist fellowship from the Department of Science and Technology, Ministry of Science, India and Research and innovation award from Galgotias University, India. She has guided 11 master's research students. Currently four doctorate students have been registered under her supervision. She has been a principal investigator in a DST-funded research project. She is guest editor in CMAM Journal from Bentham Science Publication and Member of editorial board in Der Chemica Sinicea. She is also working in the reviewer panel of many renowned journals like JSD. She has research collaboration with Prof M. A. Quraishi, King Fahd University, Saudi Arab, Dr Anjali Gupta, Galgotias University, Greater Noida, and Prof Anuradha Mishra, Gautam Buddha University, Greater Noida. She has also worked as NAAC coordinator at university level.

Dr Anjali Gupta is currently working as Associate Professor in the Department of Chemistry, School of Basic and Applied Sciences, Galgotias University, Greater Noida, India. She has research and teaching experience of approximately 9.5 years. She is a recipient of Young Scientist Award by Department of Science and Technology and Dr D.S. Kothari Postdoctoral Fellowship by University Grants Commission and Senior Research Fellowship from CSIR, India. Her research area is in-silico screening and synthesis of naturally occurring bioactive analogues. She is involved in research collaborations with Dr Divya Bajpai, Galgotias University, Dr Fahmina Zafar, Jamia Millia Islamia, Delhi, and Dr Anujit Ghosal, Jawaharlal Nehru University, Delhi. She has 15 research publications in reputed journals/book chapters/conference proceedings and six patents to her credit. She has three PhD and seven master's research students under her guidance. She has been a principal investigator in DST-sponsored research projects and received her graduate, postgraduate and doctorate degrees from University of Delhi, Delhi.

Prof. Arvind Kumar Jain is working as Dean of Students Welfare and Dean of School of Basic and Applied Sciences in Galgotias University. He has completed his PhD from IIT Roorkee in 2002. Then he worked in France as a CNRS post-doctoral fellow in ICMSB CNRS at Bordeaux. His areas of expertise are nanotechnology, analytical chemistry organic synthetic chemistry. He has delivered many invited talks at national and international level and published a number of research papers in national and international journal and conference proceedings, also guided three PhD students, eight under progress, and published six patents. He has visited France, Germany and Holland for research activity. He has been a recipient of various national and international fellowships such as travel grant $ 1345 from NSF, USA to participate and present paper in the Third Annual Conference of the Society for the Study of Nanoscience and Emerging Technologies at Tempe, Arizona, USA, November 7–10, 2011, full travel grant $ 2000 from NSF, USA to participate and deliver a lecture in the Second Annual Conference of the Society for the Study of Nanoscience and Emerging Technologies at Technische Universität Darmstadt, Germany, September 29 to October 2, 2010. He was also CSIR Senior Research Fellow at Department of Chemistry Indian Institute of Technology Roorkee, Roorkee (May 2001 to June 2002), DST Project Fellow in the Department of Chemistry Indian Institute of Technology Roorkee, Roorkee (February 1996 to December 1999).

Prof. Anuaradha Mishra is currently Full Professor in the Department of Applied Chemistry, School of Vocational Studies & Applied Sciences, Gautam Buddha University, Greater Noida, India. She has been the dean of the School of Vocational Studies & Applied Sciences for almost five years. She has also been the Dean Academics for three years and the Dean Planning & Research for more than two years. She has been the Nodal Officer of her university for Skill Development

Courses in collaboration with NSDC, Govt. of India and for National Assessment & Accreditation Council (NAAC), University Grants Commission, Government of India, a body responsible for grading of Indian university. She has research and teaching experience of over 25 years and academic administrative experience of over 15 years on various senior positions. She is a recipient of coveted commonwealth fellowship award, United Kingdom and research award for teachers by University Grants Commission, India. Her research area is synthesis of polysaccharides-based biomaterials (green materials) for water remediation. She has worked as Commonwealth Fellow, Green Chemistry Centre of Excellence, Department of Chemistry, University of York, York, UK, and Guest Faculty, Department of Agriculture & Environmental Sciences, University of Newcastle, Newcastle upon Tyne, UK. She has one US patent on polysaccharide agents and methods of their use for removing solids from water; a part of the technology has been licensed by a Company Pristana LLC, California. Prof. Anuradha has 172 research publications in reputed journals/book chapters/conference proceedings to her credit. She has also authored/edited five books on polymers, and green and sustainable chemistry. She has guided 21 PhD and 44 master's research students. She has been principal investigator in many sponsored research projects. She has been a member of many academic/research bodies of government organizations in India and abroad.

Contributors

Meenu Aggarwal
Department of Chemistry
Aggarwal College Ballabgarh
Faridabad, Haryana, India

Pooja Agrawal
Department of Chemistry
School of Basic and Applied Sciences
Galgotias University
Greater Noida, Uttar Pradesh, India

Deepali Ahluwalia
Department of Applied Chemistry
Delhi Technological University
Delhi, India

Gunjan Chauhan
Department of Chemistry
Maharishi Markandeshwar (Deemed to be
    University)
Mullana, Haryana, India

Priyanka Chhabra
Department of Bio-Sciences
School of Basic and Applied Sciences
Galgotias University
Greater Noida, Uttar Pradesh, India

Anindita De
Department of Chemistry and Biochemistry
Sharda University
Greater Noida, Uttar Pradesh, India

Priyanka Dhingra
Department of Chemistry
JECRC University
Jaipur, Rajasthan, India

Mridula Guin
Department of Chemistry and Biochemistry
Sharda University
Greater Noida, Uttar Pradesh, India

Anujit Ghosal
Department of Food & Human Nutritional
    Sciences
The University of Manitoba
Winnipeg, Canada

Anjali Gupta
Department of BioSciences
School of Basic and Applied Sciences
Galgotias University
Greater Noida, Uttar Pradesh, India

Shipra Mital Gupta
University School of Basic and Applied
    Sciences
Guru Gobind Singh Indraprastha University
New Delhi, India

Preeti Jain
Department of Chemistry and Biochemistry
Sharda University
Greater Noida, Uttar Pradesh, India

Sonali Kesarwani
School of Basic and Applied Sciences
Galgotias University
Greater Noida, Uttar Pradesh, India

Ritika Kubba
Department of Applied Chemistry
Delhi Technological University
Delhi, India

Anil Kumar
Department of Applied Chemistry
Delhi Technological University
Delhi, India

Ashutosh Kumar
Department of Basic Sciences
School of Basic and Applied Sciences
Galgotias University
Greater Noida, Uttar Pradesh, India

Manoj Kumar
Division of Chemistry
School of Basic and Applied Sciences
Galgotias University
Greater Noida, Uttar Pradesh, India

Mayuri Kumari
School of Basic and Applied Sciences
Galgotias University
Greater Noida, Uttar Pradesh, India

Anuradha Mishra
Department of Applied Chemistry
School of Vocational Studies and Basic and
    Applied Sciences
Gautam Buddha University
Greater Noida, Uttar Pradesh, India

Shilpi Mishra
Biological and Chemical Science Department
Montgomery College
Rockville, Maryland

Subhalaxmi Pradhan
Division of Chemistry
School of Basic and Applied Sciences
Galgotias University
Greater Noida, Uttar Pradesh, India

Nidhi Puri
Department of Applied Science & Humanities
I.T.S Engineering College
Greater Noida, Uttar Pradesh, India

M. A. Quraishi
Center of Research Excellence in Corrosion
Research Institute
King Fahd University of Petroleum and
    Minerals
Dhahran, Saudi Arabia

Chandreyee SahaDivision of Chemistry
School of Basic and Applied Sciences
Galgotias University
Greater Noida, Uttar Pradesh, India

Nisha Saini
Department of Chemistry
Gargi College
University of Delhi
New Delhi, India

Anil K. Sharma
Department of Biotechnology
Maharishi Markandeshwar (Deemed to be
    University)
Mullana, Haryana, India

S. K. Sharma
University School of Chemical Technology
Guru Gobind Singh Indraprastha University
New Delhi, India

Shashank Sharma
Department of Chemistry
SBAS, Galgotias University
Greater Noida, Uttar Pradesh, India

Anurag Singh
Department of Food Science and Technology
National Institute of Food Technology
    Entrepreneurship and Management
Sonipat, Haryana, India

Kuldeep Singh
Department of Chemistry
Maharishi Markandeshwar (Deemed to be
    University)
Mullana, Haryana, India

Raman Singh
Department of Chemistry
Maharishi Markandeshwar (Deemed to be
    University)
Mullana, Haryana, India

Rajni Srinivasan
Department of Chemistry, Geosciences and
    Environmental Science
College of Science and Technology
Tarleton State University
Stephenville, Texas

Divya Bajpai Tripathy
Department of BioSciences
School of Basic and Applied Sciences
Galgotias University
Greater Noida, Uttar Pradesh, India

Sudhir G. Warkar
Department of Applied Chemistry
Delhi Technological University
Delhi, India

Priyanka Yadav
University School of Basic and Applied
    Sciences
Guru Gobind Singh Indraprastha University
New Delhi, India

Vandana Yadav
The Bhopal School of Social Science
Bhopal, Madhya Pradesh, India

1 Surfactants Based on Renewable Raw Materials

An Introduction

Divya Bajpai Tripathy[1], Anjali Gupta[1] and Anuradha Mishra[2]

[1]Department of BioSciences, School of Basic and Applied Sciences, Galgotias University, Greater Noida, Uttar Pradesh, India

[2]Department of Applied Chemistry, School of Vocational Studies and Basic and Applied Sciences, Gautam Buddha University, Greater Noida, Uttar Pradesh, India

CONTENTS

1.1 INTRODUCTION

1.1.1 SURFACTANTS

Surfactants are organic compounds composed of moieties of different polarities within a molecule. Their head groups are hydrophilic in nature whereas their tail parts show affinity towards non-polar solvents (Figure 1.1). This inexplicable structural characteristic of surfactants makes them a potential candidate to be exploited in the reduction of surface tension and interfacial tension between two or more dissimilar phases in terms of polarity. Self-assembly is the unique property of surfactants which results in the creation of micelles – a hollow sphere with a diameter ranging from nanometers to microns. Amphiphilic character and capability of forming critical micelle concentrations (CMCs) make surfactants suitable for a vast range of commercial

DOI: 10.1201/9781003144878-1

1

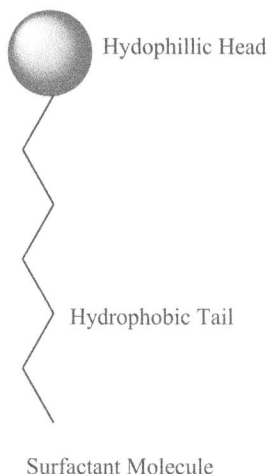

FIGURE 1.1   Surfactant molecule.

applications. Various industries, such as detergents, cleaning agents, disinfectants, emulsifiers, dispersants, adhesives, coating materials, cosmetics, corrosion inhibitors and many more, are exploiting surfactants. Moreover, surfactants have also been reported to overcome the limitations associated with nanotechnology. For instance, surfactants help to form stable dispersions of lyophilic inorganic nanomaterials like transition metal dichalcogenides, graphene, carbon nanotubes and black phosphorus [1].

In pharmaceutical industries, surfactants play a vital role in various transdermal topical compositions, medicinal formulations and gene and drug delivery systems. The accelerating requirements of a diverse range of surfactants in an infinite range of application sectors demand rigorous research to explore new and sustainable raw materials for the production of promising surfactants [2].

1.1.2   MECHANISM OF SURFACTANT MICELLE FORMATIONS

Surfactants when added to the solutions of different polarities often display strange physical behavior. When added in less quantity, a solvent surfactant behaves like any other solute and forms dilute solutions. Unexpectedly, when added in higher concentrations surfactant molecules alter the physical properties of solutions in terms of values of osmotic pressure, electrical conductance, turbidity and surface tension etc. This unpredicted behavior of surfactants is due to the formation of micelles, which are formed above the CMC. Hence, the micelle formation or micellization can be recognized as an alternate mechanism of the adsorption through which the interfacial energy of a surfactant solution gets reduced (Figure 1.2). CMC of the surfactant solution can be evaluated by investigating any of the micelle-governed physical characteristics of the solution as a function of surfactant concentration [3] (Figure 1.3).

1.1.3   CLASSIFICATION OF SURFACTANTS

1.1.3.1   On the Basis of Charge on Polar Moiety

Surfactants encompass a vast group of molecules, so there are not any standard criteria of classification of surfactants; although the most common and widely acceptable classification is based on the type of polar head group to which the hydrophobic chain is attached. On this basis, surfactants have been classified into four categories as shown in Figure 1.4 [4] and Table 1.1.

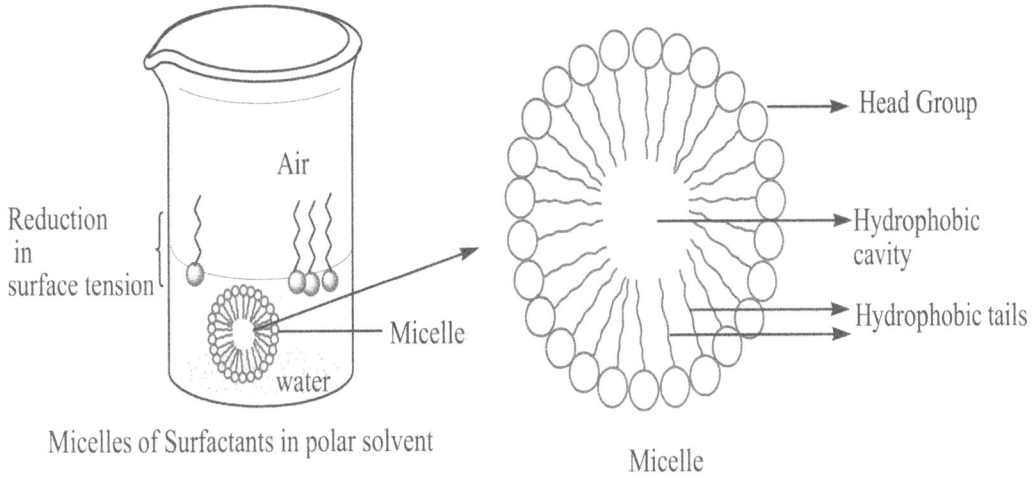

Micelles of Surfactants in polar solvent

Head Group

Hydrophobic cavity

Hydrophobic tails

Micelle

FIGURE 1.2   Micelle formation.

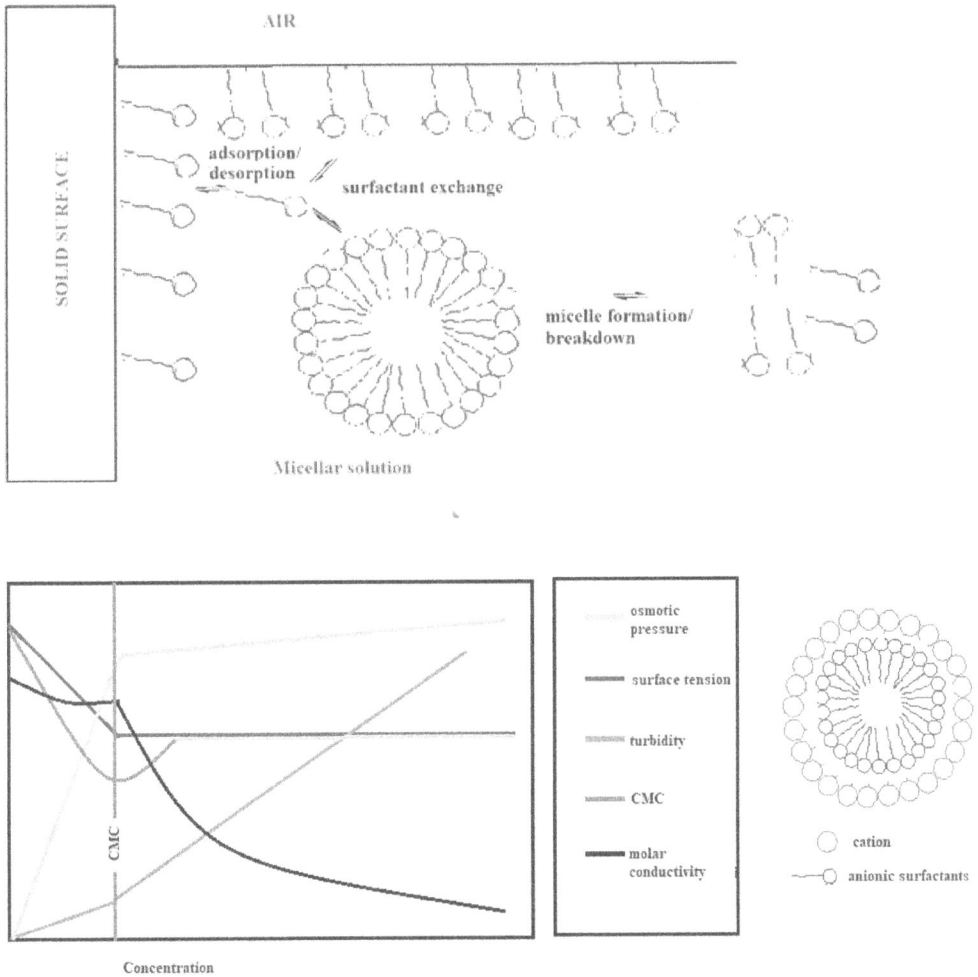

FIGURE 1.3   Effect of critical micelle concentration on different physical properties of surfactants [3].

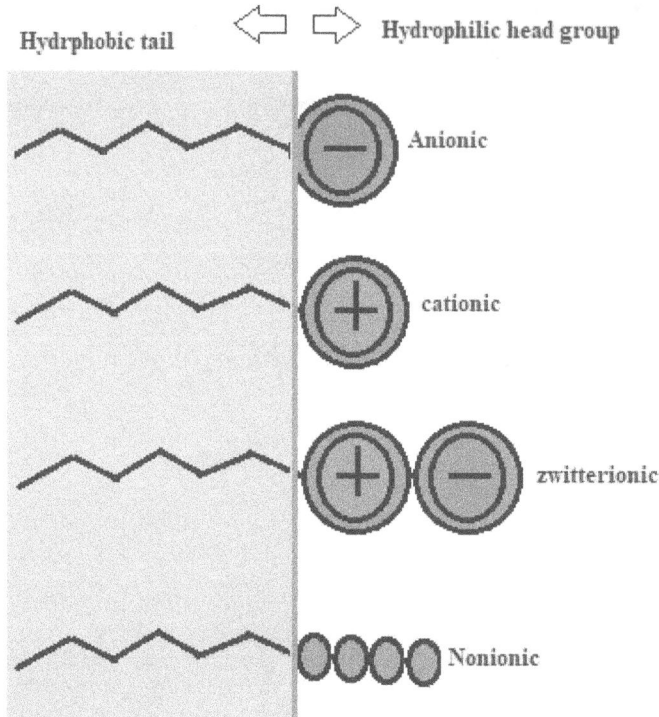

Hydrphobic tail   Hydrophilic head group

Anionic

cationic

zwitterionic

Nonionic

FIGURE 1.4   Classification of surfactants.

1.1.3.2   On the Basis of Number of Monomeric Moieties

Another approach to classify the surfactants is the number of monomeric units present in the surfactant molecules (Figure 1.5). Monomeric surfactants belong to the primitive type of this classification system that has just one hydrophilic group and a single hydrophobic tail. Examples of this class are conventionally used surfactants like sodium dodecyl sulfonate, quaternary ammonium salts, betaines and polyethylene glycols. Second but advanced category is gemini surfactants, which are gaining more attraction from the researchers of this field. In gemini surfactants, two monomeric surfactants are joined together via a spacer group. Spacer groups may be of varying length, different polarity and may be present either in between head groups or in between tails of the two monomers. These surfactants were found superior to their monomeric counterparts in terms of better surface-active properties. Examples are gemini imidazolium surfactants, alkanediyl-*a*,*x*-bis(dodecyldimethyl-ammonium bromide) [6–8].

The next type in this classification is the polymeric surfactants. This class embodies macromolecules that contain hydrophilic and hydrophobic parts. This common approach encompasses many natural polymers like polysaccharides. Best known examples of this class are emulsan and chitosan [9].

1.1.3.3   On the Basis of Source and Environmental Acceptability

Nowadays, sustainability and environmental acceptability are the major concerns of the researchers, industrialists and environmentalists. This is the most advanced and acceptable ground of classification on which the surfactants have been further classified into two categories.

1.1.3.3.1   Conventional Petrochemical-Based Surfactants

This class of surfactants is made from petrochemicals and is not eco-friendly at the end of their consumption.

TABLE 1.1

Classification of Surfactants on the Basis of Charge, Their Major Application and Common Examples [5]

Categories	Definition	Major Applications	Examples
Anionic surfactants	Anionic surfactants are the biggest class of surfactants and bear negative charge on its polar hydrophilic moiety	Wetting agent – coatings, toothpaste, laundry detergents, dishwasher detergents, shampoos, bath products, concrete plasticizer, plasterboard, DMSO, hand soap, HI&I products	Sodium alkylsulfates, other counterions, sodium alkanesulfonates, other counterions, sodium alkylarenesulfonates, sodium alkylnaphthalenesulfonates, lignosulfonates, succinates etc.
Cationic surfactants	Cationic surfactants have positive charge on their head group and ionize in water into the positively charged washing active cations and the non-washing active anions	Fabric softeners, conditioners, cosmetics, disinfectants, dispersants etc.	Imidazolinium salts, alkyltrimethylammonium salts, alkylpyridinium salts, benzalkonium chloride (BAC), cetylpyridinium chloride (CPC), benzethonium chloride (BZT) etc.
Non-ionic surfactants	Non-ionic surfactants neither form cations nor anions in water as they don't bear any charge on the molecule	Wetting agent – coatings, spermicide, food ingredients, polish, cleaner, fragrance carrier, drug delivery agents etc.	Polyoxyethylene glycol octylphenol ethers, alkylpolyoxyethylene compounds, alkylphenol ethoxylate compounds, phosphoxides, polyoxyethylene glycol alkylphenol ethers, sorbitan alkyl esters etc.
Zwitterionic surfactants	Also known as amphoteric surfactants and bear both negative and positive charges on its head group	Paints, adhesives, coagulating agents etc.	Betaines, sultaines, amino acid surfactants etc.

1.1.3.3.2 Natural Surfactants

Surfactants based on renewable raw materials belong to the class of natural surfactants. These surfactants have superior environmental acceptability. Less or no toxicity is an additional advantage of this class of surfactants.

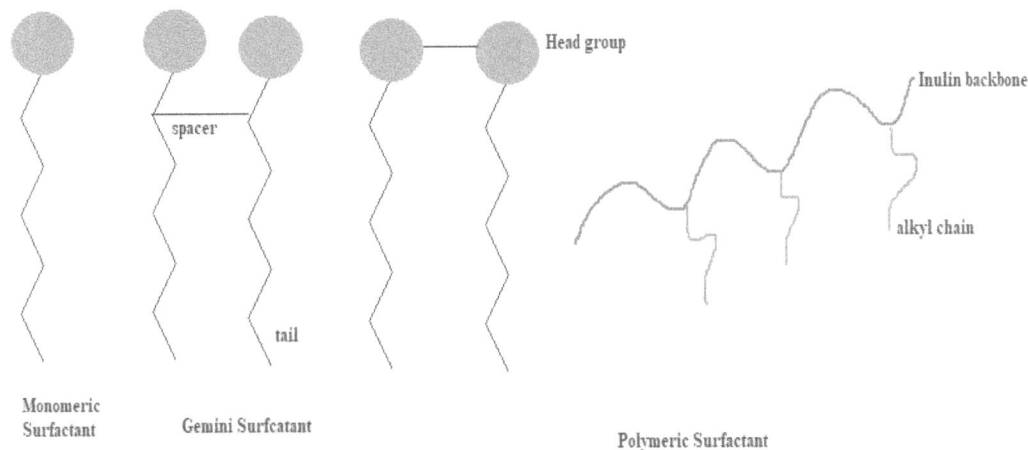

FIGURE 1.5 Representation of monomeric, gemini and polymeric surfactants.

1.2   PETROCHEMICAL-BASED SURFACTANTS, THEIR LIMITATIONS AND NEED FOR NATURAL SURFACTANTS

Surfactants can be derived from both the petrochemicals, which are type of non-renewable raw materials obtained from feedstock as well as natural sources, and plant-based raw materials, a renewable source. Earlier, the use of petrochemical as substrate for the production of surfactants was preferable due to their low cost, but the increasing environmental and health issues demand replacement of these conventional petrochemical-based surfactants with natural surfactants due to their desirable biocompatibility, biodegradability and or no toxicity.

Emerging trends to prefer non-conventional surfactants, irrespective of their relatively higher cost, attracted the researchers to go beyond the boundaries and further explore the scope of exploitation of plant-based materials as a base reactant in surfactants production.

Fatty alcohols are the most common raw materials used for the synthesis of commercial surfactants like ethoxysulfates, alcohol ethoxylates and fatty alcohol sulfonates. Fatty alcohols are the types of precursors that can be obtained both from the petroproducts and plant-based fatty acids. Nowadays, efforts have been made to promote the use of fatty alcohol obtained from natural resources (Figure 1.6).

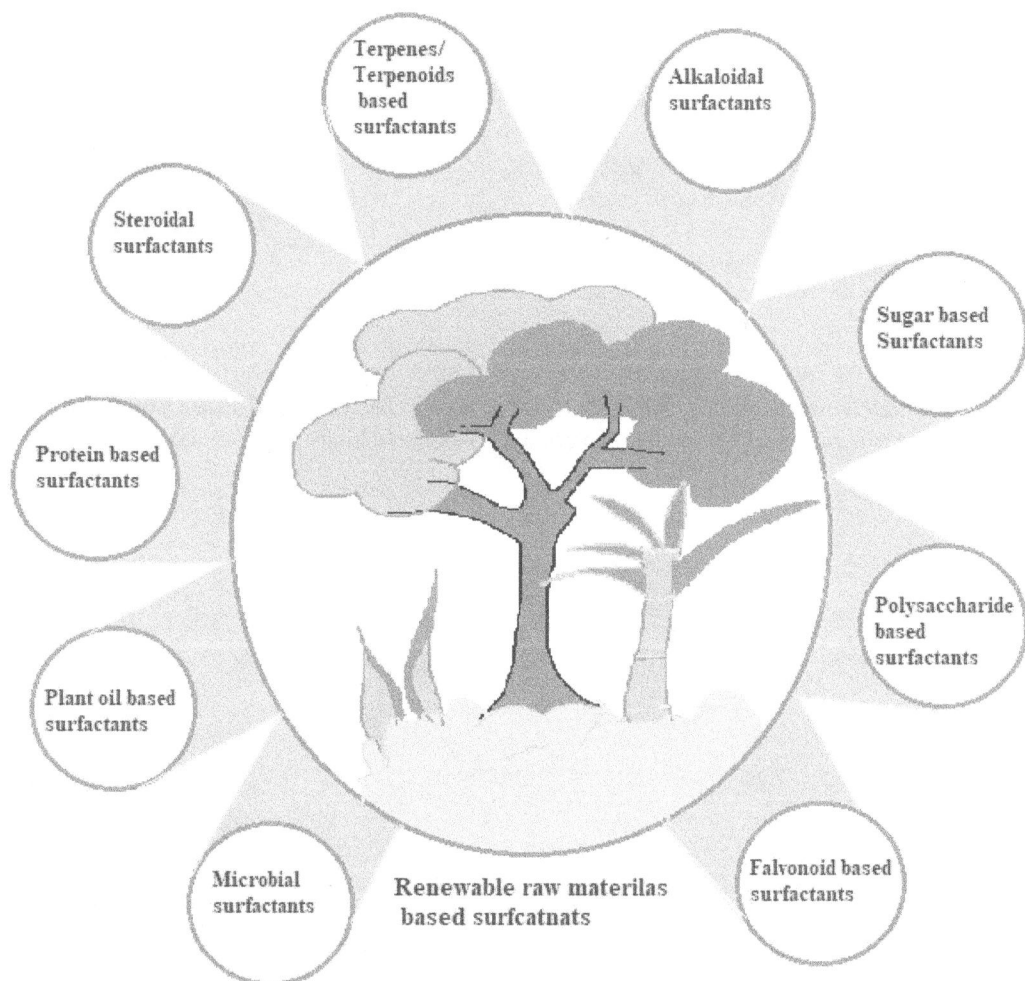

FIGURE 1.6   Surfactants based on renewable raw materials.

Vigorous investigation in the same direction emerged with various plant-based raw materials as substrates to produce surfactants. Plant oils, proteins, sugars, steroids, terpenoids, flavonoids are few examples of such sources. A wide variety of surfactants synthesized from different precursors also results in the difference in a wide range of properties thus enabling them to be exploited in various industrial applications.

1.3 RENEWABLE RAW MATERIALS-BASED SURFACTANTS TYPES

1.3.1 PLANT OIL-BASED SURFACTANTS

Vegetable oils are chemically the triglycerides mostly obtained from the seed or the fruit part of the plants. They may be edible or non-edible depending on its source. Extraction of the oil has been achieved at room temperature to retain the properties of oils. Vegetable oils differ from each other in terms of length of fatty alkyl chain, number of double bonds present in the chain, the presence or absence of conjugation and other structural properties like linearity and branched structure of chain. Different fatty acid compositions of different oils make them suitable for different applications. For instance, short-to-medium saturated alkyl chain length (C_{12}–C_{14}) of coconut and palm oil make them the best raw materials to be used in the cosmetics industry. In contrast, palm, soybean, sunflower and rapeseed oil due to long fatty alkyl chains were found best suited for the polymer and lubricants industry (Figure 1.7).

In order to be exploited as a raw material for surfactants, first the vegetable oil needs to be converted into free fatty acids via hydrolysis [10]. Fatty acids act as a source of hydrophobic chains, which on further reaction with suitable reactants yield surfactants [11].

Hence, as an abundant source of better hydrophobe, vegetable oil fascinated the surfactant manufacturers and provided nature supplies to them. Straight chains present in most of the fatty acids obtained from the natural oil also ensure better biodegradability and less toxicity.

1.3.2 PROTEIN-BASED SURFACTANTS

Protein-based surfactants [12] are another class of surfactants, earning the vast interest of researchers as natural surfactants. These surfactants are based on the material rich in amino acids (AAs) and polypeptides. Their amphoteric nature makes them applicable in the field of microbiology and pharmaceuticals and vast availability make them a cheaper base material for surfactant synthesis.

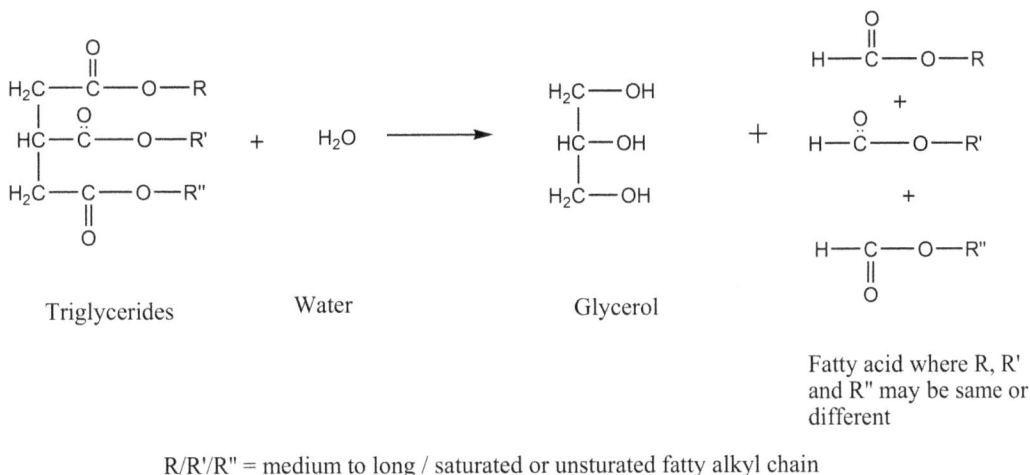

R/R'/R" = medium to long / saturated or unsturated fatty alkyl chain

FIGURE 1.7 Hydrolysis of triglycerides to produce fatty acids.

Proteins are the linear or branched molecules with peptide linkage within its polymeric structure. Basic unit in protein molecules is AA. Protein may exist in simple linear primary and secondary form as well as complex tertiary and quaternary branched structure. Structural variations of proteins validate them as potential raw material for surfactants.

1.3.3 STEROIDS-, TERPENES- AND ALKALOIDS-BASED SURFACTANTS

Steroids are the natural product naturally available in abundance in the animal and plant kingdom. Sterols are broadly classified into phytosterol (plant origin) and zoosterol (originated from animal source). Ease in availability makes them suitable as a raw material in various industries. Recently, these products gained interest from researchers for the expansion of surfactants class. Chemically, these are the cyclopentenophenanthrene nucleus having side chain at C_{17} position. The difference in the side chain at C_{17} results in the variation of their properties and applicability. Common treasure for phytosterols is soya bean, peanut, maize, rapeseed, tall oil, avocado etc.

Saponins, due to lyophilic aglycone and hydrophilic sugar moiety, exhibit amphiphilic properties. These surfactant properties of saponins enable them to produce form when interacted with the liquid-gas interface, to act as a potential emulator in liquid-liquid interfaces and to achieve dispersibility in liquid-solid interfaces. Single sugar chains containing saponins were claimed to have superior foaming ability. Increases in the number of sugar chains eventually decreased their foaming properties [13].

Terpenes are the class of plant products which have not yet received much interest from researchers in the field of surfactants hence less literature is available on this class of surfactants. Very recently the hydroaminomethylation of terpenes for the surfactants synthesis has been reported [14]. However, by-products and requirements of high pressure conditions were the major limitations found in this investigation. Moreover, hydroamination of terpenes can be a potential means to form long-chain amines that can be consequently turned into surfactants.

Alkaloids are the naturally occurring molecules with nitrogen atoms in cyclic rings, though the number of rings and position of nitrogen atoms in the ring may vary from one type to another. Additionally, the properties of the molecules are governed by the type and location of nitrogen atoms in the ring.

Some alkaloids like solenopsin alkaloids were reported to have surface-active properties and show decrease in interfacial tension due to the characteristics of forming micelles, microemulsions and property to adhere to the surfaces.

1.3.4 POLYSACCHARIDE-BASED SURFACTANTS

Polysaccharide-based surfactants belong to the class of polymeric surfactants that exhibit essential properties required for the formation of various dispersant systems. This is due to owing carbohydrate molecules as an active moiety in the polymeric chain. These surfactants were claimed to have good wettability and dispersibility [15].

1.3.5 SUGAR-BASED SURFACTANTS

Another big class of renewable raw materials available for the production of surfactants are sugars. Sugar-based surfactants are designed in such a way in which the polar group consists of carbohydrate moiety like glucose, sucrose etc., whereas a hydrophobic tail was obtained from fatty materials [16]. Most known members of this family are alkyl polyglycosides, sorbitol esters and sucrose esters. Sugars are playing an important role as a promising alternative to provide polar part of the amphiphilic molecule in the form of polyhydroxy cyclic/acyclic skeleton. These can be obtained from renewable resources and appeared as an ideal option as substrate for surfactant synthesis. Although some challenges like solubility in fatty substrates are still a major concern in high-molecular-weight sugars, which also occur in high yield of products, this limitation can be avoided by increasing the miscibility through acylation of sugar molecules prior to esterification by acetylating the sugar molecules before esterification followed by deacetylation. In terms of their performance, safety and price, these surfactants are found superior to their conventional counterparts.

1.3.6 FLAVONOID-BASED SURFACTANTS

Flavonoids, is one of the prominent natural products with a polyphenolic structure which are present in food items such as fruits, vegetables and beverages [17]. Exploitation of flavonoids obtained from the roots, stems, bark and flower extracts of many plants by the human race is well known since earlier times in the treatment of countless diseases. This practice encouraged scientists to explore, isolate and evaluate these molecular species for medical purposes. Flavonoids that are responsible to impart color to the various plant parts also act as a natural surfactant. Best-known examples of flavonoid-rich compounds are Reetha and Shikakai that have detergent-like properties and can be used as mild cleaning agents with milder action on color and texture of surfaces [18].

1.3.7 MICROBIAL SURFACTANTS (BIOSURFACTANTS)

Biosurfactants also known as microbial surfactants are amphiphilic compounds that are obtained through biological processes taking place in various microorganisms e.g., yeast, fungi or bacteria. The inquisitiveness to write on biosurfactants is their exceptional properties, viz. biodegradable, eco-friendly nature, less toxicity, diversified and selective nature, large-scale manufacture probability and many more. These renewable molecules showed great potential under grave conditions that fortified their use and attracted the attention of the researcher's community.

Biosurfactants are found to be dominant in almost all fields as antimicrobial agents, possessing dispersion, wetting, foaming, solubilization, emulsification and detergent competence. Biosurfactants minimize the surface tension almost 10–40 times lower than chemical surfactants and are required in very less amount for solubilization. Biosurfactants have great ability to emulsify/de-emulsify the emulsion and are found to be stable at different pH, salinity and temperatures. They are water soluble and simply biodegradable as compared to chemical surfactants. Chemical surfactants may also show biomagnification and accumulate in the food chain so research should be boosted towards the development of such sustainable biosurfactants.

The wide range of applications of biosurfactants also includes soil remediation with around 70–80% efficiency, hence show great potential in oil industry and other metal and bioremediation processes as well. Other applications of biosurfactants are in agriculture, ceramics, textile, pulp, dyes, food processing and cosmetic industries and they are also known to play an important role in reducing carbon dioxide emission [19–21].

1.4 ENVIRONMENTAL ACCEPTABILITY OF RENEWABLE RAW MATERIALS-BASED SURFACTANTS

Environmental acceptability of surfactants can be calculated as the extent of their harmful effect on the environment during their use and disposal after being exploited. The less harmful the effects, the more likely the environmental acceptability of the surfactant. These detrimental effects are also considerable during their risk assessment. The first step in the risk assessment of surfactants is the approximation of the surfactant's concentration in the environmental boxes of interest, like effluents obtained from wastewater treatment plants, soils, sediments and surface waters [22]. Estimation can be done either through exact measurement or via modeling and predictions. Then after, comparison has been made between the concentrations found after measurements and predictions of the surfactants under testing with the concentrations of reference surfactant which is toxic to living beings in that particular environmental sector. If the measured or predicted concentration of the tested surfactants is relatively less toxic to the reference surfactants, then the surfactants are claimed as relatively safer [23].

Renewable raw materials available for the production of surfactants also help to reduce the CO_2 emissions as these naturally occurring products once degrade, releasing only the measurable amount of carbon to the environment. The amount of carbon is normally found almost equal to the amount of carbon used by the plant in the manufacture of that particular plant part. In contrast, petrochemical resources used in surfactant productions surely release the entrapped carbon, responsible for the increase of greenhouse gases [24]. In disparity to the general stated earlier, there are some contradictory interpretations related to the sustainable expansion of surfactants category exploiting renewable building blocks. Tropical oils like coconut and palm kernel oils with fatty alkyl chain length C_{12}–C_{18} are habitually considered as renewable building blocks for manufacturing surfactants; however, these tropical plants are cultured after removing the natural rain forests and their wild populaces [25]. Resolutions taken by the General Assembly also appeal to enhance water quality through pollution reduction, dump elimination, reduction in pollution, and minimize the release of harmful chemicals to the environment [26]. Surfactants, being a most amply used chemical commodity, not only should be designed using a renewable and sustainable attitude but also need smart designing so that they can be simply degraded in the ecosystem once released after use. Surfactants with certain functional groups like esters easily undergo biodegradation which is not comparable to the synthetic chemicals and surfactants [27–30]. This is basically due to the various esterase enzymes present in the environment that ensure good biodegradation of any substrate having ester linkage [29].

1.5 CONCLUSIONS

Surfactants based on renewable raw materials are progressively becoming widespread and are being substituted for their conventional competitors, petrochemical and synthetic surfactants, in many diverse application sectors. Surfactant producers have launched various novel eco-friendly surfactant-based products into the market. Increase in the consumer awareness and commitment for sustainable growth has led to progress of several fresh varieties of natural substrate-based surfactants. Their better biodegradation accompanied with low toxicity make them the standard choice to design new formulations for commercial and consumer benefit. Demand for such surfactant is increasing at the rate of 3–4% per year and is expected to exponentially increase in coming years.

REFERENCES

1. Shaban, S. M., Kang, J., & Kim, D.-H. (2020). Surfactants: Recent advances and their applications. *Composites Communications*, 22, 100537.
2. Bhadani, A., Kafle, A., Ogura, T., Akamatsu, M., Sakai, K., Sakai, H., & Abe, M. (2020). Current perspective of sustainable surfactants based on renewable building blocks. *Current Opinion in Colloid & Interface Science*, 45, 124–135.
3. Shaw, D., & Corocoran, T. (2002). Surface activity and micelle formation. Modified by T. Corcoran.
4. Beringer, J., & Kurz, J. (2011). Hospital laundries and their role in medical textiles. In *Handbook of Medical Textiles* (pp. 360–386). Woodhead Publishing.
5. Porter, M. R. (2013). *Handbook of Surfactants*. Springer.
6. Kamal, M. S. (2016). A review of gemini surfactants: Potential application in enhanced oil recovery. *Journal of Surfactants and Detergents*, 19(2), 223–236.
7. Bajpai Tripathy, D., & Mishra, A. (2017). Convenient synthesis, characterization and surface active properties of novel cationic gemini surfactants with carbonate linkage based on C12-C18 sat./unsat. fatty acids. *Journal of Applied Research and Technology*, 15(2), 93–101.
8. Tripathy, D. B., & Mishra, A. (2017). Microwave synthesis and characterization of waste soybean oil-based gemini imidazolinium surfactants with carbonate linkage. *Surface Review and Letters*, 24(05), 1750062.

9. Raffa, P., Wever, D. A. Z., Picchioni, F., & Broekhuis, A. A. (2015). Polymeric surfactants: Synthesis, properties, and links to applications. *Chemical Reviews*, *115*(16), 8504–8563.
10. Biermann, U., Bornscheuer, U., Meier, M. A., Metzger, J. O., & Schäfer, H. J. (2011). Oils and fats as renewable raw materials in chemistry. *Angewandte Chemie International Edition*, *50*(17), 3854–3871.
11. Bajpai, D., & Tyagi, V. K. (2008). Microwave synthesis of cationic fatty imidazolines and their characterization. *Journal of Surfactants and Detergents*, *11*(1), 79–87.
12. Xia, J. (2001). *Protein-Based Surfactants: Synthesis: Physicochemical Properties, and Applications* (Vol. 101). CRC Press.
13. Oleszek, W., & Hamed, A. (2010). Saponin-based surfactants. *Surfactants from Renewable Resources*, *1*, 239–251.
14. Faßbach, T. A., Gaide, T., Terhorst, M., Behr, A., & Vorholt, A. J. (2017). Renewable surfactants through the hydroaminomethylation of terpenes. *ChemCatChem*, *9*, 1359–1362.
15. Kurečič, M., Smole, M. S., & Stana-Kleinschek, K. (2013). Use of polysaccharide based surfactants to stabilize organically modified clay particles aqueous dispersion. *Carbohydrate Polymers*, *94*(1), 687–694.
16. Foley, P., Kermanshahi-Pour, A., Beach, E. S., & Zimmerman, J. B. (2012). Derivation and synthesis of renewable surfactants. *Chemical Society Reviews*, *41*, 1499–1518.
17. Middleton, E. (1998). Effect of plant flavonoids on immune and inflammatory cell function. *Flavonoids in the Living System*, 175–182.
18. Cui, Q., Liu, J. Z., Yu, L., Gao, M. Z., Wang, L. T., Wang, W., ... & Jiang, J. C. (2020). Experimental and simulative studies on the implications of natural and green surfactant for extracting flavonoids. *Journal of Cleaner Production*, *274*, 122652.
19. Banat, I. M., Franzetti, A., Gandolfi, I., Bestetti, G., Martinotti, M. G., Fracchia, L., ... & Marchant, R. (2010). Microbial biosurfactants production, applications and future potential. *Applied Microbiology and Biotechnology*, *87*(2), 427–444.
20. Drakontis, C. E., & Amin, S. (2020). Biosurfactants: Formulations, properties, and applications. *Current Opinion in Colloid & Interface Science*, *48*, 77–90.
21. Naughton, P. J., Marchant, R., Naughton, V., & Banat, I. M. (2019). Microbial biosurfactants: Current trends and applications in agricultural and biomedical industries. *Journal of Applied Microbiology*, *127*(1), 12–28.
22. Salimon, J., Salih, N., & Yousif, E. (2012). Industrial development and applications of plant oils and their biobased oleochemicals. *Arabian Journal of Chemistry*, *5*(2), 135–145.
23. Stallings, D., Iyer, S. K., & Hernandez, R. (2018). National diversity equity workshop 2017: *Focus on underrepresented minorities in chemistry faculties*. In (pp. 109–140). American Chemical Society.
24. Stuart, L. (2017). Transforming Our World: The 2030 Agenda for Sustainable Development A/RES/70/1: Theme: statement, 'Do not leave Indigenous Australians behind'. In *Session of the Permanent Forum of Indigenous Issues (PFII): Tenth Anniversary of the United Nations Declaration on the Rights of Indigenous Peoples: measures taken to implement the Declaration*. University of the Sunshine Coast, Queensland.
25. Jordan, A., & Gathergood, N. (2015). Biodegradation of ionic liquids—A critical review. *Chemical Society Reviews*, *44*(22), 8200–8237.
26. Tehrani-Bagha, A., & Holmberg, K. (2007). Cleavable surfactants. *Current Opinion in Colloid & Interface Science*, *12*(2), 81–91.
27. Bhadani, A., Kafle, A., Ogura, T., Akamatsu, M., Sakai, K., Sakai, H., & Abe, M. (2019). Phase behavior of ester based anionic surfactants: Sodium alkyl sulfoacetates. *Industrial & Engineering Chemistry Research*, *58*(16), 6235–6242.
28. Bhadani, A., Endo, T., Sakai, K., Sakai, H., & Abe, M. (2014). Synthesis and dilute aqueous solution properties of ester functionalized cationic gemini surfactants having different ethylene oxide units as spacer. *Colloid and Polymer Science*, *292*(7), 1685–1692.
29. Bhadani, A., Shrestha, R. G., Koura, S., Endo, T., Sakai, K., Abe, M., & Sakai, H. (2014). Self-aggregation properties of new ester-based gemini surfactants and their rheological behavior in the presence of cosurfactant—Monolaurin. *Colloids and Surfaces A: Physicochemical and Engineering Aspects*, *461*, 258–266.
30. Bhadani, A., Tani, M., Endo, T., Sakai, K., Abe, M., & Sakai, H. (2015). New ester based gemini surfactants: The effect of different cationic headgroups on micellization properties and viscosity of aqueous micellar solution. *Physical Chemistry Chemical Physics*, *17*(29), 19474–19483.

2 Nonedible Oils
Potential Feedstock for Synthesis of Amino Acid and Sugar-Based Green Surfactants

Subhalaxmi Pradhan, Chandreyee Saha and Manoj Kumar
Division of Chemistry, School of Basic and Applied Sciences,
Galgotias University, Greater Noida, Uttar Pradesh, India

CONTENTS

2.1 INTRODUCTION

Amphipathic molecules having hydrophobic and hydrophilic entities which participate between the interface of two phases of fluid are known as surfactants. Generally, surfactants reduce interfacial surface tension and form microemulsions for which there is easy solubilization of hydrocarbons in water and vice versa. The demand of surfactants per annum is $9.4 billion and is increasing worldwide at the rate of 35% toward the end of the century [1]. Most surfactants used currently are derived from petrochemicals and are of vast applications as domestic detergents, in industries like food, fibre and paint and pharmaceutical agents. Mostly used surfactants are linear alkyl benzenes (LAB) which create environmental problems such as

DOI: 10.1201/9781003144878-2

eutrophication and form toxic foaming components which affect aquatic organisms [2]. Due to water solubility, there is difficulty in recovery and reusability of surfactants discharged into the water bodies if not biodegradable. Unsurprisingly, they have harmful effects to the environment when they are released in huge scale [2]. To prevent environmental pollution and preserve energy, surfactants should be reused and recycled. Surfactants of biological origin have various benefits in comparison to chemical surfactants, such as high selectivity, biodegradability, good compatibility to environment, excellent foaming action, low toxicity and high activity [3, 4]. Considering the environmental point, improved biodegradability and lower toxicity, more focus is given for the development of surfactants from natural and renewable resources. According to Indian Oil Corporation, the deficit of synthetic surfactants, primarily LAB, will be about 130 kt by 2018, so the price of petroleum products is increasing rapidly. So, nonedible seed oil-based surfactants can be used as a substitute for synthetic surfactants. Many surfactants are readily biodegradable, and by secondary treatment in wastewater plants, their amount is tremendously reduced. Due to which researchers fetch their attention toward environmentally friendly surfactants. Due to biocompatibility and biodegradability of sugar- and amino acid-based surfactants over other species, it has wide applications in detergent and pharmaceutical industries [5]. Surfactant production from edible oils is unaffordable in India, so various nonedible oil seeds such as castor, neem and jatropha are used for this purpose. Among these nonedible oil seeds, neem and castor are indigenous to India. Globally India stands first and second in neem and castor seed production, respectively. The present work will be concerned with synthesis of sugar- and amino acid-based surfactants from nonedible oils, and optimization of various reaction parameters such as enzyme type, molar ratio of substrates, solvent and temperature is also explained.

Vegetable oils are chemically the triglycerides extracted from various parts of the plants; however, seeds, buds and flowers are mostly rich in this component. On the basis of their origin and applicability, oil may be edible as well as nonedible. Properties of oils vary as per the their varying hydrophobic chains like length, saturation, type (linear or branched), which makes them suitable for various applications. Various properties also lead to their applicability in various applications. Coconut (C_{12}) and palm oil (C_{14}) are found best in cosmetics compositions, whereas palm, soybean, sunflower and rapeseed like long alkyl chains materials show good efficiency as polymer and lubricants.

Vegetable oils can be of edible as well as of nonedible type. Majority of the vegetable oils are taken from the fruits and seed with liquid physical state at normal environmental temperature.

Vegetable oils extracted from plants are chemically the triglycerides. Exact industrial applicability of oil has been governed by the fatty acid compositions of the oil. For example, coconut oil and palm kernel oil are finest to get commercially exploited in surfactants and cosmetics, which is due to the presence of short-to-medium fatty alkyl chains present in the molecule. On the contrary, oil with a long fatty alkyl chain containing fatty acids such as palm, soybean, rapeseed and sunflower are best suited as polymers and lubricants.

A vast range of citrus fruits such as lemon, grape and orange are the great sources of essential oils, thus used as flavors and fragrances. Sunflower seed oils, soya, palm kernel, rape and coconut oil have been industrially used for manufacturing of surfactants. First, these oils were chemically reformed to yield surfactants with high efficacy. For instance, in fabric softener formulations, easier and better degradability of esterquats completely substituted poorly degradable petroleum-based conventional counterparts.

Features of natural fatty acids attracted the surfactant producers seeking molecules as a source of better hydrophobe. Vast availability and good hydrophobicity made them a best alternate and a sustainable link as an interrupted supply to the surfactant's producers. Moreover, the linear tail confirms the good biocompatibility and desired hydrophobicity in the efficient surfactant molecule.

Three main varieties of hydrophobics required are:

1. Medium chain saturated (C_{10}–C_{14}) as in lauric acid
2. Long-chain (C_{16}–C_{18}) saturated as in stearic acid
3. Long-chain unsaturated as in oleic acid

All above three are present naturally in the form of plant oils. In addition, easy availability and vast variety of plant oil make them cost effective renewable raw materials for surfactants industries.

Various nonedible oils, despite having no edible use, are getting special attention of oil specialists, and this is due to their applicability as raw materials of various commercial products. In India, plants like neem, banyan, Castor are known to have many medicinal values since centuries.

Moreover, their abundance and hassle-free cultivation procedure make them potential candidates to be exploited as a raw material in the designing of many important and versatile molecules like surfactants.

2.2 POTENTIAL OF JATROPHA, CASTOR AND NEEM IN INDIA

India has a diverse agro-climate that allows for the production of various nonedible oil seed plants in wastelands. *Jatropha curcas* (L.) is a perennial spurge plant in the Euphorbiaceae family that has got a lot of attention as a possible source of vegetable oil to replace fossil fuels and synthetic surfactants. It is a fast-growing shrub that is grown in the country's tropical and subtropical regions, as well as on wastelands. In India, Jatropha is known by many names, including Ratanjyot, Jamalgota, Jangli arandi and others. The Jatropha seed oil has been discovered to be a promising and commercially viable alternative to diesel, which is a renewable energy source. It cultivates well all-around India, and Andhra Pradesh, Gujarat, Rajasthan, Karnataka, Chhattisgarh, Uttarakhand, Tamil Nadu, Maharashtra, Orissa and the North Eastern states are some of the promising states where it grows as a semi-wild bush or shrub near villages and cities, as well as hedge vegetation. It is commonly planted as a hedge to defend fields since it is not browsed by cattle and can easily be propagated by seeds and cuttings. While Jatropha prefers alkaline soils, it can thrive in a wide range of other conditions, including sandy soils [6]. Even though it survives prolonged droughts by shedding its leaves, it needs a minimum of 600 mm of rain. Jatropha produces fruit in the winter season in our nation. Unlike the majority of other trees that bear fruit during the monsoon season, with all the difficulties that entails in terms of post-harvest processing, Jatropha has the natural benefit of being able to harvest the fruit in summer [7]. Jatropha yields about 6 tonnes of fruit per hectare in poorly maintained hedges [8], but under ideal circumstances, maximum yields of up to 11 tonnes per hectare can be produced. Seeds contain 30–40% oil, resulting in up to 2200 kg of oil per hectare, depending on seed yields. Fruits are picked by hand or harvested by striking them with a long stick. Older trees/branches were shaken to harvest the fruit. All of the harvested fruits are usually dried in the sun for 1–2 days, manually decorticated to remove the seeds, and then stored for post-harvest processing [9, 10].

Neem (*Azadirachta indica*) belongs to the mahogany family Meliaceae and is widely cultivated in India. Neem thrives in a range of soil types, including clayey, acidic and alkaline soils. A mature tree can live for 150–200 years and produce 37–55 kg of fruit per year. A fully grown tree produces approximately 50 kg of fruit [11]. India is the world's leading producer of neem, with around 25 million trees producing 1.25 million tonnes of fruit, 5.4 lakh tonnes of seed, 1.07 lakh tonnes of neem oil and 4.25 lakh tonnes of neem cake each year [12]. In India, the oil content of the seed ranges from 20 to 32.61%, while the oil yield of the kernel averages 47.6% [13]. Since the pulp makes up 48% of the fresh fruit fraction, 6.0 lakh tonnes of pulp will be produced each year. In the fermentation industry and for methane gas processing, neem pulp is used as a rich source of carbohydrate. The sugar content of neem pulp is 15–18%, which can be converted to ethanol. According to stoichiometric calculations, 1000 kg of fermentable sugar yield 583 Lt of pure ethanol. Hence, about 52 million liters of ethanol will be produced annually. The oil has a heavy odor and a bitter taste.

It is yellow to dark in color and has a bitter taste. Neem is a storehouse of insecticidal azadirachtin, which possess antifeedant, growth disrupting and larvicidal properties against an array of agricultural insect pests [14]. Neem oil is commonly used in the production of soaps, skin care goods, waxes, lubricants, epoxy compounds, biosurfactants, biopesticides and lighting fuel. The fatty acid of the oil in reaction with sugar and amino acids can be converted into biosurfactants.

Castor, *Ricinus communis* L (Euphorbiaceae), is an oleaginous plant that can be found in almost every tropical and subtropical region. Following Brazil, India is the world's second-largest producer of castor seed, with an average yield of 1500–3500 kg/ha. Each year, India produces 8–12 lakh tonnes of castor seed, which yields 5.8 lakh tonnes of oil and 6.2 lakh tonnes of cake. Castor seeds come in a variety of shapes and sizes, but on average, they contain 46–55% oil by weight [15]. Since castor seeds contain ricin, ricinine and some allergens that are harmful to humans and animals, they are poisonous to both humans and animals [15]. The cake has a protein content of 34–36%. Castor bean residue is currently used as an organic fertilizer, but due to its toxicity, it is not used as a protein substitute. Despite this, the cake contains 48% starch, which can be hydrolyzed and fermented to generate alcohol. As a result, about 3.0 lakh tonnes of starch will be extracted from 6.2 lakh tonnes of cake, yielding about 175 million l of ethanol. Solid-state fermentation (SSF) of castor cake can produce a variety of enzymes. The toxic protein levels in cake are decreased during the hydrolysis and fermentation processes, allowing it to be used as animal feed. Since the seed is rich in oil, the fatty acids of the oil can be converted to sugar- and amino acid-based biosurfactants.

2.3 LAB EXTRACTION AND PHYSICOCHEMICAL CHARACTERIZATION OF SEED OIL

Amount of oil in the seed was investigated by the Soxhlet extraction method. The powdered seed was extracted using hexane in Soxhlet apparatus to find out the oil yield [16]. About 50 g of the sample was packed in a thimble and that thimble was placed inside a 500-ml capacity Soxhlet. The Soxhlet was fitted to a 500-ml round-bottomed (RB) flask and the RB was placed in the heating mantle adjusting the heating rate to give a condensation rate of two to three drops. Cold water circulation to Soxhlet was given through a condenser. About 300 ml of solvent was used for the extraction. Extraction was carried out for 8 h and then the extracted material was filtered to separate the residue and filtrate after cooling. Evaporation of solvent was done using Rotavapor (Heidolph Instruments, Laborota 4000-Efficient, Germany), the remaining oil was weighed after cooling. Then the oil was analyzed to determine various physicochemical properties. The oil yield is more in the hot extraction method using Soxhlet apparatus. Oil content and physicochemical properties of oil of various seeds are given in Table 2.1.

It is found that among the nonedible feedstocks, castor seed has high oil content and low acid value compared to other oils. The values obtained for physicochemical characterization of various oils agree with the reported data of other researchers.

TABLE 2.1
Physicochemical Characterization of Oil

Properties	Jatropha	Neem	Castor
Oil content (%)	36.4	24.3	38.25
Acid value	3.9	3.6	0.91
Iodine value (g I_2/100 g oil)	97	73.57	89
Saponification value (mg KOH/g oil)	189.2	237.2	185
Viscosity at 40°C (cSt)	34.31	25.4	231

Source: Pradhan, S., Naik, S. N., Khan, M. A. I., & Sahoo, P. K. (2012). Experimental assessment of toxic phytochemicals in *Jatropha curcas:* oil, cake, bio-diesel and glycerol. *J. Sci. Food Agric., 92,* 511–519.

2.4   HYDROLYSIS OF OIL AND ANALYSIS OF FATTY ACID BY GC

Hydrolysis of oil was carried out by conventional methods [17, 18]. The mixture of oil (100 g) and 10% NaOH solution (in 250-ml methanol) was taken and heated at 45°C for 2 h in a water bath. After that bi-distilled water (400 ml) was added with stirring to make a clear solution. Then the reaction mixture was cooled and 30% HCl solution (300 ml) was mixed into it. The addition of HCl will be done in five slots, each slot having 60 ml of HCl with stirring and the reaction is carried out for 3 h. The substrate mixture was transferred to a separating funnel after cooling to ambient temperature, and the oil phase was extracted from the aqueous phase using ether. The separated oil phase was cleaned with 200 ml of bi-distilled water three times to remove the excess HCl and dried in a rotary evaporator. Then the reaction product dissolved in 0.3 ml of petroleum ether was injected into gas chromatograph (Shimazdu 2010) using split injection and FID (flame ionization detector) at 250°C using capillary column (DB-Wax, 30 m × 0.32 mm × 0.25 µm) at column oven isothermal at 210°C. Identification of fatty acid methyl esters (FAME) of sample was done by comparative evaluation of retention time of standard FAME from Sigma-Aldrich, United States running in GC with same condition. The % composition of fatty acids was quantified by using peak areas. Fatty acid profiling of various oils is given in Table 2.2, and it is found that all the three nonedible oils have less percentage of saturated fatty acids (SFA) compared to unsaturated fatty acids (USFAs). The predominant fatty acid in Jatropha and Neem oil is oleic acid, whereas the major fatty acid in castor oil is ricinoleic acid, which is a hydroxy C18:1 USFA. Due to the presence of this hydroxy USFA castor oil is used as a better lubricant.

Hydrolysis of oil can be done in the presence of various enzymes to separate corresponding fatty acids. Instead of conventional methods, hydrolysis of oil can be carried out in subcritical and supercritical conditions to separate respective major fatty acids. The fatty acids after separation are used for esterification reactions with glucose and fructose to form sugar-based bio-surfactants.

2.5   SYNTHESIS OF AMINO ACID-BASED SURFACTANTS

Esterification of mixture of extracted fatty acids is generally done by refluxing with polyethylene glycol as per stoichiometric proportion in solvent xylene at 140°C. The reaction is continued to

TABLE 2.2
Fatty acid Profiling of Oils

Fatty Acid	Jatropha Oil	Neem Oil	Castor Oil
Myristic acid	–	0.13	–
Palmitic acid	13.94	16.85	1.3
Palmitoleic acid	0.96	0.2	–
Stearic acid	6.16	14.89	1.5
Oleic acid	42.03	53.82	3.4
Ricinoleic acid	–	–	87
Linoleic acid	35.8	10.9	5.0
Linolenic acid	0.22	<0.1	0.6
Arachidic acid	0.19	1.4	–
Ecosenoic acid	0.36	0.38	–
Docosanoic acid	0.07	<0.1	–
Saturated fatty acids	20.34	32.6	3.0
Unsaturated fatty acids	79.46	65.4	96.0

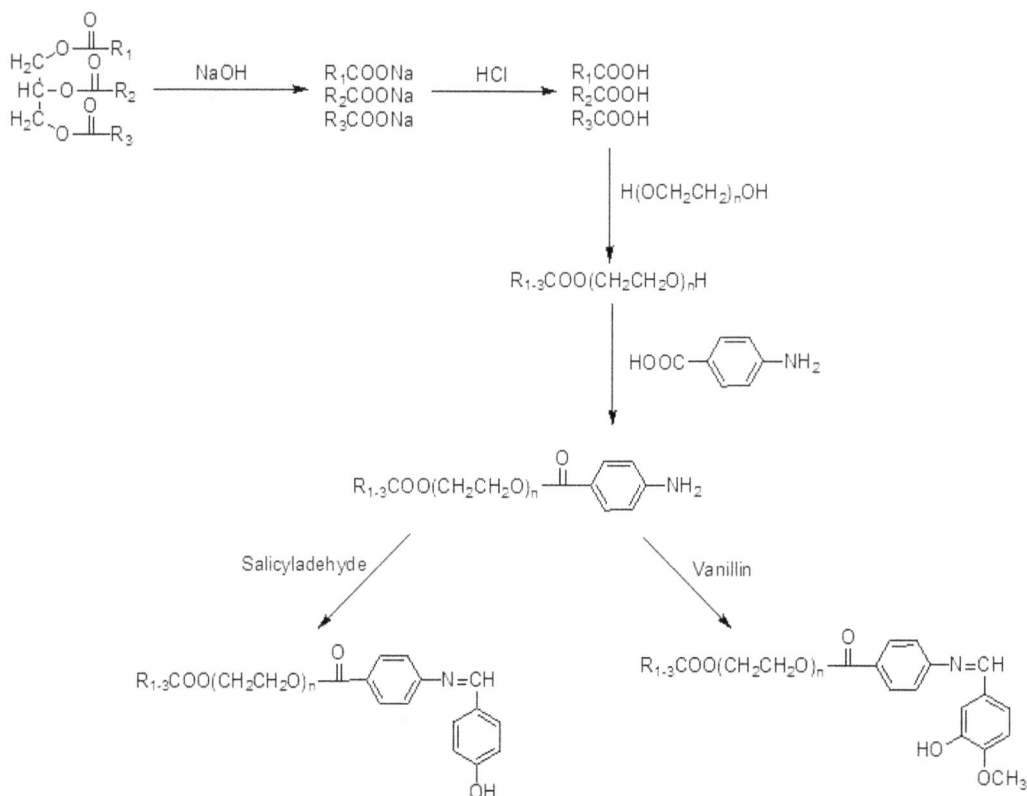

SCHEME 2.1   Synthesis of amino acid-based Surfactants [19].

remove water completely from the reaction mixture. About 27.4 g of *p*-aminobenzoic acid is mixed with the reactant and refluxed nearly at 140°C for complete removal of H_2O produced at the reaction. When 0.2 mol of H_2O is removed from the reaction, the reaction is stopped and then the reaction is done between produced diesters and carbonyl groups (vanillin and salicylaldehyde) in ethanol to obtain a different Schiff base [19]. The process of reaction is represented in Scheme 2.1.

Various properties such as emulsion stability, critical micelle concentration (CMC) and surface tension of surfactants are generally checked to determine the quality of surfactant. There are two characteristic regions obtained by changing the concentration of surfactants. Extrapolating the trends in the two characteristic regions determines CMC. Low CMC is exhibited by short PEG chain length species while improved CMC is obtained for higher chain PEG. There is a minor change in CMC value by incorporating vanillin moiety instead of o-hydroxy benzaldehyde [19]. In aqueous medium, surfactant molecules exhibit both repulsive force and attractive force canceling each other. The polar phase applies repulsive force upon hydrophobic chains, and there is attractive force among the partially charged nonionic chains and H_2O. The surface activity of the surfactant molecules at the interface is measured from surface tension at the CMC. Surfactants developed from salicylaldehyde exhibited low CMC compared to the surfactants obtained from vanillin [19].

2.6   SYNTHESIS OF SUGAR-BASED SURFACTANTS

Fatty acid esters of sugar are basically synthesized by transesterification of sugar and fatty acids in the presence of enzyme catalyst (Scheme 2.2).

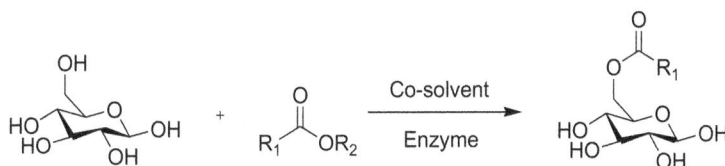

SCHEME 2.2 Synthesis of sugar-based surfactants.

Fatty acids extracted from nonedible oil and sugar are the substrates used for production of sugar-based surfactants in the presence of biocatalyst, lipase. The catalytic activity of the lipase depends upon the properties of substrates. The effectiveness of a given lipase catalyst during esterification depends upon the properties of the substrate. There are some lipases which effectively catalyze fatty acids of medium and long chain but some selectively catalyze branched and short-chain fatty acids [20]. Synthesis of sugar fatty acid esters (SFAEs) generally carried out in the presence of organic solvents and organic solvents are used as adjuvants in solid-phase enzymatic synthesis [21–23]. But the problem is finding a solvent which will solubilize both sugar and fatty acid and also that is compatible with enzymes. Organic solvents such as pyridine and dimethyl formamide can be used to solubilize both fat and sugars, but lipase activity is poor by using these solvents [24]. However, organic solvents are not used in the context of food applications as they are nonvolatile and toxic. To overcome the difficulty, ionic liquids (ILs) will provide an alternative green reaction medium over traditional organic solvents for production of SFAEs, because ILs promote the solubility of sugar, reactivity of enzymes and regioselectivity [25–31].

2.7 OPTIMIZATION OF REACTION CONDITIONS FOR SUGAR-BASED SURFACTANTS

2.7.1 MOLAR RATIO OF REACTANTS

The esterification reaction is influenced by molar proportion of the reactants (fatty acids and sugar). The amount of one substrate influences the dissolving ability of another substrate in the reaction medium as the polar nature of the reaction medium depends upon solubility. It is observed in reaction of glucose with stearic acid that excess fatty acid in the reaction medium increases the reaction efficiency, yielding a higher amount of sugar ester [32]. High yield is obtained at 1:3 proportion of sugar to fatty acids at a temperature range of 50–78°C, utilizing immobilized enzyme *Candida antarctica* lipase B. Reaction of glucose with short-chain fatty acids decreases the yield of SFAE [32], whereas fructose short-chain fatty acids favors the esterification reaction with the same enzyme *C. antarctica* lipase B [33].

This result may be obtained as long-chain fatty acids saturate the active site of the enzyme and hinder the entree of fructose to the active site of *C. antarctica* lipase B. This may have occurred due to variations in viscosities of short- and long-chain fatty acid solutions at a constant stoichiometric proportion [34]. Long-chain fatty acids enhance the conversion rate at equal stoichiometric proportion of fructose to fatty acids, but the rate of conversion reduces with higher molar proportion of fatty acid: sugar [35, 36]. During synthesis of palmitic ester of glucose, higher conversion rate is observed at equimolar ratio of glucose and palmitic acid (1:1), but rate of conversion decreases by increasing glucose concentration as glucose inhibits the enzyme activity [37]. Yan et al. [32] investigated the esterification of glucose with capric acid and observed a contrasting result that conversion rate increases by increasing sugar concentration, i.e., 18% conversion at 1:1 molar ratio, whereas 64% conversion rate at 6:1 molar proportion of glucose to capric acid. Hence, it could be concluded that for fatty acids having less than or equal to 10 carbon atoms, a high molar amount of fatty acid to sugar increases the conversion rate of synthesis of SFAEs using lipase catalysts. For fatty acids having 16 or more C-atoms, low molar amount of fatty acid to sugar exhibits high conversion rate of synthesis SFAEs, because high amount of fatty acid reduces the dissolution of sugar in the reactant mixture.

2.7.2 Nature of Co-solvent

Lipase-catalyzed synthesis of SFAEs is carried out in the presence of nonaqueous solvent. Generally, a suitable solvent is required for dissolving both the reactants during esterification reaction, but a different organic solvent is required for dissolving sugar and fatty acid. This creates complications during the synthesis of SFAEs because a high amount of sugar and fatty acid is not soluble in a single solvent. The stability, reactivity, enantioselectivity and specificity of the biocatalyst (enzyme lipase) are also affected by the solvent [38–42]. There is dehydration of the enzyme in water insoluble organic solvents, which reduces reactivity, stabilization and catalytic power of enzymes [43, 44]. Lipases that have better resistance during long interaction with organic solvents are preferred because the resistance of lipase toward organic solvent improves by immobilization of lipases. Preferably, a better solvent is that which highly solubilizes the reactants and less solubilizes the product, so that the product is easily separated after the reaction [45]. Some other researchers extensively studied the effect of organic solvents during enzyme catalytic esterification of sugar and fatty acids and interpreted that due to difference in solvent polarity, there is difference in partition coefficient of solvent, which influences reactivity and specificity of enzymes [46, 47]. The dielectric constant of solvent influences the ionic and polar interactions within the 3D structure of the enzyme, thus affecting the reactivity by changing specificity of the enzyme [48, 49].

Lipase-mediated synthesis of SFAEs is not always favored by hydrophobic solvent because glucose being hydrophilic is less soluble in hydrophobic solvent and decreases the yield of SFAEs [50–52]. Blending two or more solvents enhances the ionization and polarity of a non-hydrous reaction mixture [51, 53, 54]. Yan et al. synthesized glucose caprylate by enzymatic esterification of glucose and capric acid, and high conversion is achieved in acetone (90%) compared to butanone (66%) [32]. About 88–96% of xylitol fatty acid esters were obtained for different fatty acids in the presence of Novozyme 435-immobilized lipase using hexane as the solvent [55]. Dilauroyl maltose is synthesized using a solution of acetone-hexane as solvent in the presence of lipase [54]. Degn and Zimmermann [51] synthesized carbohydrate esters from various carbohydrates and myristic acid using immobilized *C. antarctica*, and the conversion of myristic glucose was enhanced per hour by altering the solvent from tertiary butanol to 45% (v/v) tert-butanol and pyridine.

Lipase-mediated synthesis of SFAEs is also favored using ILs as solvent because ILs are nonvolatile and composed of ions, which easily solubilize sugars as well as complex carbohydrates [56–62]. It has been observed that aqueous sodium carbonate is used as an additive to improve the solubilizing activity of ILs during lipase-catalyzed esterification [60]. The esterification reaction, which does not occur in absence of additive solvents, occurs at comparable rates in nonpolar organic solvents in the presence of additives [60]. The activity of enzymes is also affected by ILs. Lee et al. [59] examined the activity and stability of lipase B 435 (*C. antarctica*) in different ILs and their mixtures. The activity of the enzyme is stable in 1-[2-(2-methoxyethoxy)-ethyl]-2,3-dimethyl imidazolium tetrafluoro nitrate [Bmim][Tf2N], however rapidly reduced in 1-[2-(2-methoxyethoxy)-ethyl]-2,3-dimethyl imidazolium tetrafluoro oxide and enzyme is more stable in presence of [Bmim][Tf2N] as an additive to ILs [59].

2.7.3 Enzyme Type

Generally, hydrolysis of triglycerides is catalyzed by enzyme lipases. Lipases also catalyze esterification reaction, and the reaction shifts toward synthesis of sugar esters in anhydrous solvents with a little amount of water. Although water is not necessary for the production of SFAEs, a little amount of water is required for hydrating lipase. The 3D conformation of lipase confirms the stability and catalytic activity of lipases. Enzymatic activity and stability is also influenced by the hydration state of the enzyme. There is a movable lid region at the site of interaction of enzyme with substrate [63, 64]. The interaction of hydrophobic solvent and an aqueous medium causes opening of the lid of the enzyme at their interface [65–67]. The specificity, selectivity and activity of enzymes for synthesis of SFAEs are enhanced by chemical modification of enzymes [68–70]. The yield of some SFAE is improved by using detergent modified powdered lipase compared to that of the lipase without detergent treatment [68]. Immobilized lipase is always used for the synthesis of SFAE, and the rate of reaction at initial state increases by increasing the concentration of lipase, but more amount of immobilized enzyme reduces the rate of reaction due to difficulty in mixing of the reaction substrates [71]. Henceforth, excess use of enzymes is not affordable due to its high cost and reduced yield of reaction. Lipases such as *Caulerpa cylindracea, C. antarctica, Candida rugosa, Bacillus subtilis, Bacillus licheniformis* and *Rhizomucor miehei* are normally used for synthesis of SFAEs [72–75].

2.7.4 EFFECT OF TEMPERATURE

Temperature is a major factor which affects the enzyme stability, dissolution of the substrates and product, reaction rate and the equilibrium condition. Many of the lipases are thermally stable. There is no loss of catalytic activity of Novozyme 435 (*C. antarctica* immobilized lipase B) at reaction temperature 60–80°C [73, 74]. Some researchers reported less conversion of the substrate to product using Novozyme 435 as the catalyst at temperatures less than 30°C, and conversion rate decreases due to decrease in catalytic activity of the enzyme at temperature more than 70°C [75–79]. Hence, 30–70°C is a preferable temperature range for synthesis of SFAE using *C. antarctica* immobilized lipase B, and some specific reactions may be carried out up to 80°C [76]. Yu et al. observed that during enzymatic synthesis of glucose ester, the activation energy (Ea) required for acyl-enzyme complex is 52.9 kJ/mol, and the enzymatic activity is increased by increasing the temperature from 35 to 45°C [80]. High conversion was achieved during synthesis of short-chain fatty acid esters of glucose at 35–45°C using lipase as catalyst, and large fatty acid sugar esters can be synthesized at raised temperature of 60°C [32]. In this temperature range, the product is less soluble and could be easily precipitated out from the reaction medium. Hence, optimum temperature of the reaction will be influenced by the number of carbon atoms (chain length) of the fatty acids [80].

2.8 OPTIMIZATION OF ENZYMATIC SYNTHESIS OF AMINO ACID-BASED SURFACTANTS

Fatty acids of vegetable oils react with amino acids in the presence of a number of enzymes to synthesize amino acid-based surfactants. Generally, the reaction is carried out with various amino acids and fatty acids with 1:1 molar ratio in the presence of lipozymes using hexane as a co-solvent at 60°C for 3 days in an orbital shaker fixed at 150 rpm. The progress of the reaction is monitored by TLC. Molar ratio of reactants, type of enzyme, reaction temperature, enzyme concentration, nature of co-solvent and moisture content has tremendous effect on yield of the product.

2.8.1 IDENTIFICATION OF SUBSTRATE

Identification of suitable substrates is an important parameter during the synthesis. Some researchers carried out lipozyme-catalyzed reactions of numerous free amino acids with fatty acids, different fractions of oil and triglycerides [81]. They found that no product was formed with some amino acids L-glycine and L-serine, but acylation of L-arginine occurred slightly. It may happen due to steric hindrance initiated by the groups near to the massive groups in amino acids. The terminal amino group of L-lysine is isolated from the massive part by four CH_2 groups, so they react with fatty acids to give desired product which showed ninhydrin-positive spots having R_f similar to the standards N-$\mathcal{E}$-palmitoyl and N-$\mathcal{E}$-oleoyl lysine. It can be concluded that free fatty acids obtained after the hydrolysis of triglycerides will act as the substrate for synthesis of surfactants. Among six substrates such as oleic acid (C18:1), palmitic acid (C16:0), triolein, tripalmitin, palm olein (PO), palm kernel olein (PKO), oleic and palmitic acids are good substrate for the synthesis of amino acid-based surfactants using L-lysine [81].

2.8.2 Effect of Enzyme at varied Temperature

Soo et al. screened five different enzymes at equimolar ratio of fatty acid and amino acid at temperature range of 30–80°C using 3-ml hexane as solvent. They found that lipozyme exhibited highest yield for all six substrates of fatty acids and L-lysine, whereas the enzyme Novozyme 435 and Lipase AK showed moderate yield with three substrates, olein, triolein and kernel olein of palm oil with same amino acid. The enzyme *C. rugosa* and protease trypsin exhibited lowest yield at all ranges of temperature. Lipozyme exhibited excellent activity at a temperature range of 70–80°C. The activity of the enzymes reduced at lower temperatures, and they permitted the crystallization of high-melting free fatty acids, limiting mass transfer. The activity of enzymes is boosted at higher temperatures, aiding the melting of FFAs, resulting in improved substrate dissolution and intermixing. Some researchers synthesize N-$\mathcal{E}$-acyl lysines using lipozyme at 90°C and exhibited 35% yield in 24 h [82].

For such reactions, 70°C is taken as optimum temperature for subsequent reactions because of comparable yield at 80°C and also due to less energy consumption and decomposition of enzymes at high temperature. When the six reactants were compared, olein from the kernel was found to be the most effective reacting species. Triolein and kernel olein showed 80% conversion, whereas PO and other fractions exhibited 60% results. Palmitic acid and tripalmitin were relatively weak acyl donors, while oleic acid exhibited moderate results. It happens due to the effectiveness of lipozyme for long-chain fatty acids or the reaction medium's incomplete dissolution of the strong palmitic acid and tripalmitin substrates [83].

2.8.3 Nature and Effect of Solvent

Soo et al. carried out synthesis of amino acid-based surfactants from fatty acid and L-lysine in the presence of different high boiling organic solvents. Highest yield is obtained in hexane for C18:1 acid, tripalmitin, PO and PKO, whereas triolein exhibits comparable yield in the presence of toluene as solvent. The efficacy of enzymes varies according to the nature of solvent, because it depends upon the degree of hydration of enzymes by solvent rather than their direct effect on the enzyme or substrates [84]. In acetonitrile and ethyl acetate having log P < 2, no enzyme-catalyzed reaction of fatty acid and amino acid occurred. Hydrophilic solvents strip the essential water of the enzyme and cause denaturation of enzymes [24]. Dioxane and THF also have the negative impact toward the synthesis of N-$\mathcal{E}$-oleyl lysine using lipozyme [82]. Solvents with log P values greater than 2 dissolve more slowly in water, allowing the enzyme to be adequately hydrated in its effective configuration and active for synthesis [24]. This is in accordance with the lipozyme manufacturer's recommendations, which claimed that the enzyme is compatible with hydrophobic solvents such as hexane, heptane and ether. Montet et al. [82] announced that a lipozyme-catalyzed synthesis of N-$\mathcal{E}$-oleyl lysine in hexane, heptane and pentane gave good yields of 35% at 90°C.

2.8.4 Effect of Molar ratio of Substrates

Soo et al. carried out the production of bio-surfactants at different molar proportions of fatty substrate and amino acid. They observed that oleic acid and palmitic gave better yield at molar ratios of 3:1 and 5:1, respectively. The result obtained is comparable with the result of Valivety et al. [85]

who stated that good yield is obtained at 1:2 molar ratio of l-*O-(NCbz-*l*-phenylalanyl)glycerol* and myristic acid. When more amino acid (1:0.5 and 1:0.75 of amino acid to fatty substrate) is taken, the rate of conversions is less at the same reaction time [85]. When palm oil triglyceride is used, good conversion is obtained at equimolar ratio of substrates, but for triolein, good yield is obtained at 1:3 proportion of fatty acid to amino acid. Excess triglycerides may have been hydrolyzed, resulting in high amounts of FFAs which fall in such a range that they serve as reaction inhibitors. The adverse effect of FFAs on enzymatic reactions catalyzed by lipase is due to pH change that disturbs the ionization states of substrates and ionizable groups in the active site of lipase [86]. FFAs are fixed onto macroporous anion exchange resin of the enzyme, resulting in a lipophilic configuration that avoids L-lysine from reaching the reaction site [82]. An abundance of L-lysine resulted in a more pronounced drop in yield; one potential explanation is that the existence of high quantities of the undissolved amino acid intensified mass transfer limitations of the system.

2.8.5 EFFECT OF TIME PERIOD

Some researchers carried out various reactions at 1:1 molar proportion of fatty substrate and amino acid at 70°C using lipozyme in the presence of co-solvent hexane to find out good yield in less time [81]. It is observed that good yield is obtained for FFAs and triglycerides with L-lysine at a time period of 96 h, but more time (144 h) is required for the reaction between L-lysine and palm oil fractions. Since the oil contains a mixture of fatty acids and some fatty acids have less reactivity toward enzyme-catalyzed synthesis, it requires more time for completion of reaction. When palm oil fractions are used directly, it resulted in higher yields in most situations, allowing for the production of mixed-chain compounds which were expected to have improved surfactant activity than single-chain-length compounds at less cost of feedstocks [87].

2.9 SYNTHESIS OF SURFACTANTS FROM WASTE COOKING OILS

Waste cooking oil (WCO) belongs to a nonedible oil variety that generates in large amounts and its improper disposal may have a detrimental effect on human life and the environment. WCO, on the other hand, is unquestionably a green bio-based material feedstock. To resurrect WCO, researchers developed a simple and high-yield method for converting WCO to bio-based zwitterionic surfactants with excellent surface and interfacial properties [88].

2.9.1 PRODUCTION OF ZWITTERIONIC SURFACTANTS FROM WCO

Zhang et al. [88] used WCO and *N*, *N*-dimethyl-1,3-propanediamine to make zwitterionic bio-based surfactants. In a RB flask, 20 g of WCO is placed and hydrolyzed with 30 ml of NaOH (methanol: water = 2:1) by refluxing for 6 h at 70°C. The reaction mixture is neutralized after the reaction, and 6-M HCl is added to keep the pH around 2. After that, separation of fatty acids and glycerol is done by adding water and transferring the mixture to a separating funnel. Upper oily layer is washed by boiled H_2O for five times to maintain the pH of the aqueous layer at 6. Then drying of the oily phase is done by cyclohexane. The efficiency of hydrolysis was 98%, which was confirmed from the acid value of oil (1.37 mg KOH/g) and hydrolysis product (174.33 mg KOH/g). Fatty acids were converted to phenyl fatty acids (PFA) by refluxing mixture of fatty acids (18 mmol) and 2.40 g of $AlCl_3$ (18 mmol) in 25 ml of C_6H_6 in a RB flask at 65°C. During reflux, anhydrous $CaCl_2$ is kept in a drying tube above the reactor. The reaction is carried out for 6 h, the product is washed by 15 ml of 1-M HCl thrice, then surplus C_6H_6 is evaporated on a Rotavapor to give PFA. Then synthesis of *N*-phenyl fatty amido propyl-*N*,*N*-dimethylamine (PFAPMA) is done by reaction of PFA with thionyl chloride and treating product with *N*,*N*-dimethyl-1,3-propanediamine. About 5.9-g PFA (18 mmol) is dissolved in 25 ml of chloroform and added dropwise to a drying flask containing 1.45-ml thionyl chloride (20 mmol), then stirring of reaction mixture is carried out at 40°C. After 2 h, the reaction mixture is evaporated to eliminate the solvent and surplus thionyl chloride. The remaining material after evaporation is solubilized in 10 ml of acetone, and 2.30 ml of *N,N*-dimethyl-1,3-propanediamine (20 mmol) is slowly mixed in it upon cooling at zero temperature. Then the reactant is stirred for 2 h at 40°C. The additional diamine and solvent are evaporated on a Rotavapor to yield *N*-phenyl fatty amido propyl-*N,N*-dimethylamine (PFAPMA). After that, the product is quaternized using ClCH$_2$COONa with a molar proportion of 1:1.25 in mixture of CH$_3$OH and H$_2$O (1:4) and refluxed at 75°C for half a day. After reaction solvent was evaporated under reduced pressure and residue dissolved in ethanol and filtered. After that, ethanol was removed by distillation, and the product was recrystallized in ethyl acetate to produce *N*-phenyl fatty amido propyl-*N,N*-dimethylcarboxylbetaine (PFAPMB) with 84.9% yield. The synthesized compound is characterized by various spectroscopic techniques, GC-MS, ESI-HRMS and ^{1}H NMR. Surface active properties such as surface tension, interfacial tension, contact angle, emulsification property, foaming property and biodegradation study of *N*-phenyl fatty amido propyl-*N,N*-dimethylcarboxylbetaine is carried out to proof it as a biosurfactant.

2.9.2 BIOSURFACTANT SYNTHESIS FROM WCO USING NaOH AND H$_2$O$_2$

WCO is pretreated by heating at 60°C and filtered to remove solid material and other contaminants present in the oil [89]. A mixture of 10-ml WCO and NaOH (1M) solution is heated on a hot plate with a magnetic stirrer at 40°C. After continuing the reaction for a certain time, the reaction mixture is treated with 5-ml H$_2$SO$_4$ (3M) to maintain the required pH for further reaction. Then 5-ml of H$_2$O$_2$ is added to the mixture, and the reaction is continued with stirring till the foam is diminished. After that the reaction mixture is cleaned using a saturated NaCl solution and then filtered and kept in the oven at a temperature of 60°C for 1 day for drying. The researchers carried out the above-mentioned reaction at different reaction operating parameters such as temperatures of 40, 50, 60, 70 and 80°C, different concentrations of NaOH, 1M, 2M, 3M and 4M for varied time periods of 20, 30, 40, 50 and 60 min to find out the optimum reaction condition. The physicochemical properties of biosurfactants, such as pH, hard water interaction and emulsification properties, are analyzed and calculated according to ASTM D460, and the properties are found to be comparable to those of commercialized detergent.

2.10 CONCLUSION

Fatty acids of nonedible oils could be the best source for synthesis of amino acid and sugar-based surfactants to replace synthetic surfactants. Sugar esters and amino acid esters produced from the fatty acids of nonedible oils can be widely used as green and eco-friendly biosurfactants. Synthesis of sugar esters using biocatalysts such as lipase has great advantages compared to traditional methods of production of sugar esters using chemical catalysts. SFAEs can be prepared by the nonenzymatic chemical synthesis, but the enzymatic route has several advantages over conventional chemical methods. Industrial-scale synthesis of simplest carbohydrate (glucose, fructose and sucrose) esters of fatty acid is carried out using basic metal hydroxide catalysts at a temperature greater than 100°C by transesterification of fatty acid methyl esters with sugars. Amino acid-based surfactants are a form of bio-based surfactant that has good surface properties, a wide range of bioactivity, a low propensity for toxicity and a low ecotoxicity. Furthermore, chemical and enzymatic catalysis can be used to effectively prepare them. These characteristics make them an excellent clean and healthy alternative to petroleum-based surfactants. As a result, these new amino acid-based surfactant productions will meet the growing need of eco-friendly surfactants in pharma and food industries. This chemical synthesis is not economical as it involves multistep separations and use of high temperature makes the overall production process costly. Compared to conventional methods, enzymatic synthesis of sugar esters proceeds in a single step, no protection/deprotection of the –OH groups, and involvement of less energy as the synthesis occurs at intermediate reaction temperature 40–60°C. India being the importer of edible oils, use of edible oils for surfactant production is unaffordable. So fatty acids of nonedible oils could be the best source for synthesis of sugar- and amino acid-based green surfactants.

REFERENCES

1. Safary, A., Ardakani, M. R., Suraki, A. A., Khiavi, M. A., & Motamedi, H. (2010). Isolation and characterization of biosurfactant producing bacteria from Caspian Sea. *Biotechnology*, *9*, 378–382.
2. Okbah, M. A., Ibrahim, A. M. A., & Gamalm, M. N. M. (2013). Environmental monitoring of linear alkylbenzene sulfonates and physicochemical characteristics of seawater in El-Mex Bay (Alexandria, Egypt). *Environ. Monit. Assess.*, *185(4)*, 3103–3115.
3. Desai, J. D., & Banat, I. M. (1997). Microbial production of surfactants and their commercial potential. *Microbiol. Mol. Biol. Rev.*, *61(1)*, 47–64.
4. Kosaric, N. (1992). Biosurfactant in industry. *J. Am. Oil Chem. Soc.*, *64(11)*, 1731–1737.
5. Infante, M., Pinazo, A., & Seguer, J. (1997). Non-conventional surfactants from amino acids and glycolipids: structure, preparation and properties. *Colloid. Surf. A*, *123*, 49–70.
6. Kumar, A., & Sharma, S. (2008). An evaluation of multipurpose oil seed crop for industrial uses (*Jatropha curcas* L.): a review. *Ind. Crop. Prod.*, 28, 1–10.
7. Bringi, N. V. (1987). *Non-Traditional Oilseeds and Oils in India*. New Delhi: Oxford & IBH Publishing Co. Pvt. Ltd.
8. Henning, R. (1996). *The Jatropha Project in Mali*. Weissensberg, Germany: Rothkreuz 11, D-88-138.
9. Manurung, R., Heeres, H. J., Ratnaningsih, E., Suryatmana, P., Satyawati, M. R., Junistia, L., & Daniel, L. (2006). Paper Presented at Initial Stage on the Volarisation of *Jatropha curcas* Using Biorefinery Concept—In International Symposium on Volarisation of Indonesian Renewable Resources and Particularly *Jatropha curcas* Using the Biorefinery Concept The Agency for the Assessment and Application of Technology, Jakarta, November 20–21.V.
10. Lakshmikanthan, V. (1978). Tree Borne Oilseeds, Directorate of Nonedible Oils & Soap Industry. Khadi & Village Industries Commission, Mumbai, India, 38–39.
11. Parmar, B. S., & Ketkar, C. M. (1993). Commercialization in Neem Research and Development. (Eds) N. S. Randhawa and B. S. Parmar, *Publication No 3*. Society of Pesticide Science, India, 270–283.
12. Ogbuewu, I. P., Odoemenam, U. V., Obikaonu, H. O., Opara, M. N., Emenalom, O. O., Uchegbu, M. C., Okoli, I. C., Esonu, B. O., & Iloeje, M. U. (2011). The growing importance of neem (*Azadirachta indica* A. Juss) in agriculture, industry, medicine and environment: a review. *Res. J. Med. Plant*, *5(3)*, 230–245.
13. Kaura, S. K., Gupta, S. K., & Chowdhury, J. B. (1998). Morphological and oil content variation in seeds of *Azadirachta indica* A. Juss. (Neem) from northern and western provenances of India. *Plant. Food Hum. Nutr.*, 52 (4), 293–298.
14. Jadeja, G. C., Maheshwari, R. C., & Naik, S. N. (2011). Extraction of natural insecticide azadirachtin from neem (*Azadirachta indica* A. Juss) seed kernels using pressurized hot solvent. *J Supercrit Fluids*, *56(3)*, 253–258.
15. Ogunniyi, D. S. (2006). Castor oil: a vital industrial raw material. *Bioresour. Technol.* 97, 1086–1091.
16. AOAC. (1984). *Official Methods of Analysis*, 14th ed. Washington, DC: Association of Official Analytical Chemists.
17. Jules, J., & Paull, R. E. (2008). *The Encyclopaedia of Fruit and Nuts*. CABI Publishing Series, 71–73.
18. Sayed, G. H., Ghuiba, F. M., Abdou, M. I., Badr, E. A., Tawfik, S. M., & Negm, N. A. (2012). Synthesis, surface, thermodynamic properties of some biodegradable vanillin-modified polyoxyethylene surfactants. *J. Surf. Deterg.* 15, 735–743
19. Nabel, A., El-Tabl, N. A. S, Ismail, A., Zakareya, A. K., & Moustafa, A. H. (2013). Synthesis, characterization, biodegradation and evaluation of the surface active properties of nonionic surfactants derived from Jatropha oil. *J. Surfact. Deterg*, 16, 857–863.
20. Adachi, S., & Kobayashi, T. (2005). Synthesis of esters by immobilized-lipase catalyzed condensation reaction of sugars and fatty acids in water-miscible organic solvent. *J. Biosci. Bioeng.*, 99, 87–94.
21. Kobayashi, T. (2011). Lipase-catalyzed syntheses of sugar esters in non-aqueous media. *Biotechnol. Lett.*,33, 1911–1919.
22. Cao, L., Bornscheuer, U. T., & Schimid, R. D. (1996). Lipase-catalyzed solid phase synthesis of sugar esters. *Fett Lipid*, 98, 332–335.
23. Otto, R. T., Bornscheuer, U. T., Syldatk, C., & Schmid, R. D. (1998). Lipase-catalyzed synthesis of aryl aliphatic esters of -D(+)-glucose, n-alkyl and aryl glucosides and characterization of their surfactant properties. *J. Biotechnol.*, 64, 231–237.

24. Laane, C., Boeren, S., Vos, K., & Veeger, C. (1987). Rules for optimization of biocatalysis in organic solvents. *Biotechnol. Bioeng.*, *30*, 81–87.

25. Yang, Z., & Huang, Z. L. (2012). Enzymatic synthesis of sugar fatty acid esters in ionic liquids. *Catal. Sci. Technol.*, *2*, 1767–1775.

26. Katsoura, M. H., Katapodis, P., Kolisis, F. N., & Stamatis, H. (2007). Effect of different reaction parameters on the lipase-catalyzed selective acylation of polyhydroxylated natural compounds in ionic liquids. *Process Biochem.*, *42*, 1326–1334.

27. Rahman, M. B. A., Arumugam, M., Khairuddin, N. S. K., Abdulmalek, E., Basri, M., & Salleh, A. (2012). Microwave assisted enzymatic synthesis of fatty acid sugar ester in ionic liquid-tert-butanol biphasic solvent system. *Asian J. Chem.*, *24*, 5058–5062.

28. Galonde, N., Brostaux, Y., Richard, G., Nott, K., Jerôme, C., & Fauconnier, C. (2013). Use of response surface methodology for the optimization of the lipase-catalyzed synthesis of mannosyl myristate in pure ionic liquid. *Process Biochem.*, *48*, 1914–1920.

29. Fischer, F., Happe, M., Emery, J., Fornage, A., & Schütz, R. (2013). Enzymatic synthesis of 6- and 6'-O-linoleyl-α-d-maltose: From solvent-free to binary ionic liquid reaction media. *J. Mol. Catal. B. Enzym.*, *90*, 98–106.

30. Mai, N. L., Ahn, K., Bae, S. W., Shin, D. W., Morya, K., & Koo, Y. M. (2014). Ionic liquids as novel solvents for the synthesis of sugar fatty acid ester. *Biotechnol. J.*, *9*, 1–8.

31. Findrik, Z., Megyeri, G., Gubicza, L., Bélafi-Bakó, K., Nemestóthy, N., & Sudar, M. (2016). Lipase catalyzed synthesis of glucose palmitate in ionic liquid. *J. Clean. Prod.*, *112*, 1106–1111.

32. Yan, Y., Bornscheuer, U. T., Stadler, G., Lutz-Wahl, S., Reuss, M., & Schmid, R. D. (2001). Production of sugar fatty acid esters by enzymatic esterification in a stirred-tank membrane reactor: optimization of parameters by response surface methodology. *J. Am. Oil Chem. Soc.*, *78*, 147–153.

33. Yan, Y. (2001). Enzymatic production of sugar fatty acid esters. Germany: University of Stuttgart; Ph.D. Thesis.

34. Soultani, S., Engasser, J. M., & Ghoul, M. (2001). Effect of acyl donor chain length and sugar/acyl donor molar ratio on enzymatic synthesis of fatty acid fructose esters. *J. Mol. Catal. B Enzym.*, *11*, 725–731.

35. Humeau, C., Girardin, M., Rovel, B., & Miclo, A. (1998). Effect of the thermodynamic water activity and the reaction medium hydrophobicity on the enzymatic synthesis of ascorbyl palmitate. *J. Biotechnol.*, *63*, 1–8.

36. Cao, L., Bornscheuer, U. T., & Schmid, R. D. (1999). Lipase-catalyzed solid-phase synthesis of sugar esters. Influence of immobilization on productivity and stability of the enzyme. *J. Mol. Catal. B. Enzym.*, *6*, 279–285.

37. Berger, M., Laumen, K., & Schneider, M. P. (1992). Enzymatic esterification of glycerol .1. lipase-catalyzed synthesis of regioisomerically pure 1,3-sn-diacylglycerols. *J. Am. Oil Chem. Soc.*, *69(10)*, 955–960.

38. Tarahomjoo, S., & Alemzadeh, I. (2003). Surfactant production by an enzymatic method. *Enzyme Microb. Technol.*, *33*, 33–37.

39. Klibanov, A. M. (1990). Asymmetric transformations catalysed by enzymes in organic solvents. *Acc. Chem. Res.*, *23*, 114–120.

40. Sakurai, T., Margolin, A. L., Russell, A. J., Klibanov, A. M. (1998). Control of enzyme enantioselectivity by the reaction medium. *J. Am. Chem. Soc.*, *110*, 7236–7237.

41. Wescott, C.R., & Klibanov, A. M. (1994). The solvent dependence of enzyme specificity. *Biochem. Biophys. Acta*, *1206*, 1–9.

42. Liu, Y., Wang, F., & Tan, T. (2009). Effects of alcohol and solvent on the performance of lipase from *Candida* sp. in enantioselective esterification of racemic ibuprofen. *J. Mol. Catal. B. Enzym.*, *56*, 126–130.

43. Salem, J. H., Humeau, C., Chevalot, I., Harscoat-Schiavo, C., Vanderesse, R., & Blanchard, F. (2010). Effect of acyl donor chain length on isoquercitrin acylation and biological activities of corresponding esters. *Process Biochem.*, *45*, 382–389.

44. Hudson, E. P., Eppler, R. K., & Clark, D. S. (2005). Biocatalysis in semi-aqueous and nearly anhydrous conditions. *Curr. Opin. Biotechnol.*, *16*, 637–643.

45. Rubio, E., Fernandez-Mayorales, A., & Klibanov, A. M. (1991). Effects of solvents on enzyme regioselectivity. *J. Am. Chem. Soc.*, *113*, 695–696.

46. Lortie, R. (1997). Enzyme catalyzed esterification. *Biotechnol. Adv.*, *15*, 1–15.

47. Paula, A. V., Barboza, J. C., & Castro, H. F. (2005). Study of the influence of solvent, carbohydrate and fatty acid in the enzymatic synthesis of sugar esters by lipases. *Quimica Nova*, *28*, 792–796.

48. Affleck, R., Haynes, C. A., & Clark, D. S. (1992). Solvent dielectric effects on protein dynamics. *Proc. Natl. Acad. Sci. U.S.A.*, *89*, 167–170.

49. Affleck, R., Xu, Z. F., Suzawa, V., Focht, K., Clark, D. S., & Dordick, J. S. (1992). Enzymatic catalysis and dynamics in low water environments. *Proc. Natl. Acad. Sci. U.S.A.*, *89*, 1100–1104.

50. Akkara, J. A. (1999). Enzymatic synthesis and modification of polymers in nonaqueous solvents. *Trends Biotechnol.*, *17*, 67–73.

51. Degn, P., & Zimmermann, W. (2001). Optimization of carbohydrate fatty acid ester synthesis in organic media by a lipase from *Candida antarctica*. *Biotechnol. Bioeng.*, *74*, 483–491.

52. Oosterom, W. M., van Rantwijk, F., & Sheldon, R. A. (1996). Regioselective acylation of disaccharides in t-butyl alcohol catalysed by *Candida antarctica* lipase. *Biotechnol. Bioeng.*, *49*, 328–333.

53. Akoh, C. C., & Mutua, L. M. (1994). Synthesis of alkyl glucoside fatty acid esters: effect of reaction parameters and the incorporation of n-3 polyunsaturated fatty acids. *Enzyme Microb. Technol.*, *16*, 115–119.

54. Jia, C., Zhao, J., Feng, B., Zhang, X., & Xia, W. (2010). A simple approach for the selective enzymatic synthesis of dilauroyl maltose in organic media. *J. Mol. Catal. B. Enzym.*, *62*, 265–269.

55. Adnani, A., Basri, M., Chaibakhsh, N., Salleh, A., & Rahman, M. (2011). Lipase-catalyzed synthesis of a sugar alcohol-based nonionic surfactant. *Asian J. Chem.*, *23*, 388–392.

56. Chen, Z., Zong, M., & Gu, Z. (2007). Enzymatic synthesis of sugar esters in ionic liquids. *Chin. J. Org. Chem.*, *27*, 1448–1452.

57. Fan, Y., & Qian, J. (2010). Lipase catalysis in ionic liquids/supercritical carbon dioxide and its applications. *J. Mol. Catal. B: Enzym.*, *66*, 1–7.

58. Ganske, F., & Bornscheuer, U. T. (2005). Lipase-catalyzed glucose fatty acid ester synthesis in ionic liquids. *Org. Lett.*, *7*, 3097–3098.

59. Lee, S. H., Ha, S. H., Hiep, N. M., Chang, W. J., & Koo, Y. M. (2008). Lipase-catalyzed synthesis of glucose fatty acid ester using ionic liquids mixtures. *J. Biotechnol.*, *133*, 486–489.

60. Park, S., & Kazlauskas, R. J. (2001). Improved preparation and use of room-temperature ionic liquids in lipase-catalyzed enantio-and regioselective acylations. *J. Org. Chem.*, *66*, 8395–8401.

61. Lee, S. H., Nguyen, H. M., Koo, Y. M., & Ha, S. H. (2008). Ultrasound-enhanced lipase activity in the synthesis of sugar ester using ionic liquids. *Process Biochem.*, *43*, 1009–1012.

62. Moniruzzaman, M., Nakashima, K., Kamiya, N., & Goto, M. (2010). Recent advances of enzymatic reactions in ionic liquids. *Biochem. Eng. J.*, *48*, 295–314.

63. Sharma, R., Chisti, Y., & Banerjee, U. C. (2001). Production, purification, characterization, and applications of lipases. *Biotechnol. Adv.*, *19*, 627–662.

64. Secundo, F., Carrea, G., Tarabiono, C., Gatti-Lafranconi, P., Brocca, S., & Lotti, M. (2006). The lid is a structural and functional determinant of lipase activity and selectivity. *J. Mol. Catal. B. Enzym.*, *39*, 166–170.

65. Saifuddin, N., & Raziah, A. Z. (2008). Enhancement of lipase enzyme activity in nonaqueous media through rapid three phase partitioning and microwave irradiation. *e-J. Chem. (Online)*, *5*, 864–871.

66. Tejo, B. A., Salleh, A. B., & Pleiss, J. (2004). Structure and dynamics of *Candida rugosa* lipase: the role of organic solvent. *J. Mol. Model.*, *10*, 358–366.

67. Cajal, Y., Svendsen, A., Girona, V., Patkar, S. A., Alsina, M. A. (2000). Interfacial control of lid opening in *Thermomyces lanuginosa* lipase. *Biochemistry*, *39*, 413–423.

68. Tsuzuki, W., Kitamura, Y., Suzuki, T., & Kobayashi, S. (1999). Synthesis of sugar fatty acid esters by modified lipase. *Biotechnol. Bioeng.*, *64*, 267–271.

69. Lee, S. B., & Kim, K. J. (1995). Effect of water activity on enzyme hydration and enzyme reaction rate in organic solvents. *J. Ferment. Bioeng.*, *79*, 473–478.

70. Cabrera, Z., Gutarra, M. L. E., Guisan, J. M., & Palomo, J. M. (2010). Highly enantioselective biocatalysts by coating immobilized lipases with polyethyleneimine. *Catal. Commun.*, *11*, 964–967.

71. Sabeder, S., Habulin, M., & Knez, Z. (2006). Lipase-catalyzed synthesis of fatty acid fructose esters. *J. Food Eng.*, *77*, 880–886.

72. Contesini, F. J., Lopes, D. B., Macedo, G. A., Nascimento, M. G., & Carvalho, P. O. (2010). *Aspergillus* sp. lipase: potential biocatalyst for industrial use. *J. Mol. Catal. B: Enzym.*, *67*, 163–171.

73. Guncheva, M., & Zhiryakova, D. (2011). Catalytic properties and potential applications of Bacillus lipases. *J. Mol. Catal. B: Enzym.*, *68*, 1–21.

74. Krishna, S. H., & Karanth, N. G. (2002). Lipases and lipase-catalyzed esterification reactions in non-aqueous media. *Catal. Rev.*, *44*, 499–591.

75. Rodrigues, R. C., & Fernandez-Lafuente, R. (2010). Lipase from *Rhizomucor miehei* as an industrial biocatalyst in chemical process. *J. Mol. Catal. B. Enzym.*, *64*, 1–22.

76. Yoshida, Y., Kimura, Y., Kadota, M., Tsuno, T., & Adachi, S. (2006). Continuous synthesis of alkyl ferulate by immobilized *Candida antarctica* lipase at high temperature. *Biotechnol. Lett.*, *28(18)*, 1471–1474.

77. Oguntimein, G. B., Erdmann, H., & Schmid, R. D. (1993). Lipase catalyzed synthesis of sugar ester in organic solvents. *Biotechnol. Lett.*, *15*,175–180.

78. Schlotterbeck, A., Lang, S., Wray, V., & Wagner, F. (1993). Lipase-catalyzed monoacylation of fructose. *Biotechnol. Lett.*, *15*, 61–4.

79. Ljunger, G., Adlercreutz, P., Mattiasson, B. (1994). Enzymatic synthesis of octyl-[beta]-glucoside in octanol at controlled water activity. *Enzyme Microb. Technol.*, *16*, 751–755.

80. Yu, J., Zhang, J., Zhao, A., Ma, X. (2008). Study of glucose ester synthesis by immobilized lipase from *Candida* sp. *Catal. Commun.*, *9*, 1369–1374.

81. Soo, E., Salleh, A. B., Basri, M., Rahman, R. A., & Kamaruddin, K. (2003). Optimization of the enzyme-catalyzed synthesis of amino acid-based surfactants from palm oil fractions *J. Biosci. Bioeng.*, *95(4)*, 361–367.

82. Montet, D., Servat, F., Pina, M., Graille, J., Galzy, P., Arnaud, A., Ledon, H., & Marcou, L. (1990). Enzymatic synthesis of *N-ℰ*-acyllysines. *J. Am. Oil Chem. Soc.*, *67*, 771–774.

83. Yankah, V. V., & Akoh, C. C. (2000). Lipase-catalyzed acidolysis of tristearin with oleic or caprylic acids to produce structured lipids. *J. Am. Oil Chem. Soc.*, *77*, 495–500.

84. Hailing, P. J. (1994). Thermodynamic predictions for biocatalysis in non-conventional media: theory, tests and recommendations for experimental design and analysis. *Enzyme Microb. Technol.*, *16*, 178–206.

85. Valivety, R., Jauregi, P., Gig, I., & Vulfson, E. (1997). Chemoenzymatic synthesis of amino acid-based surfactants. *J. Am. Oil Chem. Soc.*, *74*, 879–886.

86. Maugard, T., Remaud-Simeon, M., & Monsan, P. (1998). Kinetic study of chemoselective acylation of amino-alditol by immobilized lipase in organic solvent: effect of substrate ionization. *Biochim. Biophys. Acta*, *1387*, 177–183.

87. Porter, M. R. (1994). Handbook of Surfactants. Blackie Academic and Professional, Glasgow.

88. Zhang, Q., Cai, B. X., Xu, W. J., Gang, H. Z., Liu, J. F., Yang, S. Z., & Mu, B. Z. (2015). The rebirth of waste cooking oil to novel bio-based surfactants. *Sci. Rep.*, 5, 09971. DOI: 10.1038/srep09971

89. Azman, N. A. B. (2015). Biosurfactant Synthesis from Waste Cooking Oil, B.Tech Thesis.

3 Protein-Based Surfactants

Mayuri Kumari[1], *Divya Bajpai Tripathy*[1],
Anjali Gupta[1] *and Anurag Singh*[2]

[1]School of Basic and Applied Sciences, Galgotias
University, Greater Noida, Uttar Pradesh, India

[2]Department of Food Science and Technology, National Institute of Food
Technology Entrepreneurship and Management, Sonipat, Haryana, India

CONTENTS

DOI: 10.1201/9781003144878-3

3.1 INTRODUCTION

Surfactants or surface-active agents are chemical compounds that have been continuously exploited on a large scale by humans as soaps, cosmetics, drugs, dispersants, corrosion inhibitors and many more. Vast production and use of surfactants raise the concern of industrialists as well as environmentalists towards its biodegradability, nontoxicity, mildness, production cost and recyclability. In this path of exploration of a suitable candidate, protein-based surfactants (PBS) gathered the huge interest of researchers as they are not only based on renewable raw materials like amino acids (AAs) and their polypeptides that make them environment friendly but also their easy availability and less or nontoxicity makes them multifunctional as well as cost-effective especially for the field of microbiology and pharmaceuticals where the high cost, irritability and toxicity are the main challenges associated with conventional chemical-based surfactants.

Proteins are the linear or branched polypeptide chains formed via peptide linkage in between AAs. Protein molecules are found in simple linear (primary), helical or β sheets (secondary) to complex tertiary and quaternary complex structures. The complex 3-D structure formed due to the crosslinking as well as weak hydrogen bonding between peptides. This structural diversity along with terminal and side functional groups justified proteins as a potential raw material for surfactants synthesis. Also, a wide variety of available AA (20 naturally occurring AA) allows to greatly alter the properties of surfactants by altering AAs.

PBS are made up of a polypeptide chain of AA. Many of these molecules are amphiphilic as they have a long alkyl chain linked to the central carbon atom and an amino or a peptide as the hydrophilic part. Two active functional groups of AAs/protein molecules are responsible for imparting activity to the molecule and the ionic character of the molecule has been decided by the pH of the environment in which the protein molecule is present [1]. The hydrophobic chain may or may not have acyl, ester, amide, alkyl, or ether linkage (Figure 3.1).

The activity of surfactant molecules is generally associated with smaller molecules instead of biomacromolecules such as proteins. Many of the protein molecules alone cannot be directly used as surfactants and require certain structural modifications to use them as surfactants [2]. Animal-based (such as leather waste) and plant-based proteins are being looked for as potential raw materials for the condensation reaction of protein hydrolysate with fatty acids or fatty acids derivatives Cl (chloride) to produce PBS. Very few proteins are also claimed to have a good intrinsic surface-active property and can be naturally employed as surfactants, i.e. pulmonary surfactants [3].

Certain hydrophobic modifications in protein molecules either by a chemical process or enzymatic process can enhance their surface activity. AAs may be used as starting material, dipeptide or polypeptide in such processes, depending on the type of product required. The hydrophobic moiety can be introduced chemically through acylation or enzymatically using biotechnological processes.

FIGURE 3.1 Structure of protein.

3.2 HISTORY OF PROTEIN-BASED SURFACTANTS

Proteins are usually being used as surfactants for home care and personal care products for the last 70 years [4]. The use of PBS has increased as these surfactants could be synthesized from renewable resources as well as they suggested a solution to the issue of disposal of waste produced as a result of by-products of animal and vegetable proteins [5]. Modification of proteins like soybean, casein, collagen, keratin or albumen to PBS has been firstly reported in 1937 when a US-based company (the Maywood Chemical Company) first became acquainted with commercial PBS. These surfactants were synthesized via condensation of hydrolyzed proteins with fatty acids [4]. Some commonly known trade names of commercially available PBS were Crotein, Polypeptide, Magpon, Lexein, Protolate, Super Pro and Sol-U-Twins.

Bondi et al. in the early 19th century reported that *N*-acyl-glycine and *N*-acyl-alanine were used for the very first time as hydrophilic moieties of the surfactant for the synthesis of the AA-based surfactants (AASs). Successive work encompassed the formation of lipoamino acid by exploiting glycine and alanine [6]. Hentrich et al. patented an AASs series. They also filed their first patent regarding the applications of acyl-aspartate/sarcosinate-based AASs in domestic goods like detergents, toothpaste and shampoos [7].

Afterwards, several scientists worked on the synthesizing processes, physical and chemical properties and commercial applications of acyl AAs [8–10]. Since now, an important figure of research journals has been issued on the production of surface-active properties, commercial applications, nontoxicity and biodegradability of PBS [11–17].

3.3 TYPES OF PROTEIN-BASED SURFACTANTS

PBS are broadly categorized as AASs and peptide surfactants based on the structure of the hydrophilic group of PBS whether it has peptide chain as active hydrophilic moiety as in peptide surfactants or single amino group (AAS) [18] (Figure 3.2).

3.3.1 AMINO ACID SURFACTANTS (AASs)

This kind of PBS can usually be made using animal- as well as agricultural-derived feedstocks. In the early 20th century, AAs attained the main interest of researchers as a substrate for surface-active agents. The main reasons for attraction towards AASs involved bio-based synthesis,

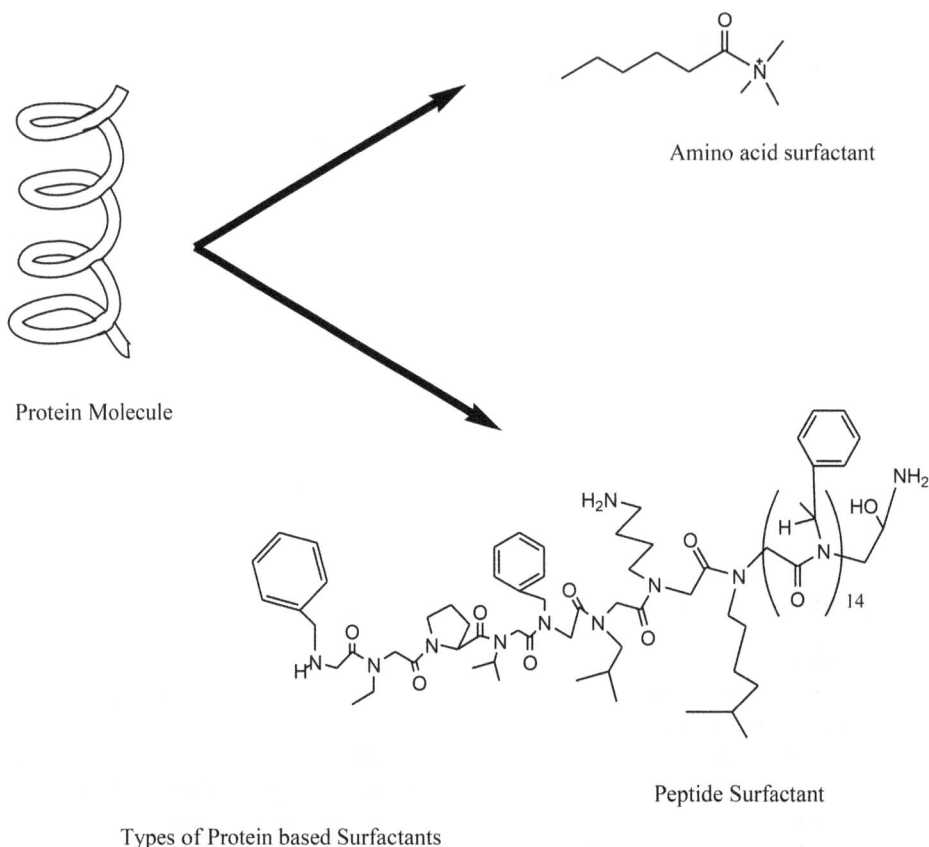

Amino acid surfactant

Protein Molecule

Peptide Surfactant

Types of Protein based Surfactants

FIGURE 3.2 Protein-based surfactant types.

nontoxicity, easy degradability and their eco-friendly behavior [18]. AASs can be explained such as surfactant groups comprising AA groups. AAs are the type of molecule that has two differently charged functional groups that increase the opportunity for deriving a vast and varied range of surfactants. To date, standard protein genic AAs available in nature are 20 in the count, which differ from each other only based on alkyl moiety. Alkyl groups also vary greatly on account of their charges and polarity [19].

Nontoxicity and easy biodegradability of AASs make them greater alternatives to their predictable equivalents in preservatives pharmaceutical compositions and cosmetic formulations [20]. These surfactants were also found to be bioactive against a big range of bad bacteria, viruses as well as in tumors. Recently, some AAs are found to be easily produced at a commercial scale through yeasts, therefore extenuating their manufacture to be more biodegradable [20]. Furthermore, AASs are categorized into different subclasses as shown in Figure 3.3.

3.3.2 PEPTIDE SURFACTANTS

Peptide surfactants are the class of PBS where the hydrophilic moiety is not a single AA type but these have been synthesized through condensation reaction between dipeptides or tripeptides. Like AAS, the hydrophobic moiety of peptide surfactants is from medium to long alkyl chains originated from the fatty source. Many researchers revealed the chemical synthesis of these types of surfactants, whereas few reports on their biosynthesis have also been reported. Many useful reports on peptide surfactants have appeared in the literature.

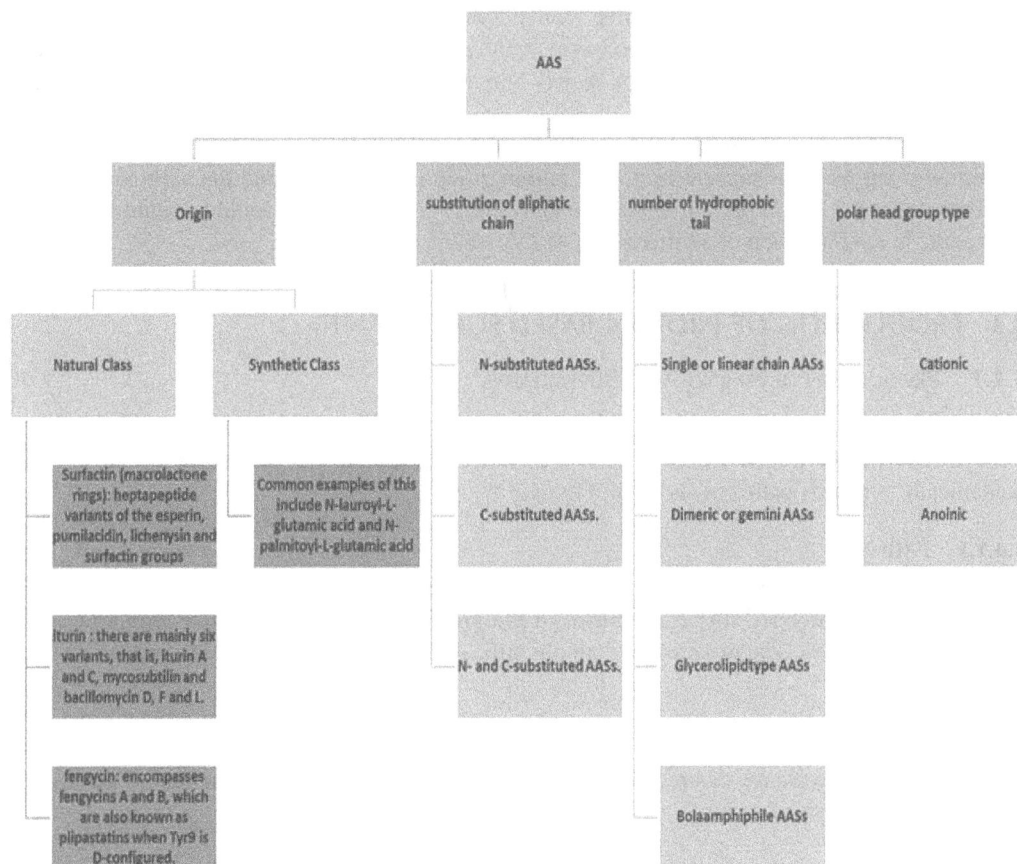

FIGURE 3.3   Brief categorization of AASs.

Mhaskar and Lakshminarayana reported the synthesis of lauroyl-dipeptides using diethanol-amides (DEA). The synthesis involved the coupling of five AAs followed by the condensation DEA in the presence of NaOMe [21]. Another recent example of peptide surfactant was reported by Infante et al. in which the researcher revealed the synthesis of acidic and basic Nα-lauroyl argi-nine dipeptides. Peptide chain incorporated in peptide surfactants may either be homologous or heterologous [22]. Cationic peptide surfactants were found to have remarkable antimicrobial prop-erties. That can be described as revealing α-helical structure or containing β-sheet elements with an α-helical domain. Rest are rich in proline or histidine residue [22].

3.3.3   CHEMISTRY OF PROTEIN-BASED SURFACTANTS

Proteins are the polymerized peptides or polypeptides, in which 20 amino acids (AAs) are linked together via a peptide linkage. AAs are the molecules that are characterized by the presence of active functional groups such as amino (–NH$_2$) group and carboxylic acid (–COOH) group linked to the common tetrahedral (α-C) atom [23]. It varies with the R-groups linked to the α-carbon (the only glycine has hydrogen in place of the R-group). R-groups are different from one another in terms of acidity/basicity, size and chemical structure. These alterations in the R-group decide the polarity of AAs.

The four different substituents on α-carbon atom imparts chirality and optical activity to the molecules of AAs (excluding glycine). Proteins contain PBs as building blocks and are also chiral in tryptophan, tyrosine, and phenylalanine. The presence of aromatic R-groups makes them able to absorb UV light with an absorbance maximum at 280 nm. Active acidic α-COOH and basic α-NH$_2$ groups impart ionic character to the AA moiety and make them able to produce ionic equilibrium as follows.

PBs comprise at least 2 weakly acidic groups, yet the (–COOH) group is much more acidic than its amino group which is present in it. The amino group is protonated and the carboxyl group is unprotonated at pH 7.4. At pH 7.4, there is no ionizable R-group AA that would be neutral in charge and results in the formation of zwitterions [24].

3.4 PRODUCTION OF PROTEIN-BASED SURFACTANTS

3.4.1 PRODUCTION OF AMINO ACID SURFACTANTS

AASs may contain the hydrophobic group each at amine moiety/carboxylic acid moiety/attached through the side chain of the PBs. Based on this method, researchers can adopt any of the four fundamental synthesis paths, as shown in Figure 3.4.

3.4.1.1 Path 1
Path 1 shows manifest reflux of alcohol and fatty acids in the existence of dehydrating agents and acid catalysts that produce surfactants through the process of amphiphilic ester amine by using esterification reactions. The processing of amphiphilic ester amine is done by using an esterification reaction.

3.4.1.2 Path 2
Path 2 conveys to create the desired amide bond; alkylamine and AA react and hence synthesis of amphiphilic amidoamine occurs.

3.4.1.3 Path 3
Path 3 reveals that the amine group of PBs reacts with fatty acid to form AA.

3.4.1.4 Path 4
Path 4 reveals the formation of long-chain alkyl PBs when the amine group reacts with alkyl halogen.

3.4.1.5 Path 5
Path 5 indicates that anhydride is formed when a (–COOH) group of aspartic and glutamic acids reacts with a fatty alcohol; it comprises the coupling of a specific function of the side group of the PB.

3.4.2 SYNTHESIS OF PBs

3.4.2.1 Formation of Peptide Surfactants or Single-Chain PBs
Peptide or PBs (N-acyl and O-acyl) can be synthesized by enzyme-catalyzed acylation of (–NH$_2$)/ (–OH) by fatty acids. Solvent-free lipase-catalyzed synthesis of methyl ester derivatives and PB amide by *Candida antarctica* has also been reported by various researchers. The outcome was in the 25–90% which was dependent on the nature of PB [25]. In certain reactions, butanone (C$_4$H$_8$O) was also used as a solvent.

The N-acylation is either of protein hydrolysates, AAs or their derivatives using binary solvents (for example dimethylformamide-water or butyl methyl ketone), catalyzed by lipase or proteases. Very low yield is usually the main problem with the enzymatic production of PBSs suggested by Vonderhagen et al. [26].

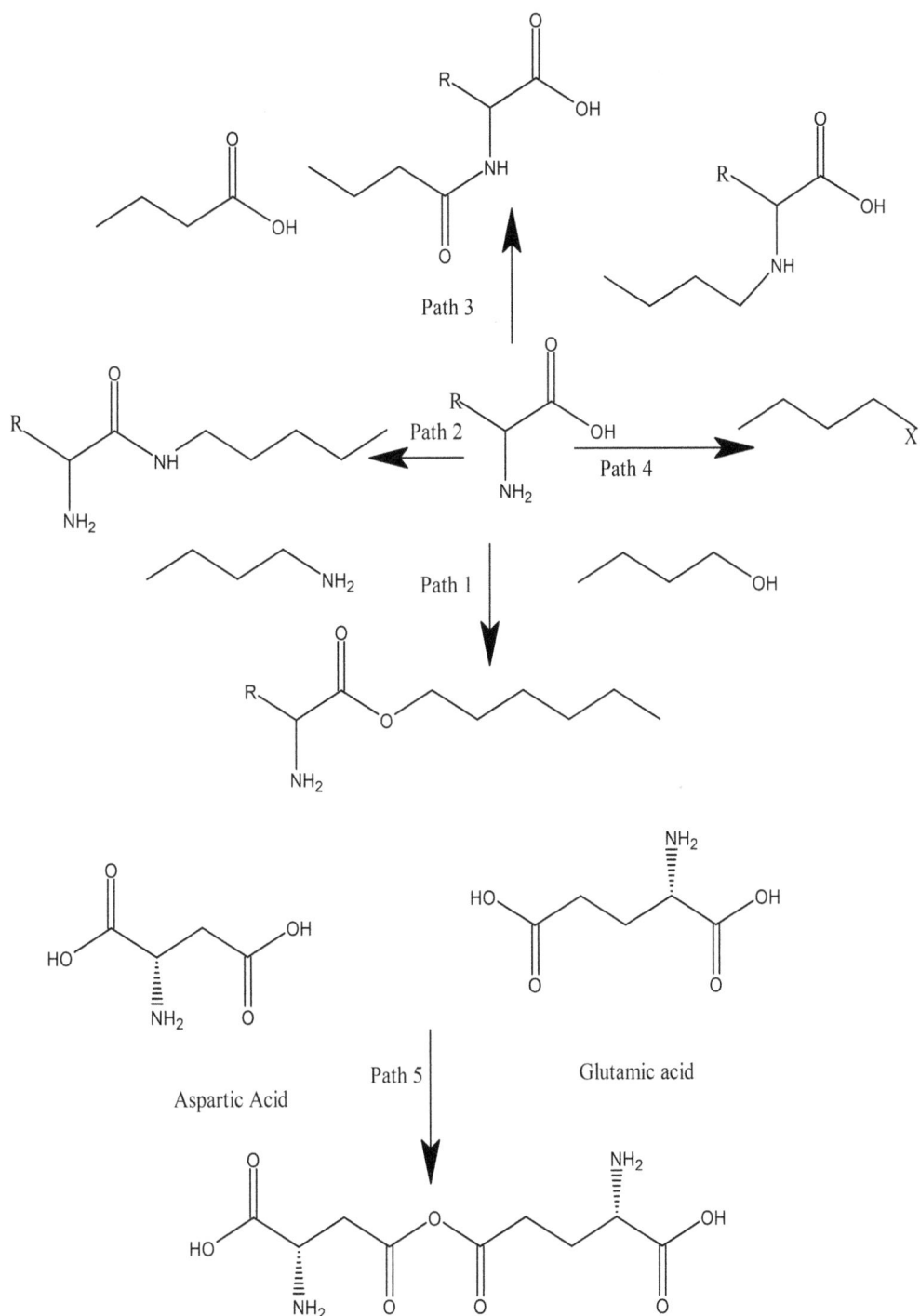

FIGURE 3.4 Fundamental production paths of (AASs).

N-Myristol PB derivatives yield 2–10% even though using different lipases for several days in incubation at 70°C. They also raised a similar problem during the synthesis of *N*-cbz-lysine methyl ester derivatives. In this research, 80% of 3-*O*-myristol-L-serine was claimed under a solvent-free environment, while using *N*-protected serine as a substrate and novo-enzyme as a catalyst [27].

Montet et al. also reported issues regarding low yields of *N*-acyl-lysine PB during synthesis through vegetable oils and fatty acids. A maximum 19% product yield was obtained under solvent-free conditions with organic solvents [28].

In an aqueous buffered medium, the outcomes of *O*-acylation of L-tyrosine (LET), L-serine, L-threonine and L-homoserine attained by *Rhizopus delemar* and *Candida cylindracea* was analyzed using lipases by Nagao and Kito [29]. Up to a certain extent, they also claimed that acylation of L-serine and L-homoserine and was obtained in lesser quantity, while on LET and L-threonine, no acylation was observed. Inexpensive and accomplishable substrates are being used by scientists to process profitable PBSs.

Soo et al. proposed that palm oil–based surfactants gave the best results with immobilized lipoenzyme, though the palm oil fractions took 6 long days to react and produce these results [30].

The manufacturing and surfactant properties of various racemic mixtures chirality and optically active of PBSs based on palmitic acids such as phenylglycine, phenylalanine, threonine, leucine, proline and methionine was acknowledged by Gerova et al. [31].

The details of dicarboxylic acid–based and amino-based monomers undergoing the copolycondensation for the synthesis of a chain of biodegradable functional PB-based polyester amides in the solution are mentioned by Pang and Chu [32].

The esterification of R-COOH groups of Boc-Asp-OH and Boc-Ala-OH with long-linkage aliphatic alcohols and Sepharose 4B as a catalyst and diols using dichloromethane as a solvent. They obtained 63% of Boc-Ala-OH with fatty alcohols having 6 and 12 C atoms which were found better than 51% yield of Boc-Ala-OH with fatty alcohols having 16 C atoms announced by Cantacuzene or Guerreiro [33].

Clapes et al. announced that when Cbz-Arg-OMe reacts with various fatty alcohols and long-chain alkyl amines leads to the formation of ester and amide bonds between them, which in turn yielded derivatives of *N*-arginine alkyl amide (58–76%) in the existence of papain (from the latex of *Carica papaya*) with 99.9% purity [34].

3.4.2.2   Gemini PB/Protein Surfactants Synthesis

Two linear PB surfactant molecules connected through a spacer at polar/hydrophilic heads formed gemini surfactants [35]. The PB-based gemini surfactants were chemoenzymatically synthesized through any of the two possibilities (Schemes 3.1 and 3.2). Scheme 3.1 indicates the reaction between two PB derivatives and the spacer. The two hydrophobic groups are then included. Scheme 3.2 indicates that a bifunctional spacer chain directly links two linear structures [36].

Yoshimura et al. acknowledged various traits of gemini PB surfactants based on *N*-alkyl bromide and cystine, including their synthesis, adsorption and aggregation, the synthesized surfactants were also compared to their monomeric counterparts [35].

The details regarding the formation of anionic urea-based monomeric PBSs and their similar derivatives, e.g. D-cystine, L-cystine and DL-cystine, L-methionine and L-lysergic acid have also been claimed. Other descriptions regarding steady-state fluorescence spectroscopy, equilibrium surface tension and electrical conductivity techniques were also mentioned. Comparative studies showed that the gemini surfactants have lower critical micelle concentration (CMC) than the monomeric form of surfactants stated by Faustino et al. [37].

3.4.2.3   Glycerol-Lipid Amino Acid/Protein Surfactants Synthesis

Structural analogues of mono-, di-acyl glycerides and phospholipids belonging to a class of lipoamino PBs are formed by glycerol-lipid PB or peptide surfactants are formed by linking together 1 or 2 aliphatic links and one PB through ester linkage in the backbone of glycerol. These surfactants are

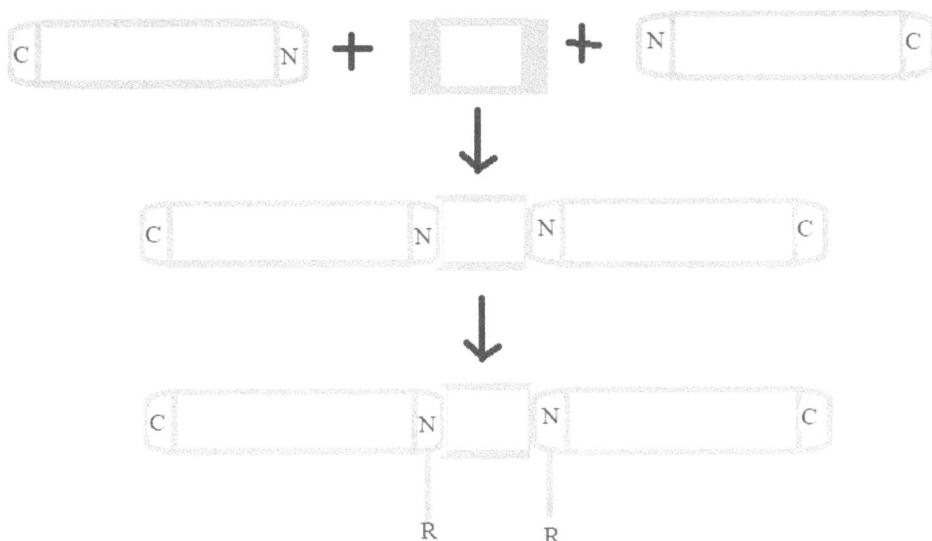

SCHEME 3.1 Synthesis of gemini AAS using AA derivatives and spacer, followed by the attachment of the hydrophobic group.

synthesized by first forming glyceryl esters of PB with an acid catalyst (BF_3) at increased temperature [27].

Scheme 3.3 is indicative of using hydrolases, proteases and lipases as a catalyst during enzymatic synthesis [38].

Conjugation of di-lauroylated arginine glyceride with papain during enzymatic processing is also observed. Studies on conjugates of di-acyl-glycerol-lipid synthesized from acetyl-arginine and their physical and chemical properties have been conducted [39–41].

3.4.2.4   Bola-Amphiphile Amino Acid/Protein Surfactants Synthesis

PB-based bola-amphiphiles contain two PBs linked to a hydrophobic chain. Franceschi and his co-authors explained the bola-amphiphiles synthesis with 2 PBs (D-/L-alanine) and an alkyl link of variable length and examined their properties of surface-active. They examined the production and mass of unique bola-amphiphiles containing PB moieties using alcohol or rare β-PBs and spacers with C 12–20. The rare β-PBs used may be an Aza-do-thymine, a sugar PB resulting in Aza-do-thymine-derived amino alcohol, PB, and a norbornene PB (Scheme 3.4) [42]. The production of

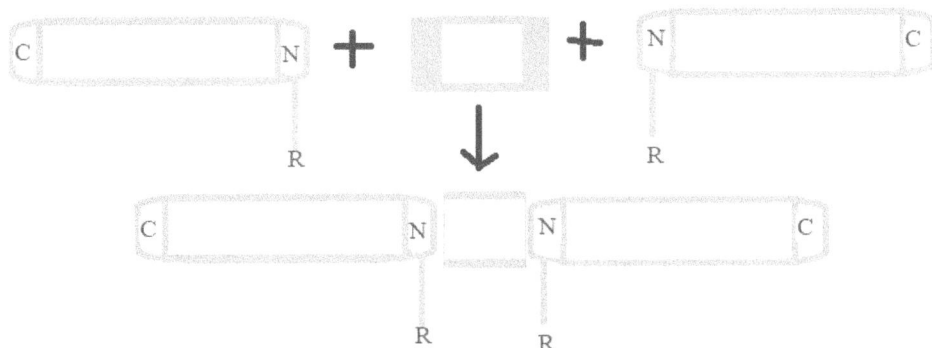

SCHEME 3.2 Synthesis of gemini AASs using bifunctional spacers and AAS.

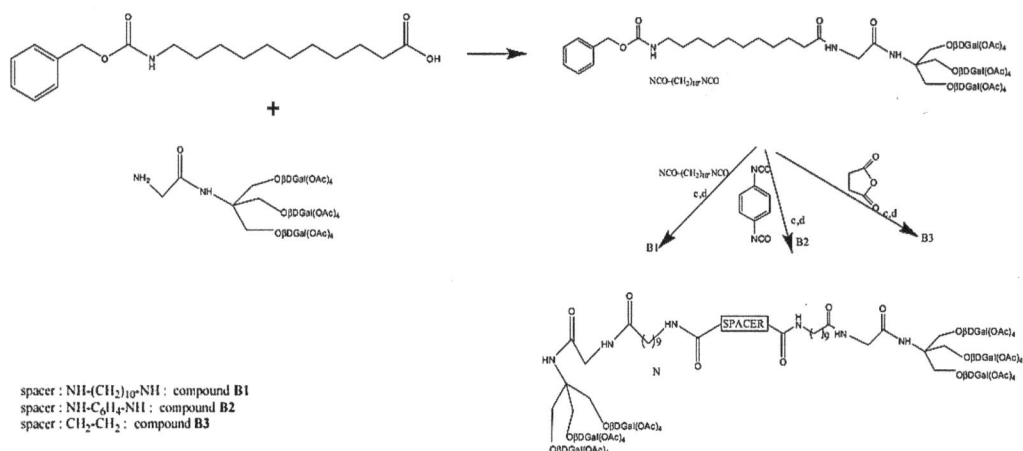

spacer : NH-(CH₂)₁₀-NH : compound **B1**
spacer : NH-C₆H₄-NH : compound **B2**
spacer : CH₂-CH₂ : compound **B3**

SCHEME 3.3   Glycerol-lipid amino acid/protein surfactants synthesis.

symmetric bola-amphiphiles resultant from tris(hydroxymethyl)aminomethane (Tris) (Scheme 3.4) has also been explained by Polidori et al. [43].

3.5   PHYSICOCHEMICAL PROPERTIES OF PBs

Surfactants derived from AAs have several advantageous properties due to which it has several applications, that is, it is a good solubilizer, good emulsifier, highly efficient, highly surface-active and good lime resistor (has tolerance for calcium).

Various scientists have analyzed other characteristics of AASs, for example Krafft temperature, phase behavior, surface tension and CMC, and inferred that the surface activity of AAs is superior to their conventional counterparts.

The characteristics of PBs are as follows:

3.5.1   Critical Micelle Concentration (CMC)

The CMC of surfactants is a vital occurrence and responsible for controlling various properties of surface active like–interaction with biological membranes, lytic action and solubilization. Elongation of chain length towards the hydrocarbon tail usually reduces the CMC values of surfactant solution and eventually enhances its surface action [44]. The value of CMC is lesser in PBSs as compared with conventional surfactants [40, 41].

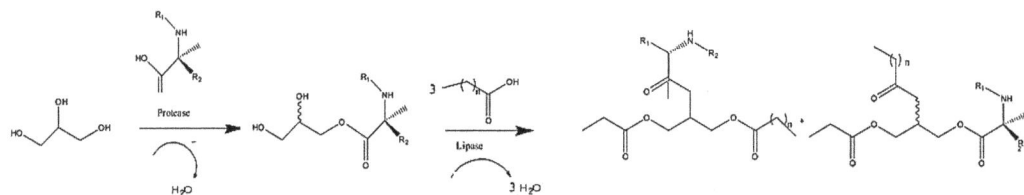

SCHEME 3.4   Scheme 4 Synthesis of symmetric bola-amphiphiles derived from tris(hydroxymethyl) aminomethane (Tris).

TABLE 3.1

CMC and Surface Tension of Nα-Acyl Arginine Surfactants

Properties	NAE	LAE	CAM	MAE
CME (mg L^{-1})	820 ± 50	410 ± 10	>1500	350 ± 30
γ (mN L^{-1})	26.1 ± 0.5	25.5 ± 0.5	27.0 ± 0.5	24.0 ± 0.5

CAE, N-α-octanoyl-L-arginine ethyl ester; LAE, ethyl-N-α-lauroyl-L-arginate; MAE, N-α-myristoyl-L-arginine ethyl ester; NAE, N-α-nonanoyl-L-arginine ethyl ester.

Infante et al. [27] stated the synthesis of three arginine-based PBSs by making variations in the hydrophilic head and hydrophobic tail groups. They considered their CMC and γ CMC values, and the consequences revealed a reduction in CMC and that increase in the values of γ CMC, which increases the length of a hydrophobic tail. The CMC reduces with increasing in C atoms in a hydrophobic tail of Nα-acyl arginine surfactants (Table 3.1) [45].

Yoshimura and co-authors [35] carried out an analysis of cysteine-derivative PBSs for its CMC value revealing that how the CMC value decreased on increasing the carbon length within the hydrophobic chain beyond 10 or 12, which again increased if the length was increased beyond 14 carbons. They also ensured that gemini surfactants have a low aggregation tendency.

Faustino et al. [37] announced that cystine-based anionic gemini surfactants' aqueous solution resulted in the development of mixed micelles. The gemini surfactants are also often compared with their monomeric counterparts, C8 Cys. The observation was made that lipid-surfactant mixtures had CMC values lower than the pure ones. A micelle which is a water soluble forming–phospholipid (1,2-diheptanoyl-sn-glycero-3-phosphocholine) and gemini surfactants have CMC in the mm range.

The aqueous medium of mixed type PBSs of both anionic and cationic nature results in the development of viscoelastic worm-shaped micelles when no salt was present which is considered by Shrestha and Aramaki [46]. It was observed during this study that N-dodecyl glutamic acid had a higher Krafft temperature. Also when L-lysine, an alkaline PB, was added to neutralize these surfactants, it leads to micelle formation and the solution showed Newtonian behavior at 25°C.

3.5.2 Water Solubility

The characteristics of the PBSs are water-soluble, which is caused by the existing additional CO–NH chain. It makes them eco-friendly and more readily biodegradable as compared to their conventional counterparts. The two carboxylic acids present in these surfactants make N-acyl-glutamic acid more soluble in water. The two ionic R-groups are present in one molecule of Cn(CA)$_2$, and increases solubility in water too, which results in more actual diffusion and adsorption on the cell interface and showing actual antimicrobial activity even at lesser amounts [23].

3.5.3 Krafft Temperature/Krafft Point

The rare solubility behavior of surfactants and rapid rise in their solubility overhead precise temperature is defined as Krafft temperature [47]. It is the minimum temperature at which micelle formation starts. Ionic surfactants synthesize solid hydrates through precipitation from the aqueous medium. The common observation will be that the solubility of the surfactants will go through a harsh and intermittent rise at a certain distinct/Krafft temperature [48].

The ionic surfactants usually exhibit this solubility feature that can be described depending on the low solubility of nondispersed surfactants below Krafft temperature which gradually increases until it approaches Krafft point. The surfactants should be prepared above the Krafft point to gain 100% solubility [49].

Various scientists have learnt about the Krafft temperature of PBSs which were contrasted with that of conventional synthetic surfactants. The Krafft temperature of arginine-based surfactants observed that premicellar aggregation CMC values more than $2–5 \times 10^{-6}$ M and later showed the normal micellization at $3–6 \times 10^{-4}$. N-Hexadecanoyl PBSs were varied in six types and the link between the Krafft temperature and PB residue was observed by Ohta et al. [50].

An experiment conducted showed that with the reduction in the PB residue size, the Krafft temperature of the N-hexadecanoyl PBSs elevated. However, this was not observed with phenylalanine. Furthermore, the size reduction of PB residues showed that the rise in the solution's enthalpy was endothermic. Glycine and phenylalanine were found exceptional. The inference was made from these investigations that the D–L interaction of N-hexadecanoyl PBS salts in the solid-state in both phenylalanine and alanine systems was superior to the L–L interaction.

Brito et al. [51] checked the three sequences of novel PBSs for their Krafft temperatures by differential scanning microcalorimetry. It was observed that the rise in Krafft temperature was comparatively higher (47–53°C, i.e., ≈6°C) due to the variation from trifluoroacetate to iodide. The Krafft temperature noticeably decreases if a cis- = chain and unsaturation in long-chain Ser derivative existed. For N-dodecyl glutamic acid, Krafft temperature was found to be more. Though micelle developed in the solution acted like a Newtonian fluid at 25°C when alkaline AA and L-lysine were added to neutralize.

3.5.4 SURFACE TENSION

Elongation of the bond length of the hydrophobic moiety provided more tension to the surfactants. The surface tension value of sodium cocoyl glycinate was detected by Zhang et al. [52] through the Wilhelmy plate method using a DCAT 11 tensiometer at 25 ± 0.2°C. The value of surface tension was calculated as s33 mN m^{-1} at the CMC of 0.21 mmol L^{-1}.

The surface tension values for certain 2C n Cys-type PBSs were detected by Yoshimura et al. [53]. It was detected that the value of surface tension at CMC diminished when the chain was elongated up to $n = 8$, while it was found increasing for surfactants having chain length up to $n = 12$ or above.

Certain studies were conducted to observe what impact is exerted on the surface tension of dicarboxylic PB type of surfactants by CaCl$_2$ [54]. Experiments were performed by adding CaCl$_2$ to the 3-dicarboxylic PBS (C 12 MalNa$_2$, AspNa$_2$ and GluNa$_2$). The result of the impact caused by Ca ions on the stuffing of surfactants of anionic nature at the interface of water and air media was observed that the lower quantity of CaCl$_2$ showed a lower value for surface tension in contrast with the beyond plateau value.

The surface tension was found to be nearly constant up to a 10-mm concentration of CaCl$_2$ of the salts of N-dodecyl-aspartic acid and amino-malonic acid. Above which, the surface tension was found to increase very fast as the surfactant formed calcium salt precipitate. The surface tension noticeably decreased on moderate addition of CaCl$_2$ which was constant for a wide range of CaCl$_2$ concentrations of the disodium salt of N-dodecyl glutamic acid.

Adsorption kinetics detected at the interface of water and air for gemini PBSs, the dynamic surface tension was calculated via a maximum bubble pressure method. The dynamic surface tension for 2C 12 Cys has observed no changes even after a long period. The number, length and concentration of the hydrocarbon tail are the only factors affecting the dynamic surface tension. When the surfactant was concentrated, and the length and numbers of the chain were reduced, it was seen to decay faster. It was observed that observations made at greater concentrations of C n Cys ($n = 8–12$) were quite near to the γ CMC acquired through Wilhelmy surface tension.

Some other investigations conducted informed that the dynamic surface tension of sodium di-dec-amino cysteine and sodium di-dec-lauroyl-cystine (SDLC) was detected through Wilhelmy plate technique and Khan DCA-315 Tensiometer, and the drop volume method determined the equilibrium surface tension of their aqueous solution. The reaction of the disulfide link by another process was also considered [55].

The surface tension considerably elevated from 34 to 53 mN m^{-1}, when 0.1-mmol L^{-1} SDLC solution and mercaptoethanol were mixed. 0.1-mM L^{-1} disulfide groups of SDLC get oxidized with 5-mmol L^{-1} NaClO$_3$ without the formation of aggregates in the solution, which is confirmed by the results obtained using electron transmission microscopy and differential light scattering techniques. After around 20 minutes, it was shown that the surface tension of SDLC elevated from 34 to 60 mN m^{-1}.

3.5.5 BINARY SURFACE INTERACTION

Important characteristics of cationic PBSs, for instance diacyl-glycerol-arginine–based surfactant and phospholipids at the interface of water and air, were claimed by many researchers. They inferred that such nonideal characteristics attribute to the common electrostatic interactions [56].

3.5.6 AGGREGATION PROPERTIES

The gemini surfactants and PB-based monomeric properties to form aggregates are identified by differential light scattering measurement techniques at concentrations above the CMC. The diffusion coefficient so calculated is changed into the visible hydrodynamic diameter D H = 2R H. C n Cys and 2C n Cys formed comparatively larger aggregates with a wide size distribution compared with other surfactants. Aggregates of about 10 nm are formed by non-2C 12 Cys surfactants. Gemini surfactants formed bigger micelles than their monomeric counterparts [37]. More is the length of the hydrocarbon chain; bigger is the size of micelles [57].

The property of mass formation of three different aqueous forms of stereoisomers of tetramethylammonium N-dodecyl-phenyl-alanyl-phenyl-alaninate was explained by Ohta et al. [58]. It disclosed the similar critical accumulation concentrations of diastereomers in aqueous solutions.

The development of optically active chiral aggregates of N-lauroyl-L-valine, N-lauroyl-L-glutamic acid, and its methyl ester in various solvents such as acetonitrile, tetrahydrofuran, 1,2-dichloroethane and 1,4-dioxane by calculating vapor pressure osmometry, nuclear magnetic resonance chemical shift of N–H proton, circular dichroism was analyzed by Iwahashi et al. [59].

3.5.7 INTERFACIAL ADSORPTION

Interfacial adsorption is also a characteristic of AASs that distinguish them from their synthetic conventional counterparts. l-tyrosine (LET) and l-phenylalanine (LEP) derived dodecyl esters of aromatic AAs were tested for their interfacial adsorption properties which helped to determine that LET and LEP have the lowest surface area at the hexane interface, followed by water and air.

The adsorption and solution behavior of 3-dicarboxylic PB-based surfactants, dodecyl aspartic acid and the disodium salts of dodecyl glutamic acid, and dodecyl amino-malonic acid with 3, 2 and 1 C atoms, separately, between the COOH groups at the interface of water and air was stated by Bordes et al. [54]. They stated that the value of CMC for dicarboxylic surfactants was four to five times more than mono-COOH groups containing dodecyl glycinate. It qualified the occurrence of H bonds between molecules over the amide groups in the dicarboxylic surfactants.

3.5.8 PHASE BEHAVIOR

The sealed ampoules having required amounts of reagents can be employed for the detection of surfactants' phase behavior. These reagents were agitated at high temperatures and allowed to homogenize using repeated centrifugation in a vortex mixer. Once properly agitated, the samples were stored for some days at 25°C in a thermostatically controlled water bath to ensure equilibrium. Normal and crossed polarizers are used to visualize the equilibrated phases. The isotropic discontinuous cubic phase was seen for the high concentration of surfactants. The discrete, small aggregates of positive curvature were triggered by a bigger hydrophilic head group of the surfactant molecule [60].

Marques et al. observed the phase behavior of 12 Lys12/12 and 8 Lys 8/16 Seer Systems and mentioned the phase separation for 12 Lys12/12 Seer Systems amid the micellar and vesicle solution. The coexistence of small micelles existed midst of the vesicles in the previous, while 8 Lys 8/16 Seer Systems exhibited change process continuously, which is the elongated micelles existed between the vesicles and small micelles [61].

3.5.9 EMULSIFICATION

The emulsification potency of L-glutamate, N-[3-lauryl-2-hydroxypropyl]-L-arginine, including other PBSs, for properties like viscosity, interfacial tension and disperse ability was analyzed by Kouchi et al. [62]. Synthesized surfactants were compared with their conventional amphoteric and nonionic counterparts. The inference was made that the emulsification potency of PBSs was more than conventional surfactants. Novel PB-based anionic surfactants were processed by Baczko et al. [63] and their utility as chiral oriented nuclear magnetic resonance spectroscopy solvents was also analyzed.

Wu and co-authors mentioned that by using pupa oil Na N-fatty acyl PBSs was synthesized [64]. The pupa protein hydrolysates were extracted as a waste material of the silk industry. The emulsifying power of Na N-fatty acyl PB and pupa protein hydrolysates in water/oil emulsion was observed. It was inferred that these surfactants possessed higher emulsifying properties with ethyl acetate than the oil phase over n-hexane.

3.5.10 LIME TOLERANCE

The endurance of the surfactant to precipitate such as lime soap when magnesium and calcium ions exist in hard water is known as lime tolerance. Surfactants having high tolerance against the hardness of water were effective in preparing personal care products and detergents. It can be detected by calculating the solubility change and surface activity in surfactants in the occurrence of ions of calcium [54].

It can also be detected by calculating the percentage of the number of a gram of the surfactant necessary for the dispersal of lime soap formed from 100 g sodium oleate in water [65]. In certain areas, mineral content, magnesium and calcium found in high concentrations cause issues in a few practical applications. In the making of anionic surfactants, sodium ions find their use as counterions. The ions of bivalent calcium get linked to two surfactant molecules causing easy precipitation of surfactants from solution and eventually reduce its detergent properties.

Lime tolerance properties of PBSs were studied, which revealed that lime and acid tolerance increases significantly when another carboxyl group is introduced and tends to rise farther when the spacer group length between 2 carboxyl groups is increased.

The lime and acid tolerance observed was to be lesser in C 12 glycinate when compared to C 12 glutamate and C 12 aspartate. Di-carboxylic-amide linkage surfactants were compared with di-carboxyl amino linkage and it was found that the pH range of surfactants with di-carboxyl amino linkage was broader and its surface activity was raised with the adequate addition of acid. Di-carboxyl N-alkyl PBs exhibit a chelating effect and C 12 aspartate forms a white gel when calcium ions exist. C 12 glutamate exhibits greater surfactant property when calcium ions are highly concentrated. They can be effectively used in the purification of seawater [66].

3.5.11 Dispersibility

The surfactants' capability is to stop the accumulation and resolving of dispersed surfactant molecules in the solution is called dispersibility. It is an essential property that makes surfactants suitable to be used as cosmetics, detergents and pharmaceutical products [67].

The dispersing agents should possess amide or amine bond, ester and ether within the hydrophilic and the hydrophobic groups and a straight-chain of a hydrophobic group [68]. The anionic surfactants such as the zwitterionic surfactants and sulfated alkanol-amide such as amido-sulfobetaine are usually useful as dispersing agents in lime soaps [69–71].

The dispersibility was detected through several studies where *N*-lauroyl lysine was poorly compatible with water. Cosmetic formulations were not prepared with ease. *N*-Acyl basic AA possesses high dispersion ability and is incorporated to be used in cosmetic industries for improving interpretations [72].

3.6 TOXICITY

Conventional surfactants, such as cationic surfactants, are found to be severely toxic for aquatic life (including algae, fish and mollusk). The ionic interaction of the conventional surfactant at the interface of cell and water tends to disrupt the integral membrane and is therefore responsible for their acute toxicity. The higher adsorption property of surfactants on the interfaces is due to its lowered CMC value, which enhances its acute toxicity. The acute toxicity of a surfactant is also enhanced due to elongation in the hydrophobic chain length. Most PBSs being less toxic or nontoxic to humans as well as the surroundings (including marine life) can be used as ingredients in cosmetics, food and medicines [72–78]. These are found to be mild and are nonirritant to the surface of the skin. [76]. Arginine-based surfactants are also not as lethal as compared with their conventional counterparts.

The physical, chemical and toxic characteristics of PB-based amphiphiles and their strong cationic vesicles derived from lysine (Lys), serine (Ser), hydroxyproline (Hyp) and tyrosine (Tyr) were acknowledged by Brito et al. [57]. The data of their acute toxicity was presented to Daphnia magna (IC 50). The cationic vesicles of do-decyl-trimethyl-ammonium-bromide/Lys-derivative mixtures were formed and analyzed for their eco-toxicity and hemolytic potential. They also mentioned that all PBSs and their vesicles containing mixtures exhibited lesser eco-toxicity as compared to the conventionally used surfactant DTAB (dodecyltrimethylammonium bromide).

The link between the stable cationic PB-based vesicles and DNA was acknowledged by Rosa et al. [79]. They suggested that the cationic surfactants being used conventionally are usually more toxic, while the interaction between the cationic amino PBSs was concluded to be harmless. The spontaneously stable vesicles are obtained from this arginine-based PBS with some anionic surfactants. PB-based corrosion inhibitors were also observed to be nontoxic [80–85]. Such surfactants were easily synthesized with high purity (=99%), inexpensive, easily biodegradable and were soluble in the liquid material. PBSs containing sulfur were more competitive as corrosion inhibitors [80, 86–90].

The desired toxicological properties of rhamnolipids were stated by Perinelli et al. [91] in their recent studies. Rhamnolipids enhance permeability. They also affect the epithelial absorbency of macromolecular drugs.

3.7 ANTIMICROBIAL ACTIVITY

The minimal inhibitory concentration of the surfactants best explains its antimicrobial activity [22]. A detailed study on the antimicrobial properties of surfactants based on arginine-AA was made [16, 20, 34, 53, 92, 93]. The resistance of Gram-negative bacteria to arginine-based surfactants was more as compared to Gram-positive bacteria. The unsaturation or cyclopropane, hydroxyl group within the acyl chain existing, enhances the antimicrobial action of surfactants.

The acyl chains' length and the (+) charge that detects the HLB (hydrophile-lipophile balance) exert no effect on the membrane disrupting property of the surfactants. One more essential class of cationic surfactants is $N\alpha$-acyl arginine methyl ester which has a broad-spectrum antimicrobial activity and is easily biodegradable suggested by Castillo et al. [94]. It is also lightly toxic or nontoxic [14, 95]. The interaction among 1,2-dimyristoyl-sn-glycero-3-phosphocholine, 1,2-dipalmitoyl-sn-glycero-3-phosphocholine, $N\alpha$-acyl arginine methyl ester-based surfactants, living organisms, including or else excluding external barriers, and model membranes also focused on their worthy antimicrobial activity [45].

3.8 BIODEGRADABILITY

The biodegradable property of PBSs was investigated by Kamimura [95], Shida et al. [96, 97] and Kubo et al. [98] and observed that N-acyl PBs undergo simply biodegradation by undergoing breakdown into fatty acids and PBs. Single hydrophobic chain surfactants get more easily degraded than their branched counterparts, for example Args (bis). More hydrophobic surfactants normally have poor biodegradability.

The fatty acids–based PBSs and their physical and chemical properties along with biodegradability were examined and processed by Abe et al. [99]. The microbial degradation of these surfactants was observed between 57 and 73% after 14 days upon conducting the biodegradation studies on them.

The supramolecular hydrogel mixtures of the bio-surfactants based on PBs and $C_{24}H_{39}NaO_4$ (sodium deoxycholate) like arginine, lysine, alanine, glycine, etc., with various buffered solutions and explained their distinctive sensitivity to multi stimuli environments, their simplistic biodegradability and pH sensitivity make them a likely and useful way for drug delivery and it was synthesized by Zhang et al. [100].

3.9 HEMOLYTIC ACTIVITY

The five anionic lysine-based PBSs vary amongst them based on their counterion. Their capability of breaking the cell under different incubation periods, concentration and pH was studied by Nogueira et al. [101]. The pH-sensitive hemolytic activity results and better kinetics of such surfactants at the endosomal pH range were verified. Surfactants interact with the lipid bilayer of cell membranes.

The surfactant membrane interaction mechanisms were acknowledged by Pérez et al. [41]. The activity of the three arginine-based cationic and five lysine-based anionic AASs on hypotonic hemolysis was controlled. The behaviors of unalike anti-hemolytic amongst AASs, both connected to the maximal protective concentration, were confirmed. The physical, chemical and structural properties of these are monitored for protection in contrast to hypotonic hemolysis. The concentrations of cationic surfactants and CMC causing the highest prevention as compared to hypotonic hemolysis were well correlated. On the other hand, there was no correlation between anionic surfactants. Lysine-based surfactants only vary in their counterions, which is a cause for the anti-hemolytic potency and the hemolytic activities.

The gemini surfactants and arginine-based monomeric ability to interfere with the erythrocyte membranes depends on hydrophobicity and size was disclosed by conducting toxicological studies [102].

The interaction of erythrocyte with $N\varepsilon$-dioctyl lysine salts, $N\alpha$ with several counterions such as Na^+, $Tris^+$, K^+, Li^+ and Lys^+, was described by Najjar [103]. Surfactants act together with erythrocytic membranes in a biphasic way, at a lower concentration by prevention besides hypotonic hemolysis and the high amount by introducing hemolysis.

3.10 RHEOLOGICAL PROPERTIES OF SURFACTANTS

Rheological properties of surfactants have a vital part to identify their applications in various industries like oil extraction, medicinal drugs, eatable food as well as in-skin products [104–106]. Many types of research explained the relation between the CMC of AASs and viscoelastic characteristics [46, 107, 108].

3.11   APPLICATIONS IN INDUSTRIES

Biodegradability, nontoxicity and special structural characteristics of the PBSs make them appropriate for applications in many industries.

3.11.1   USED IN AGRICULTURE

PBSs find their application as plant growth inhibitors, herbicides and insecticides for agricultural purposes. Betaine ester surfactants can be used as 'temporary biocides' which are classified under cationic surfactants and simply get hydrolyzed into nontoxic components [96, 109, 110].

US patent claimed that a lawn pesticide was produced with the mixture of refined oils obtained from PBS solution and the Cupressaceae plants. The PBS added 20–50% through the weight of the earlier mentioned solution. Herbicidal actions of the nonionic-type PB surfactants were also talked about [97].

3.11.2   USED IN WASHING DETERGENTS

The need for washing detergent preparations based on PB is rising worldwide. PBSs have good foaming, cleaning and fabric softening characteristics making them fit for use as household shampoos, detergents, body washes etc. From aspartic acid, AAS obtained is found to be a real detergent having ampholytic and chelating properties. Irritation on the skin was less frequently seen with compositions of detergents made up of N-alkyl-β-amino-ethoxy acids [108]. The liquid detergent composition of N-coco-β-amino-propionate seemed to be a good cleaning solution for oil mark on metal surfaces [111, 112]. $C_{14}CHOH\ CH_2CHCH_2COONa$, an amino-carboxylic surfactant, was also found to possess well abilities of detergency and utilized for cleaning hair, glass, textiles, carpets etc. [113]. Products formed by reacting propionic acid and ethylenediamine acetic acid are also known as a permanent component in blenching formulations with good complex-forming capability.

The composition of detergent is dependent on N-β-alanine and was stated to have well-capability and permanency, better-softening ability of foam [113]. The compositions of acyl PB–based detergent were synthesized and patented by Keigo and Tatsuya [114]. Kao Corporation designed N-acyl-1-N-hydroxy-β-alanine–based detergent compositions and reported it's less irritating on the skin surface, excessive detergent activity and high resistance for water [115]. Ajinomoto Company [116] utilized less noxious and simply biodegradable L-lysine-, L-arginine-, and L-glutamic acid-based PBSs as a basic component in cosmetics, shampoo and detergents (Figure 3.5). The capability of having enzymatic additives in compositions of detergent to eliminate protein-based has also been confined [117]. N-Acyl PBSs obtained from methyl glycine, aspartic acid, serine, alanine and glutamic acid were known for their good cleaning fluid activity in aqueous solutions. All these surfactants exhibit no rise even at very low temperatures in viscosity and can get transferred simply from containers, thus imparting even froth [118].

3.11.3   LUBRICANTS

PBSs of amphoteric nature usually find their applications as lubricants. It has a tremendous adhesive nature for hydrophilic surfaces and low-friction coefficients make them suitable for use as lubricants. The lubricating properties of PBSs such as glutamic acid have been understood by many researchers [119–121].

3.11.4   USED IN MAKING MEDICINE

3.11.4.1   For Preparation of Functional Liposomes and Drug Carriers

Recently, many research groups reported the ability of synthetic acyl PBs or peptides to be utilized as a carrier of drug and for the preparation of functional liposomes with lipopeptide ligands.

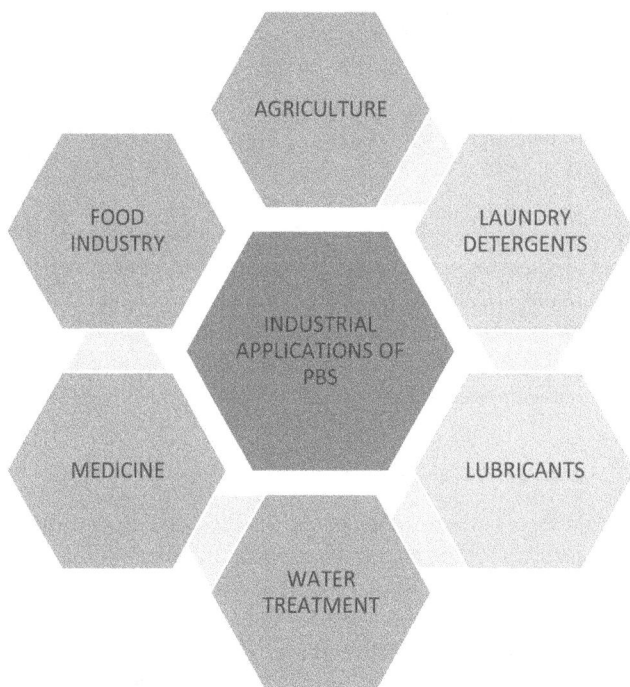

FIGURE 3.5 Weight per cent of AASs as the main ingredients in cosmetics, detergent and hair wash.

The long aliphatic linkage of $N\alpha$-acyl PBs vesicles revealed more efficient encapsulation for solutes than the conventional liposomes of lecithin [121–123].

3.11.4.2 For DNA Transfection and Gene Therapy

The crucial method presently used in the field of life science to carefully include particular genes into the living cells is gene therapy. Gemini surfactants can be potentially used as the main medium for transporting bioactive molecules. 2,4-Diamino-butyric acid- and lysine acid-based polycationic gemini surfactants are simply produced by regular peptide chemistry [124].

The novel class of AA-based gemini surfactants as a carrier for the transfer of genes into cells was synthesized by Peña et al. [125]. Initial outcomes showed that mixing these α-based gemini surfactants by di-oleoyl-phosphate-di-ethanol-amine (DOPE) allowed the synthesis of the liposomes of varied sizes and lipid constituents. It was observed that DOPE surfactant mixtures suspension in the aqueous medium resulted in a lipid vesicles' mixture with many complex structures equivalent to particles with diameter of 500 nm. Surfactants (50/50, 60/40 and 70/30) and various molar ratios of DOPE impacted luciferase expression in Chinese hamster ovary (CHO) cells at distinguishable concentrations. The molecular composition and colloidal size of the gemini surfactants should be considered while providing optimal gene expression in living standards. The DNA transfection of cysteine-based AASs with low MW (molecular weight) has low efficiency and its equivalent gemini surfactant could be considered [126]. No cytotoxicity was exhibited by these surfactants and was known for their greater efficacy when compared with their counterparts available commercially that transfect CHO K1 cells.

3.11.4.3 Antiviral Agents

Scientists are attracted towards the lipo-PBs because of their considerable antiviral activity. A few derivatives of acyl-PB are subjected to inhibit influenza neuraminidase [127]. AA cationic surfactants extracted esterified dibasic PB like arginine and lauric acid through the condensation process that might be overused as protection against microorganisms and were found effective against certain viral infections. Adding ethyl-N-α-lauroyl-L-arginate into the culture of bovine parainfluenza 3 virus and herpes virus type 1 vaccinia virus and results in the nearly whole reduction of the virus in these cultures. These influences were seen from 5 to 60 minutes [128, 129].

3.11.5 Used in the Food Processing Industry

Globally, PBSs are being utilized as emulsifiers in the food industries as low-calorie spread, margarine, dressing and milk items [130]. New perceptions in eating habits and health phases also need to be changed when formulating the food products periodically for the vitamins, minerals, fat content and calories which may cause a continuous request of product preparations optimization. These surfactants even involve the optimum selection of the raw materials. Industries were focusing more on the different glycerides earlier, especially, on mono-glyceride derivatives. However, keeping in mind the health awareness makes arginine-based PBS having antimicrobial action and antibacterial, broad spectra to be used as an alternative for pure glycerides found commercially in food formulations. These surfactants also exhibit noticeable antimicrobial activity against *Salmonella* and *Escherichia coli*, most communal food-borne pathogens which impose severe health risk as a result of elevated drug challenge [131].

3.11.6 Used in the Cosmetic Industry

PBSs find their use in several household care product formulations. Potassium N-cocoyl glycinate found in face cleansers for the removal of soil and make-up which is mild on the skin. N-Acyl-L-glutamic acid contains two COOH groups, providing it has the property to be soluble in water. C 12 fatty acid–based PBSs are widely being used in cleaners to remove make-up and dirt from the face. PBSs with C 18 were present, which are used as emulsifiers in cosmetic products. N-Dodecanoyl alaninate is well known for the capability of making nonirritating smooth froth for skin and so used to prepare products for babies. N-Lauroyl-based PBS used in toothpaste shows adequate froth alike to soap and strong enzyme-inhibiting value [68]. From the earlier decades, the important factors include fear of gentleness, less toxicity, safety, mildness for the use of surfactants in pharmaceutical products, beauty products and personal home care products. Customers are honestly aware of the toxicity, environmental factors, potential irritation and toxicity of these products. Currently, PBSs are used for preparing body soaps, many shampoos, and rinse as of their plentiful qualities over their conventional corresponding item for use in-skin products and cosmetics. PBSs have the necessary qualities in a home care product [69]. Some PBSs are good for the film-forming capability, while others have the good frothing capability.

PBs are essential in ensuring natural nourishing factors appear in the stratum corneum. The dead epidermal cells are present in the stratum corneum, and the proteins in the cells regularly change into PBs. Then PBs are transferred into the stratum corneum, and the fatty substances are absorbed into the skin and therefore increase the skin flexibility. Around 50% of the natural moisturizing factors available in the epidermis of the human body are composed of pyrrolidone and PBs [132]. Collagen is a normal cosmetic part that also includes PBs, which makes skin soft. Many problems such as dark coloration and skin roughness are due to lack of PBs. A research study showed that mixing PB with ointment gives aid to burn on the skin and the damaged skin turns into the actual condition and does not convert into a keloid [133].

They are also very beneficial for the repair of impaired cuticles. Dry hair and lacklustre can show a reduction in the concentration of PBs in deeply impaired cuticles. It can enter over the cuticle into the shaft hair and drag in moistness from the surface of the skin. The proficiency of PBSs becomes beneficial in hair treatment agents like hair softeners/rinses, shampoos, hair softeners and the occurrence of PBS gets brittle less hair [22].

3.11.7   Used in Microbial Enriched Oil Recovery

PBSs have been discovered effective for microbial enriched oil recovery. A *Brevibacterium aureum* MSA13, the strain of bacteria, synthesized a surfactant derivative of AAs that were potentially used for oil recovery of microbes. That surfactant has octadecanoic acid such as a hydrophobic moiety and a group of hydrophilic contains a tetrapeptide which is a short sequence of 4 PB pro-leu-gly-gly. An action bacterium, MSA13-derived surfactant, also possesses the feature of microbial enriched oil recovery in sea environments [134, 135].

3.11.8   Used in the Synthesis of Nanomaterials

Protein-based polymerizable surfactants were also observed beneficial in chiral nanoparticles synthesis. The production of PB-based chiral surfactants through polymerizable moieties and then misused to make nanoparticles by chiral surface functionality was stated by Preiss et al. [133].

Their potential such as nucleating agents in the enantioselective crystallization of PB conglomerate systems, taking rac-asparagine as a model, has been checked. The comparisons were made in the synthesis of particles from chiral surfactants containing dissimilar tail groups and outcomes showed that only the chiral nanoparticles prepared of polymerizable surfactants were observed that the enantioselective crystallization had the capability as a nucleation agent.

3.11.9   Another Application of PBSs

PBSs also find their utility in the formation of [poly(3,4-ethylene-di-oxy-thiophene)] PEDOT films [136–145]. PEDOT films made by the direct anodic oxidation of (3,4-ethylene-di-oxy-thiophene) EDOT in water constituted the sodium *N*-lauroyl-sarcosinate and was found to be eco-friendly. They are too used in optimizing laundering processes through CO_2 and even as chiral solvents [146–148].

REFERENCES

1. Li, J., Wang, J., Zhao, Y., Zhou, P., Carter, J., Li, Z., … & Xu, H. (2020). Surfactant-like peptides: From molecular design to controllable self-assembly with applications. *Coordination Chemistry Reviews*, *421*, 213418.
2. Otzen, D. (2011). Protein–surfactant interactions: A tale of many states. *Biochimica et Biophysica Acta (BBA)-Proteins and Proteomics*, *1814*(5), 562–591.
3. Notter, R. H., & Finkelstein, J. N. (1984). Pulmonary surfactant: An interdisciplinary approach. *Journal of Applied Physiology*, *57*(6), 1613–1624.
4. Halliday, H. L. (2005). History of surfactant from 1980. *Neonatology*, *87*(4), 317–322.
5. Halliday, H. L. (2008). Surfactants: Past, present and future. *Journal of Perinatology*, *28*(1), S47–S56.
6. Bondi, S. (1909). Lipoprotein and the analysis of degenerative adiposis lipopeptides, their meaning, synthesis and characteristics (laurylglycin and laurylalanin). *Biochemische Zeitschrift*, *17*, 543.
7. Hentrich, W., Keppler, H., Hintzmann, K. (1936). German Patent 635,522; Sept. 18; British Pats, 459(461,328).
8. Heitmann, P. (1968), Reactivity of sulfhydryl groups in micelles. *European Journal of Biochemistry*, *5*, 305–315.
9. Heitmann, P. (1968), A model for sulfhydryl groups in proteins. Hydrophobic interactions of the cysteine side chain in micelles. *European Journal of Biochemistry*, *3*, 346–350.
10. Presenz, P. (1996). Lipoamino acids and lipopeptides as amphiphilic compounds. *Pharmazie*, *51*(10), 755–758.
11. Pérez, L., Pinazo, A., García, M. T., Lozano, M., Manresa, A., Angelet, M., … & Infante, M. R. (2009). Cationic surfactants from lysine: Synthesis, micellization and biological evaluation. *European Journal of Medicinal Chemistry*, *44*(5), 1884–1892.
12. Lundberg, D., Faneca, H., Morán, M. D. C., Pedroso De Lima, M. C., Miguel, M. D. G., & Lindman, B. (2011). Inclusion of a single-tail amino acid-based amphiphile in a lipoplex formulation: Effects on transfection efficiency and physicochemical properties. *Molecular Membrane Biology*, *28*(1), 42–53.

13. Piera, E., Comelles, F., Erra, P., & Infante, M. R. (1998). New alquil amide type cationic surfactants from arginine. *Journal of the Chemical Society, Perkin Transactions*, 2(2), 335–342.
14. Seguer, J., Infante, M. R., Allouch, M., Mansuy, L., Selve, C., & Vinardell, P. (1994). Synthesis and evaluation of non-ionic amphiphilic compounds from amino acids: Molecular mimics of lecithins. *New Journal of Chemistry*, 18(6), 765–774.
15. Seguer, J., Molinero, J., Manresa, A., & Caelles, J. (1994). Physicochemical and antimicrobial properties of N″-acyl-L-arginine dipeptides from. *Journal of the Society of Cosmetic Chemists*, 45, 53–63.
16. Dana, V., Aurelia, P., Irina, E. C., Mihai, C. C. (2011). Aspects regarding the synthesis and surface properties of some glycine based surfactants *Scientific Bulletin, Series B*, 73(3), 147–154.
17. Franklin, T. J., & Snow, G. A. (2013). *Biochemistry of antimicrobial action*. Springer.
18. Xia, J. (2001). *Protein-based surfactants: Synthesis: Physicochemical properties, and applications* (Vol. 101). CRC Press.
19. Then, A., Mácha, K., Ibrahim, B., & Schuster, S. (2020). A novel method for achieving an optimal classification of the proteinogenic amino acids. *Scientific Reports*, 10(1), 1–11.
20. Xia, J., Xia, Y., & Nnanna, I. A. (1995). Structure-function relationship of acyl amino acid surfactants: Surface activity and antimicrobial properties. *Journal of Agricultural and Food Chemistry*, 43(4), 867–871.
21. Mhaskar, S. Y., & Lakshminarayana, G. (1992). Synthesis of diethanolamides of N-lauroyl dipeptides and correlation of their structures with surfactant and antibacterial properties. *Journal of the American Oil Chemists' Society*, 69(7), 643–646.
22. Infante, M., Molinero, J., Bosch, P., Juláa, M. R., & Erra, P. (1989). Lipopeptidic surfactants: I. Neutral N-lauroyl-I-arginine dipeptides from pure amino acids. *Journal of the American Oil Chemists' Society*, 66(12), 1835–1839.
23. Morán, M. C., Pinazo, A., Pérez, L., Clapés, P., Angelet, M., García, M. T., Vinardell, M. P., & Infante, M. R. (2004). "Green" amino acid-based surfactants. *Green Chemistry*, 6, 233.
24. Tamarkin, D., Eini, M., Friedman, D., Berman, T., & Schuz, D. (2014). U.S. Patent 8,795,693.Washington, DC: U.S. Patent and Trademark Office.
25. Godtfredsen, S. E., & Bjoerkling, F. (1990). An enzyme-catalyzed process for preparing N-acyl amino acids and N-acyl amino acid amides. World Patent, 90/14429.
26. Vonderhagen, A., Raths, H. C., & Eilers, E. (1999). Enzymatic catalyzed N-acylation of amino acids, protein hydrolysates and/or their derivatives. *Ger. Offen. DE*, 19, 555.
27. Valivety, R., Jauregi, P., Gill, I., & Vulfson, E. (1997). Chemo-enzymatic synthesis of amino acid-based surfactants. *Journal of the American Oil Chemists' Society*, 74(7), 879–886.
28. Montet, D., Servat, F., Pina, M., Graille, J., Galzy, P., Arnaud, A., … & Marcou, L. (1990). Enzymatic synthesis of N-ε-acyllysines. *Journal of the American Oil Chemists' Society*, 67(11), 771–774.
29. Nagao, A., & Kito, M. (1989). Synthesis of O-acyl-L-homoserine by lipase. *Journal of the American Oil Chemists' Society*, 66(5), 710–713.
30. Soo, E. L., Salleh, A. B., Basri, M., Rahman, R. N. Z. R. A., & Kamaruddin, K. (2003). Optimization of the enzyme-catalyzed synthesis of amino acid-based surfactants from palm oil fractions. *Journal of Bioscience and Bioengineering*, 95(4), 361–367.
31. Gerova, M., Rodrigues, F., Lamère, J. F., Dobrev, A., & Fery-Forgues, S. (2008). Self-assembly properties of some chiral N-palmitoyl amino acid surfactants in aqueous solution. *Journal of Colloid and Interface Science*, 319(2), 526–533.
32. Pang, X., & Chu, C. C. (2010). Synthesis, characterization and biodegradation of functionalized amino acid-based poly (ester amide)s. *Biomaterials*, 31(14), 3745–3754.
33. Cantacuzene, D., & Guerreiro, C. (1989). Optimization of the papain catalyzed esterification of amino acids by alcohols and diols. *Tetrahedron*, 45(3), 741–748.
34. Clapés, P., Morán, C., & Infante, M. R. (1999). Enzymatic synthesis of arginine-based cationic surfactants. *Biotechnology and Bioengineering*, 63(3), 333–343.
35. Yoshimura, T., Sakato, A., & Esumi, K. (2013). Solution properties and emulsification properties of amino acid-based gemini surfactants derived from cysteine. *Journal of Oleo Science*, 62(8), 579–586.
36. Piera, E., Infante, M. R., & Clapés, P. (2000). Chemo-enzymatic synthesis of arginine-based gemini surfactants. *Biotechnology and Bioengineering*, 70(3), 323–331.
37. Faustino, C. M., Calado, A. R., & Garcia-Rio, L. (2009). New urea-based surfactants derived from α, ω-amino acids. *The Journal of Physical Chemistry B*, 113(4), 977–982.
38. Mitin, Y. V., Braun, K., & Kuhl, P. (1997). Papain catalyzed synthesis of glyceryl esters of N-protected amino acids and peptides for the use in trypsin catalyzed peptide synthesis. *Biotechnology and Bioengineering*, 54(3), 287–290.

39. Morán, C., Infante, M. R., & Clapés, P. (2001). Synthesis of glycero amino acid-based surfactants. Part 1. Enzymatic preparation of rac-1-O-(N α-acetyl-l-aminoacyl) glycerol derivatives. *Journal of the Chemical Society, Perkin Transactions, 1*(17), 2063–2070.

40. Pérez, L., Infante, M. R., Pons, R., Morán, C., Vinardell, P., Mitjans, M., & Pinazo, A. (2004). A synthetic alternative to natural lecithins with antimicrobial properties. *Colloids and Surfaces B: Biointerfaces, 35*(3–4), 235–242.

41. Pérez, L., Pinazo, A., García, M. T., del Carmen Moran, M., & Infante, M. R. (2004). Monoglyceride surfactants from arginine: Synthesis and biological properties. *New Journal of Chemistry, 28*(11), 1326–1334.

42. Franceschi, S., de Viguerie, N., Riviere, M., & Lattes, A. (1999). Synthesis and aggregation of two-headed surfactants bearing amino acid moieties. *New Journal of Chemistry, 23*(4), 447–452.

43. Polidori, A., Wathier, M., Fabiano, A. S., Olivier, B., & Pucci, B. (2006). Synthesis and aggregation behaviour of symmetric glycosylated bolaamphiphiles in water. *Arkivoc, 4*, 73–89.

44. Mukerjee, P., & Mysels, K. J. (1971). *Critical micelle concentrations of aqueous surfactant systems.* National Standard Reference Data System.

45. Singare, P. U., & Mhatre, J. D. (2012). Cationic surfactants from arginine: Synthesis and physicochemical properties. *American Journal of Chemistry, 2*, 186–190.

46. Shrestha, R. G., & Aramaki, K. (2009). The study of salt induced viscoelastic wormlike micelles in aqueous systems of mixed anionic/nonionic surfactants. *Journal of Nepal Chemical Society, 23*, 65–73.

47. Pilemand, C. (2002). *Surfactants: Their abilities and important physico-chemical properties.* Arbejdsmiljøinstituttet.

48. Malik, N. A., & Ali, A. (2016). Krafft temperature and thermodynamic study of interaction of glycine, diglycine, and triglycine with hexadecylpyridinium chloride and hexadecylpyridinium bromide: A conductometric approach. *Journal of Molecular Liquids, 213*, 213–220.

49. Puerto, M. C. (2001). Surfactants: Fundamentals and applications in the petroleum industry. *Chemical Engineering Journal*, 1(83), 63. Cambridge University Press.

50. Ohta, A., Ozawa, N., Nakashima, S., Asakawa, T., & Miyagishi, S. (2003). Krafft temperature and enthalpy of solution of N-acyl amino acid surfactants and their racemic modifications: Effect of the amino acid residue. *Colloid and Polymer Science, 281*(4), 363–369.

51. Brito, R. O., Silva, S. G., Ricardo, M. F., Fernandes, R. M. F., Marques, E. F., Enrique-Borges, J., & do Vale, M. L. C. (2011). Enhanced interfacial properties of novel amino acid-derived surfactants: Effects of headgroup chemistry and of alkyl chain length and unsaturation. *Colloids and Surfaces B: Biointerfaces, 86*(1), 65–70.

52. Zhang, G., Xu, B., Han, F., Zhou, Y., Liu, H., Li, Y., Cui, L., Tan, T., & Wang, N. (2013). Green synthesis, composition analysis and surface active properties of sodium cocoyl glycinate. *American Journal of Analytical Chemistry, 4*, 2013.

53. Yoshimura, T., Sakato, A., Tsuchiya, K., Ohkubo, T., Sakai, H., Abe, M., & Esumi, K. (2007). Adsorption and aggregation properties of amino acid-based N-alkyl cysteine monomeric and N,N'-dialkyl cystine gemini surfactants. *Journal of Colloid and Interface Science, 308*(2), 466–473.

54. Bordes, R., Tropsch, J., & Holmberg, K. (2009). Counterion specificity of surfactants based on dicarboxylic amino acids. *Journal of Colloid and Interface Science, 38*(2), 529–536.

55. Fan, H., Han, F., Liu, Z., Qin, L., Li, Z., Liang, D., … & Fu, H. (2008). Active control of surface properties and aggregation behavior in amino acid-based Gemini surfactant systems. *Journal of Colloid and Interface Science, 321*(1), 227–234.

56. Hines, J. D., Thomas, R. K., Garrett, P. R., Rennie, G. K., Penfold, J. (1997). Investigation of mixing in binary surfactant solutions by surface tension and neutron reflection: Anionic/nonionic and zwitterionic/nonionic mixtures. *Journal of Physical Chemistry B, 101*(45), 9215–9223.

57. Brito, R. O., Marques, E. F., Silva, S. G., do Vale, M. L., Gomes, P., Araújo, M. J., … Mitjans, M. (2009). Physicochemical and toxicological properties of novel amino acid-based amphiphiles and their spontaneously formed catanionic vesicles. *Colloids Surfaces B, 72*, 80.

58. Ohta, A., Shirai, M., Asakawa, T., & Miyagishi, S. (2008). Effect of stereochemistry on the molecular aggregation of phenylalanine dipeptide-type surfactants. *Journal of Oleo Science, 57*(12), 659–667.

59. Iwahashi, M., Matsuzawa, H., Minami, H., Yano, T., Wakabayashi, T., Ino, M., & Sakamoto, K. (2002). Solvent effect on chiral aggregate formation of acylamino acids. *Journal of Oleo Science, 51*(11), 705–713.

60. Johnsson, M., & Edwards, K. (2001). Phase behavior and aggregate structure in mixtures of dioleoylphosphatidylethanolamine and poly (ethylene glycol)-lipids. *Biophysical Journal, 80*(1), 313–323.

61. Marques, E. F., Brito, R. O., Silva, S. G., Rodriguez-Borges, J. E., Vale, M. L. D., Gomes, P., ... & Soderman, O. (2008). Spontaneous vesicle formation in catanionic mixtures of amino acid-based surfactants: Chain length symmetry effects. *Langmuir*, *24*(19), 11009–11017.

62. Kouchi, J., Tabohashi, T., Yokoyama, S., Harusawa, F., Yamaguchi, A., Sakai, H., & Abe, M. (2001). Emulsifying potency of new amino acid-type surfactant (1). O/W emulsions. *Journal of Oleo Science*, *50*(11), 847–855.

63. Baczko, K., Larpent, C., & Lesot, P. (2004). New amino acid-based anionic surfactants and their use as enantiodiscriminating lyotropic liquid crystalline NMR solvents. *Tetrahedron: Asymmetry*, *15*(6), 971–982.

64. Wu, M. H., Wan, L. Z., & Zhang, Y. Q. (2014). A novel sodium N-fatty acyl amino acid surfactant using silkworm pupae as stock material. *Scientific Reports*, *4*(1), 1–8.

65. Holmberg, K., Jönsson, B., Kronberg, B., Lindman, B. (2004). Surfactants and polymers in aqueous solution. *Journal of Synthetic Lubrication*, *20*(4), 367–370.

66. Li, Y. (2012). *Synthesis and physicochemical study of novel amino acid based surfactants* (Master's thesis).

67. Satyanarayana, T., Johri, B. N., & Prakash, A. (Eds.). (2012). *Microorganisms in sustainable agriculture and biotechnology*. Springer Science & Business Media.

68. Rosen, M. J., & Kunjappu, J. T. (2012). *Surfactants and interfacial phenomena*. John Wiley & Sons.

69. Linfield, W. M. (1978). Soap and lime soap dispersants. *Journal of the American Oil Chemists' Society*, *55*(1), 87–92.

70. Parris, N., Weil, J. K., Linfield, W. M. (1973). Soap based detergent formulations. V. Amphoteric lime soap dispersing agents. *Journal of the American Oil Chemists Society*, *53*(12), 509–512.

71. Weil, J. K., Parris, N., & Stirton, A. J. (1970). Synthesis and properties of sulfated alkanolamides. *Journal of the American Oil Chemists' Society*, *47*(3), 91–93.

72. Sagawa, K., Gesslein, B. W., Popova, K., Oshimura, E., Ikeda, N., & Kamidoi, T. (2015). U.S. Patent 9,034,924. Washington, DC: U.S. Patent and Trademark Office.

73. Berger, C., & Gacon, P. (1992). Preparation of N-acyl derivatives of amino acid mixtures obtained from cereal protein hydrolyzates for cosmetics. Patent no. FR9221318.

74. Davila, A. M., Marchal, R., & Vandecasteele, J. P. (1997). Sophorose lipid fermentation with differentiated substrate supply for growth and production phases. *Applied Microbiology and Biotechnology*, *47*(5), 496–501.

75. George, A., Modi, J., Jain, N., Bahadur, P. (1998). A comparative study on the surface activity and micellar behaviour of some N-acylamino acid based surfactants. *Indian Journal of Chemistry – Section A Inorganic, Physical, Theoretical and Analytical Chemistry*, *37*(11), 985–992.

76. Ito, S., Inoue, S. (1982). Sophorolipids from *Torulopsis bombicola*: Possible relation to alkane uptake. *Applied and Environmental Microbiology*, *43*(6), 1278–1283.

77. Myers, D. (2020). *Surfactant science and technology*. John Wiley & Sons.

78. Obata, Y., Suzuki, D., & Takeoka, S. (2008). Evaluation of cationic assemblies constructed with amino acid based lipids for plasmid DNA delivery. *Bioconjugate Chemistry*, *19*(5), 1055–1063.

79. Rosa, M., del Carmen Morán, M., da Graça Miguel, M., Lindman, B. (2007). The association of DNA and stable catanionic amino acid-based vesicles. *Colloids and Surfaces, A*, *301*(1) 361–375.

80. Morad, M. S. (2008). Inhibition of iron corrosion in acid solutions by Cefatrexyl: Behaviour near and at the corrosion potential. *Corrosion Science*, *50*(2), 436–448.

81. De Souza, F. S., & Spinelli, A. (2009). Caffeic acid as a green corrosion inhibitor for mild steel. *Corrosion Science*, *51*(3), 642–649.

82. Abiola, O. K., & James, A. O. (2010). The effects of Aloe vera extract on corrosion and kinetics of corrosion process of zinc in HCl solution. *Corrosion Science*, *52*(2), 661–664.

83. Barouni, K., Kassale, A., Albourine, A., Jbara, O., Hammouti, B., & Bazzi, L. (2014). Amino acids as corrosion inhibitors for copper in nitric acid medium: Experimental and theoretical study. *Journal of Materials and Environmental Science*, *5*(2), 456–463.

84. Singh, A. K., & Quraishi, M. A. (2010). Effect of Cefazolin on the corrosion of mild steel in HCl solution. *Corrosion Science*, *52*(1), 152–160.

85. Umoren, S. A., Obot, I. B., & Obi-Egbedi, N. O. (2009). Raphia hookeri gum as a potential eco-friendly inhibitor for mild steel in sulfuric acid. *Journal of Materials Science*, *44*(1), 274–279.

86. Abiola, O. K. (2005). Adsorption of methionine on mild steel. *Journal of the Chilean Chemical Society*, *50*(4), 685–690.

87. Abd-El-Nabey, B. A., Khalil, N., & Mohamed, A. (1985). Inhibition by amino acids of the corrosion of steel in acid. *Surface Technology*, *24*(4), 383–389.

88. Rahim, M. A. A., Hassan, H. B., & Khalil, M. W. (1997). Naturally Occurring Organic Substances as Corrosion Inhibitors for mild steel in acid medium. *Materialwissenschaft und Werkstofftechnik, 28,* 98–102.

89. Özcan, M., Karadağ, F., & Dehri, I. (2008). Investigation of adsorption characteristics of methionine at mild steel/sulfuric acid interface: An experimental and theoretical study. *Colloids and Surfaces A: Physicochemical and Engineering Aspects, 316*(1–3), 55–61.

90. Chandra, N., & Tyagi, V. K. (2013). Synthesis, properties, and applications of amino acids based surfactants: A review. *Journal of Dispersion Science and Technology, 34*(6), 800–808.

91. Perinelli, D. R., Vllasaliu, D., Bonacucina, G., Come, B., Pucciarelli, S., Ricciutelli, M., … Casettari, L. (2017). Rhamnolipids as epithelial permeability enhancers for macromolecular therapeutics. *European Journal of Pharmaceutics and Biopharmaceutics, 119,* 419–425.

92. Infante, R., Dominguez, J. G., Erra, P., Julia, R., & Prats, M. (1984). Surface active molecules: Preparation and properties of long chain nα-acyl-l-α-amino-ω-guanidine alkyl acid derivatives. *International Journal of Cosmetic Science, 6*(6), 275–282.

93. Holmberg, K. (Ed.). (2003). *Novel surfactants: Preparation applications and biodegradability, revised and expanded* (Vol. 114). CRC Press.

94. Castillo, J. A., Infante, M. R., Manresa, À., Vinardell, M. P., Mitjans, M., & Clapés, P. (2006). Chemoenzymatic synthesis and antimicrobial and haemolytic activities of amphiphilic bis (phenylacetylarginine) derivatives. *ChemMedChem: Chemistry Enabling Drug Discovery, 1*(10), 1091–1098.

95. Kamimura, A. (1973). Colorimetric determination of long-chain n-acylglutamic acids with pinacyanol. *Agricultural and Biological Chemistry 37*(3), 457–464.

96. Shida, T., Homma, Y., & Misato, T. (1973). Bacterial degradation of N-lauroyl-L-valine. *Agricultural and Biological Chemistry, 37*(5), 1027–1033.

97. Shida, T., Homma, Y., Kamimura, A., & Misato, T. (1975). Studies on the control of plant diseases by amino acid derivatives. V. Degradation of N-lauroyl-L-valine in soil and the effect of sunlight and ultraviolet rays on N-lauroyl-L-valine. *Agricultural and Biological Chemistry, 39*(4), 879–883.

98. Kubo, M., Yamada, K., & Takinami, K. (1976). Effects of chemical structure on biodegradation of long chain N-acyl amino acids. *Journal of Fermentation Technology (Japan), 54,* 323–332.

99. Abe, A., Asakura, K., & Osanai, S. (2004). Synthesis and characterization of novel amphiphiles containing amino acid and carbohydrate. *Journal of Surfactants and Detergents, 7*(3), 297–303.

100. Zhang, M., Strandman, S., Waldron, K. C., & Zhu, X. X. (2016). Supramolecular hydrogelation with bile acid derivatives: Structures, properties and applications. *Journal of Materials Chemistry B, 4*(47), 7506–7520.

101. Nogueira, D. R., Mitjans, M., Infante, M. R., & Vinardell, M. P. (2011). The role of counterions in the membrane-disruptive properties of pH-sensitive lysine-based surfactants. *Acta Biomaterialia, 7*(7), 2846–2856.

102. Tavano, L., Infante, M. R., Riya, M. A., Pinazo, A., Vinardell, M. P., Mitjans, M., … & Perez, L. (2013). Role of aggregate size in the hemolytic and antimicrobial activity of colloidal solutions based on single and gemini surfactants from arginine. *Soft Matter, 9*(1), 306–319.

103. Najjar, R. (Ed.). (2017). *Application and characterization of surfactants.* BoD–Books on Demand.

104. Lin, Z., Zheng, Y., Talmon, Y., Maxson, A., & Zakin, J. L. (2016). Comparison of the effects of methyl- and chloro-substituted salicylate counterions on drag reduction and rheological behavior and micellar formation of a cationic surfactant. *Rheologica Acta, 55*(2), 117–123.

105. Yang, J. (2002). Viscoelastic wormlike micelles and their applications. *Current Opinion in Colloid & Interface Science, 7*(5–6), 276–281.

106. Saavedra, L. C. C., Gómez, E. M. P., Oliveira, R. G., & Fernández, M. A. (2017). Aggregation behaviour and solubilization capability of mixed micellar systems formed by a gemini lipoamino acid and a nonionic surfactant. *Colloids Surfaces, A, 2*(3), 1144–1148.

107. Singh, J., Michel, D., Getson, H. M., Chitanda, J. M., Verrall, R. E., & Badea, I. (2015). Development of amino acid substituted gemini surfactant-based mucoadhesive gene delivery systems for potential use as noninvasive vaginal genetic vaccination. *Nanomedicine, 10*(3), 405–417.

108. Leonard, E. O. (1976). U.S. Patent 3,960,742. Washington, DC: U.S. Patent and Trademark Office.

109. Solans, C., Infante, R., Azemar, N., & Wärnheim, T. (1989). Phase behavior of cationic lipoaminoacid surfactant systems. In: Bothorel, P., & Dufourc, E. J. (Eds.) *Trends in colloid and interface science III. Progress in colloid & polymer science* (Vol. 79). Steinkopff.

110. Walther, W., & Netscher, T. (1996). Design and development of chiral reagents for the chromatographic e.e. determination of chiral alcohols. *Chirality, 8*(5), 397–401.

111. Cooper, L. A., Simon, J., & Wilson, D. A. (1988). U.S. Patent 4,786,440. Washington, DC: U.S. Patent and Trademark Office.

112. Kennedy, R. R., Lindemann, M. K., & Verdicchio, R. J. (1980). U.S. Patent 4,181,634. Washington, DC: U.S. Patent and Trademark Office.

113. Miyamoto, N., Ikeuchi, T., & Shinjo, Z. (1988). U.S. Patent 4,749,515. Washington, DC: U.S. Patent and Trademark Office.

114. Keigo, S., & Tatsuya, V. (1966). Detergent Composition U.S. 5529712 A.

115. Nozaki, T. (1995). U.S. Patent 5,417,875. Washington, DC: U.S. Patent and Trademark Office.

116. Keshwani, A., Malhotra, B., & Kharkwal, H. (2015). Natural polymer based detergents for stain removal. *World Journal of Pharmacy and Pharmaceutical Sciences*, 4(4), 490–508.

117. Moriyama, M., Tanabe, H., Hanazawa, H., & Kajihara, Y. (1998). U.S. Patent 5,712,232. Washington, DC: U.S. Patent and Trademark Office.

118. Jiang, H., Tao, D., & Wang, B. (2005). Synthesis of the multifunctional additive N-acyl glutamic acid and application in water-based lubricating fluids. Proceedings *of the World Tribology Congress* III – 2005 (pp. 667–668).

119. Jiang, H., Zhu, C., & Lei, L. (2006). Measuring the overall heat transfer coefficient of Wei-Jing oil pipeline by the method of soil thermal conductivity. *Oil & Gas Science and Technology*, 25(6), 48–51.

120. Marqués, A. M., Pérez, L., Farfán, M., & Pinazo, A. (2021). Green surfactants: production, properties, and application in advanced medical technologies. *Biosurfactants for a sustainable future: Production and applications in the environment and biomedicine* (pp. 207–243).

121. Boeckler, C., Frisch, B., & Schuber, F. (1998). Design and synthesis of thiol-reactive lipopeptides. *Bioorganic & Medicinal Chemistry Letters*, 8(15), 2055–2058.

122. Yagi, N., Ogawa, Y., Kodaka, M., Okada, T., Tomohiro, T., Konakahara, T., & Okuno, H. (2000). Preparation of functional liposomes with peptide ligands and their binding to cell membranes. *Lipids*, 35(6), 673–680.

123. Kirby, A. J., Camilleri, P., Engberts, J. B., Feiters, M. C., Nolte, R. J., Söderman, O., … van Eijk, M. C. (2003). Gemini surfactants: New synthetic vectors for gene transfection. *Angewandte Chemie International Edition*, 42(13), 1448–1457.

124. McGregor, C., Perrin, C., Monck, M., Camilleri, P., & Kirby, A. J. (2001). Rational approaches to the design of cationic gemini surfactants for gene delivery. *Journal of the American Chemical Society*, 123(26), 6215–6220.

125. Peña, L. C., Argarañá, M. F., De Zan, M. M., Giorello, A., Antuña, S., Prieto, C. C., Veaute, C. M. I., & Müller, D. M. (2017). New amphiphilic amino acid derivatives for efficient DNA transfection in vitro. *Advances in Chemical Engineering and Science*, 7(2), 191–205.

126. Kondoh, M., Furutani, T., Azuma, M., Ooshima, H., & Kato, J. (1997). Acyl amino acid derivatives as novel inhibitors of influenza neuraminidase. *Bioscience, Biotechnology, and Biochemistry*, 61(5), 870–874.

127. Bonvila, X. R., Roca, S. F., & Pons, R. S. (2006). U.S. Patent Application No. 12/375,774.

128. Flack, E. (2018). Butter, margarine, spreads, and baking fats. In *Lipid technologies and applications* (pp. 305–327). Routledge.

129. Lynde, C. W. (2001). Moisturizers: What they are and how they work. *Skin Therapy Letter*, 6(13), 3–5.

130. Cruz, N. I., & Korchin, L. (1994). Inhibition of human keloid fibroblast growth by isotretinoin and triamcinolone acetonide in vitro. *Annals of Plastic Surgery*, 33(4), 401–405.

131. Kiran, G. S., Thomas, T. A., Selvin, J., Sabarathnam, B., & Lipton, A. P. (2010). Optimization and characterization of a new lipopeptide biosurfactant produced by marine *Brevibacterium aureum* MSA13 in solid state culture. *Bioresource Technology*, 101(7), 2389–2396.

132. Zhou, C., Liu, Z., Yan, Y., Du, X., Mai, Y. W., & Ringer, S. (2011). Electro-synthesis of novel nanostructured PEDOT films and their application as catalyst support. *Nanoscale Research Letters*, 6(1), 1–6.

133. Preiss, L. C., Wagner, M., Mastai, Y., Landfester, K., & Muñoz-Espí, R. (2016). Amino-acid-based polymerizable surfactants for the synthesis of chiral nanoparticles. *Macromolecular Rapid Communications*, 37(17), 1421–1426.

134. Ahuja, T., Mir, I. A., & Kumar, D. (2007). Biomolecular immobilization on conducting polymers for biosensing applications. *Biomaterials*, 28(5), 791–805.

135. Garrigues, S., Armenta, S., & de la Guardia, M. (2011). Challenges in green analytical chemistry. *Encyclopedia of Inorganic and Bioinorganic Chemistry*, 66, 1–9.

136. Moral-Vico, J., Carretero, N. M., Pérez, E., Suñol, C., Lichtenstein, M., Casañ-Pastor, N. (2013). Dynamic electrodeposition of aminoacid-polypyrrole on aminoacid-PEDOT substrates: Conducting polymer bilayers as electrodes in neural systems. *Electrochimica Acta*, 111, 250–260.

137. Lowe, C. R. (1989). Biosensors (review). *Philosophical Transactions of the Royal Society of London. Series B, Biological Sciences*, 324(1224), 487–496.

138. Andreescu, S., & Sadik, O. A. (2004). Trends and challenges in biochemical sensors for clinical and environmental monitoring. *Pure and Applied Chemistry*, *76*(4), 861–878.
139. Guimard, N. K., Gomez, N., & Schmidt, C. E. (2007). Conducting polymers in biomedical engineering. *Progress in Polymer Science*, *32*(8–9), 876–921.
140. Malhotra, B. D., Chaubey, A., & Singh, S. P. (2006). Prospects of conducting polymers in biosensors. *Analytica Chimica Acta*, *578*(1), 59–74.
141. Malhotra, B. D., Singhal, R., Chaubey, A., Sharma, S. K., & Kumar, A. (2005). Recent trends in biosensors. *Current Applied Physics*, *5*(2), 92–97.
142. Sackmann, E., Meiboom, S., & Snyder, L. C. (1968). Nuclear magnetic resonance spectra of enantiomers in optically active liquid crystals. *Journal of the American Chemical Society*, *90*(8), 2183–2184.
143. Sarfati, M., Lesot, P., Merlet, D., & Courtieu, J. (2000). Theoretical and experimental aspects of enantiomeric differentiation using natural abundance multinuclear NMR spectroscopy in chiral polypeptide liquid crystals. *Chemical Communications*, (21), 2069–2081.
144. Bordes, R., & Holmberg, K. (2015). Amino acid-based surfactants—Do they deserve more attention?. *Advances in Colloid and Interface Science*, *222*, 79–91.
145. Sarfati, M., Aroulanda, C., Courtieu, J., & Lesot, P. (2001). Enantiomeric recognition of chiral invertomers through NMR in chiral oriented phases: A study of cis-decalin. *Tetrahedron: Asymmetry*, *12*(5), 737–744.
146. Van Roosmalen, M. J. E., Woerlee, G. F., & Witkamp, G. J. (2004). Amino acid based surfactants for dry-cleaning with high-pressure carbon dioxide. *The Journal of Supercritical Fluids*, *32*(1–3), 243–254.
147. Weiss-López, B. E., Azocar, M., Montecinos, R., Cassels, B. K., & Araya-Maturana, R. (2001). Differential incorporation of L-and d-N-Acyl-1-phenyl-d 5-2-aminopropane in a cesium N-dodecanoyl-l-alaninate cholesteric nematic lyomesophase. *Langmuir*, *17*(22), 6910–6914.
148. Liu, Y., Jessop, P. G., Cunningham, M., Eckert, C. A., & Liotta, C. L. (2006). Switchable surfactants. *Science*, *313*(5789), 958–960.

4 Steroid-, Terpene- and Alkaloid-Based Surfactants

Meenu Aggarwal[1], Nisha Saini[2] and Pooja Agrawal[3]

[1]Department of Chemistry, Aggarwal College Ballabgarh, Faridabad, India

[2]Department of Chemistry, Gargi College,
University of Delhi, New Delhi, India

[3]Department of Chemistry, School of Basic and Applied
Sciences, Galgotias University, Greater Noida, India

CONTENTS

DOI: 10.1201/9781003144878-4

4.1   INTRODUCTION

The term surface active agents condensed into the word surfactants have the habit to adsorb at interfaces because of their bifold nature [1]. Because of the unique amphiphilic nature of surfactant molecules, the hydrophile and lipophile show substantial affinity toward polar and nonpolar solvents, respectively [2]. The surface tension of water is drastically reduced by surfactants, thereby increasing its cleansing capacity as a main component of household detergents. Also, they are utilized as emulsifiers in cosmetics, agricultural chemicals, pharmaceuticals, textiles, food and paper processing and many more [3]. Self-assembled drug delivery system is better than traditional drug carriers due to high drug load, low degradation with high target specificity. It is very important to check the stability of self-assembled prodrug and its rate of degradation while selection, but it is difficult to find the ideal prodrug. Drugs which are water insoluble require formulation for injection purpose, and to overcome this problem, surfactants are commonly used. Due to amphiphilic nature, surfactants have tendency to undergo self-aggregation to form micelles at critical micelle concentration (CMC). For any drug to be soluble, it is essential to have contact of micelle with biological membrane [4, 5]. Now lipophilic drugs can be solubilized as micelles consist of a hydrophilic part surrounding a hydrophobic core of drug. Most of the surfactants are cytotoxic in nature. Research is now focused on the development of surfactants with minor toxicity and biocompatibility apart from usage of solubilizing water insoluble drugs [6]. Surfactants below CMC increase drug solubility by hydrophobic interaction with drug. While above CMC value, they self-aggregate to develop micelle with hydrophobic tail inwards and hydrophilic head outwards [7–9].

In spite of good properties, synthetic surfactants like alkyl phenols pose a threat to the environment; thus, biodegradability of surfactant is also important factor due to safety concern of environment [10–12]. It motivates the research for the expansion of environmentally friendly surfactants which can be synthesized from so many plant-based natural products. Surfactants can be formed by utilizing the essential raw materials like fatty acids, monoglycerides and glucosides [13–15]. However, natural product-based surfactants are better than conventional surfactants, but other factors which are also taken into consideration while selection are toxicity, biodegradability, cost of sewage treatment, amount of poisonous gases produced, nature of degradation products, easy availability of raw materials and its transport cost [16–18].

4.2   CLASSIFICATION OF NATURAL SURFACTANTS

Natural surfactants, i.e., derived directly from nature based on their origin, are chiefly classified into two types: (i) plant-derived and (ii) bio-surfactants produced by the microbes such as bacteria or yeast. The biosurfactants are formed by requiring expensive carbon sources as raw materials. So, plants continued to be predominant source of natural surfactants, thereby opening area in exploration, extraction and isolation of natural surfactants from plants distributed extensively in nature [19–22]. Among the numerous plants-derived bioactive compounds, saponins exhibit surface-active properties which belong to a class of nonionic surfactants bearing amphiphilic nature [23–25]. They get their name 'saponin' from Latin word sapo, which means soap as they form soapy lather when agitated with water [26–29].

4.3 COMPOSITION OF PLANT-BASED SURFACTANTS

Plant-based surfactants or saponins is a class of high molecular weight amphiphilic compounds having sugars as hydrophilic moiety and steroid/triterpenoid aglycon as lipophilic moiety. Its large molecule has two parts: the hydrophobic sapogenin as aglycones part (steroids or triterpenoids) and the hydrophilic glycone part (sugars) as glycone part, bound by a glycosidic bond (Figure 4.1). The carbohydrate part might consist of several oligosaccharide chains as glucose, galactose, pentose, etc. Based on the aglycone part of saponins, they are classified as (i) steroidal and (ii) triterpenoidal saponins.

4.3.1 GENERAL CLASSIFICATION OF SAPONINS

Sterols which are a subgroup of steroids occur naturally in plant and animals and are the most common natural raw materials for the development of surfactants (Figure 4.2). Sterols have perhydro cyclopentenophenanthrene nucleus with a side chain at position C-17, and it is derived from gonane by the replacement of hydrogen at C-3 by hydroxyl group. Sterols derived from plants are called phytosterols like campesterol, stigmasterol and sitosterol, whereas derived from animals are called zoo steroids like cholesterol (Figure 4.3). These sterols differ in the nature of the side chain

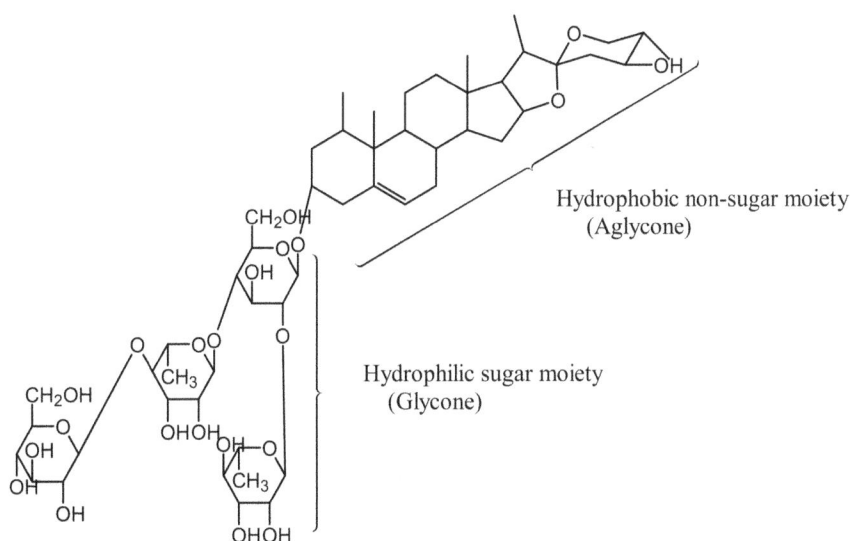

FIGURE 4.1   Structure of a saponin molecule.

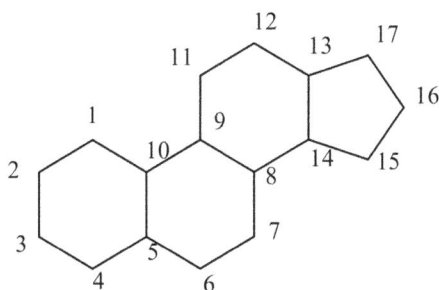

FIGURE 4.2   Gen structure of a sterol.

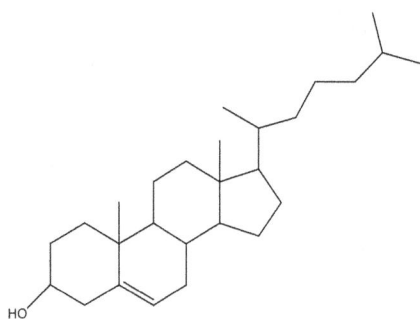

Cholesterol

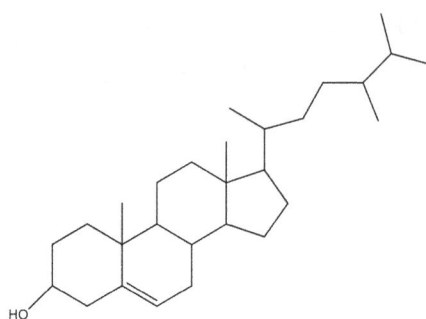

Campesterol

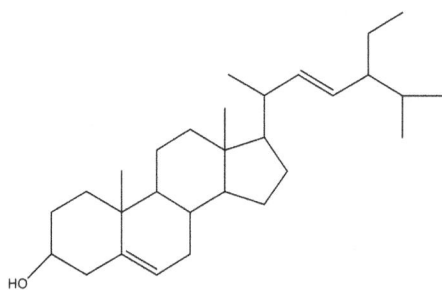

Stigmasterol

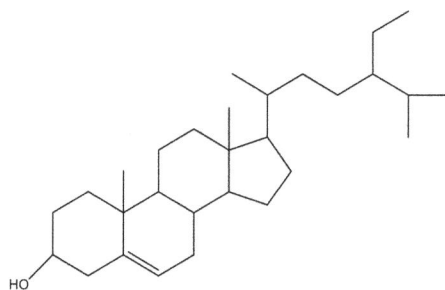

Sitosterol

FIGURE 4.3   Structures of few sterols.

present at position C-17 in the structure. The most prevalent sources for phytosterols are soyabean, maize, coconut, peanut, sesame, rapeseed, sunflower, tall oil and avocado, etc. The fatty acids and tocopherols are used in the saponification-extraction process to isolate sterols. Steroidal saponins are further divided as (i) spirostanols and (ii) furostanols glycosides (Figure 4.4) [30–32].

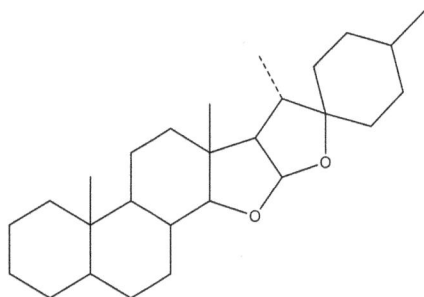

(a)

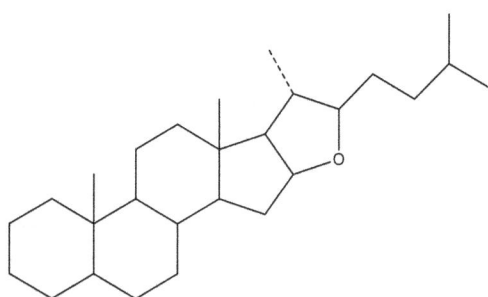

(b)

FIGURE 4.4   Steroidal saponins: (a) spirostanols and (b) furostanols.

4.3.2 STEROIDAL SURFACTANTS

Steroidal surfactants like saponins have been traditionally used as natural detergents from a long time. This class of surfactants are also industrially important due to having its several applications. Saponins word comes from Latin word 'sapo,' which means soap; therefore, the plants having saponin can form lathery foam in water. Saponins are having aglycone moiety and sugar unit as hydrophobic and hydrophilic fragments, respectively, which are responsible for its surfactant-like characteristic of it. Due to biologically active nature, triterpenoid and saponins are having several applications as anti-inflammatory, hypocholesterolemic and immune-enhancing medicines in pharmaceuticals The class of saponins which have one sugar moiety displays better foaming properties compared to the saponins compounds with two or three sugar chains.

The mechanism of reducing surface tension and foam formation involves alignment of saponin molecules vertically on aqueous surface as shown in Figure 4.4. As it is clear from the figure that saponin molecules arrange themselves, so that its hydrophilic part is inside and hydrophobic part is outside of the aqueous surface. Above CMC surfactants can form micelles which have a lipophile center in which it can trap the fat molecules (Figure 4.4) [33].

Emulsifiers in bile are basically steroidal surfactant compounds called bile salts (BSs) where they do the emulsification of dietary fats, so that these can be absorbed by the intestine. BSs have steroidal as amphiphilic moiety rather than the sugar molecule, therefore are stiffer than the conventional surfactants. BSs have special characteristics like self-assembly, solubility and permeation capabilities. Derivatives of BSs can be synthesized using natural BSs as initiators to achieve improved properties like protein solubilization, effective receptors and carriers. New category of nonconventional molecules can be prepared by changing the hydrophobic and hydrophilic arrangement in the molecule. These molecules form helical like supramolecular tubules with the inner diameters between 3 and about 500 nm. These are very applicable substances at various conditions and environment. Moreover, oligomer of BSs can be prepared by linking them together which provides different manner of structure also provides possible applications due to exclusive properties [34].

CHAPSO and CHAPS are derived from BSs and has specific structure with the head of dipolar sulfobetaine. Therefore, these are zwitterionic surfactants and used for purification and crystallization of protein, along with the formation and solubilization of bicelle and liposome.

CHAPSO and CHAPS have long chains which provide a vast isoelectric range which shows mild detergent behavior. Characteristics of BSs are shown due to the hydroxyl groups in sterol rings like cholic acid and mainly taurocholate (TC). Micelles formation in BSs is a nonconventional process as they show a wide range of CMC. Hydrophobic interactions are shown in the mechanism which includes back-to-back aggregation. Slow size growth involves aggregation of BSs surfactants due to this reason there is a wide range of CMCs values have been reported. Furthermore, a wide range of CMCs have been confirmed using Nuclear Magnetic Resonance (NMR), chromatographic and fluorescence techniques. High sensitivity and simultaneous assessment of molecular interaction makes the fluorescence measurement more suitable. By the means of characterization, the interaction behavior of micelles acting as a hosting system and entrapping capacity of guest molecules along with the nature of guest molecules can be identified. Pyranine (HPTS) found to be a suitable dye for detection of CMC and formation of aggregates of surfactants. Responses of this dye for different aggregates are distinct from each other. Barnadas et al. have performed similar research to study the aggregation mechanism and formation of pre-micellar aggregates for CHAPSO and CHAPS. This research explores the interaction of HPTS with the surfactants and found a wide range of surfactant concentration with three critical concentrations. Results show that hydroxyl groups of sterol ring and sulfate group of the dye are involved due to the formation of hydrogen bonds and for this reason negative dye incorporated in CHAPSO and CHAPS micelles. Hosting nature of negative BSs TC for a negatively charged molecule was also identified in this study [35].

4.3.3 TERPENOIDAL SURFACTANTS

Terpenes, major ingredients of essential oils isolated from plants in the form of terpenoidal aglycone part (Figure 4.5) as bifurcated hydrophobic rear end, were used in the preparation of sustainable surfactants, the surface activity of which was comparable to conventional quaternary ammonium compounds. Farnesol, a 15-carbon acyclic sesquiterpene alcohol present in numerous plant species such as *lemon grass, rose, neroli, citronella,* etc, is a precursor of many terpenes. Terpenoidal saponins can also classified according to the number of sugar units as (i) monodesmosidic: one sugar unit (ii) bidesmosidic saponins: two sugar units, (iii) tridesmosidic saponins: three sugar units (Figure 4.6). These sugar units attached to the aglycone may be linear or branched chains, that are chiefly D-glucose, L-rhamnose, D-xylose, L-arabinose, D-galactose, D-fructose or D-glucuronic acid.

Use of terpene-based surfactants is very uncommon. Recent researches showed that surfactants from terpenes can be synthesized by hydro-aminomethylation reaction of it. Although the reaction involves the high pressure and formation of by-products, to overcome these problems, a better way has been explored by the researchers for the formation of long-chain amines through the process of hydroamination. Hydroamination generally involves the addition of amines (except tertiary amine) on C=C double bond [36].

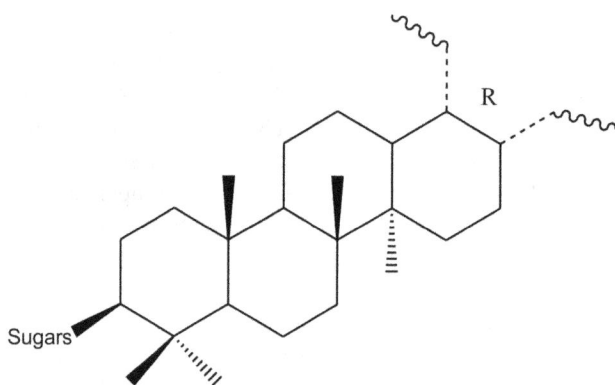

FIGURE 4.5   Terpenoidal aglycone.

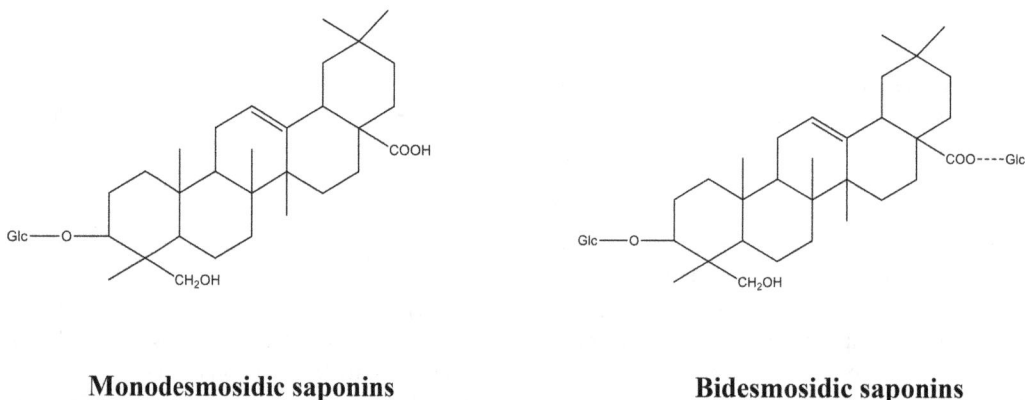

Monodesmosidic saponins   **Bidesmosidic saponins**

FIGURE 4.6   Terpenoidal saponins.

Vorholt et al. designed a prominent catalytic method to achieve the basic units for surfactants by utilizing the industrially-available terpenes through hydroaminomethylation. This study has shown hydroaminomethylation of 1,3-dienes with the highest regioselectivity (S = 97%) for the first time. The Rh/dppe catalytic system developed in this study and found to be suitable for the amines and can produce large amounts of product even in very less times. Reported reaction involves the hydroformylation and an enzyme condensation after this hydrogenation takes place. To achieve the surfactant properties, the obtained amine is converted to quaternary ammonium compounds. Therefore, the developed process for amine is a sustainable, eco-friendly and green approach to obtain surfactants from renewable feedstock. Rh/dppe catalytic system is also appropriate for other amines and terpenes. It also shows surfactants properties similar to conventional cationic surfactants [37].

4.3.4 ALKALOIDAL SURFACTANTS

Alkaloids generally contain a nitrogen atom in the carbon ring, and the number of rings can be varied having different locations of the nitrogen atom as for different alkaloids of distinct families of plants. Also, the properties of the alkaloids are altered by the specific location of the nitrogen atom (Table 4.1) [38, 39].

Glycoalkaloids are commonly found in potato and tomato. Potato tubers are a potent source of vitamin B6, vitamin C and carbohydrates (88% starch). They also contain glycoalkaloids, focused primely in the skin layer, the standard concentration of which in tubers should not be higher than 20 mg/100 g fresh weight. The mixture of two glycosides of solanidine, α-solanine and a trisaccharide of α-chaconine are present in glycoalkaloids of the potato tubers (Figure 4.7) [40].

Global tomato annual production is around 160 million tons, some of which is exploited for the processing industries comprising dried red, ketchup, pomace industries, which give rise to the tomatoes as prominent vegetables for transformation. Two glycoalkaloids – dehydrotomatine and α-tomatine – are present in tomatoes (Figure 4.8) [40].

Legumes, kidney beans, onion, garlic, asparagus, oats, spinach and tea have saponins. Medicinal plants having saponins are Mojave yucca (*Yucca schidigera*), fenugreek (*Trigonella foenum-graecum*), licorice (*Glycyrrhiza* species), ginseng (*Panax* species), alfalfa (*Medicago sativa*), soapwort (*Saponaria officinalis*) etc.

Surfactants properties of the solenopsin alkaloids were identified and found to be substantial by researchers due to its amphipathic character. As the amphipathic compounds can decrease the interfacial tension due to physical interactions. These compounds can generate the micelles and microemulsions also help to adsorb to existing surfaces. Formation of micelles can also inhibit the microbial tolerance and formation of biofilm formation. For hydrophilic surfaces, value of total free energy (ΔG) should be positive and value of water contact angle should be less than 65°. Rhamnolipid extract obtained in this study shows outstanding surfactant behavior on water surface

TABLE 4.1
Few Plant Sources and Their Alkaloids

Plant Name	Alkaloid	Uses
Papaver somniferum	Morphine	To treat severe pain
Piper nigrum	Piperine	Anti-inflammatory effects
Coffea arabica	Caffeine	To treat drowsiness, headache
Nicotiana tabacum	Nicotine	A stimulant to ease tension
Vinca minor	Vinpocetine	To increase brain metabolism
Salsola collina	Salsoline	As hypertensive agent
Peganum harmala	Harmine	To treat diabetes

Solanidine   Demissidine

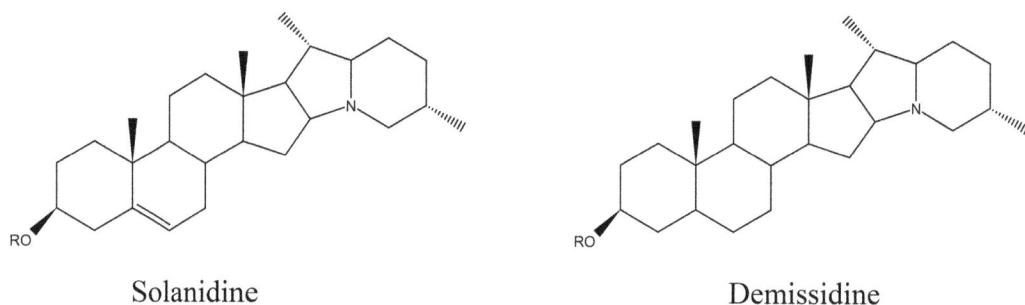

FIGURE 4.7 Glycoalkaloids of potato tubers.

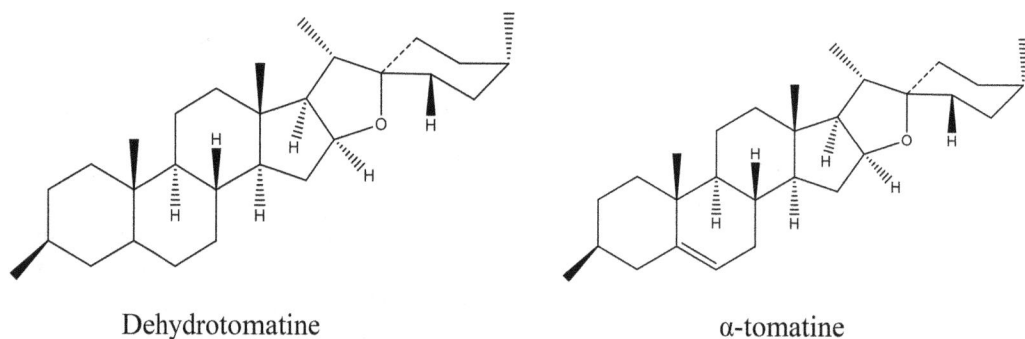

Dehydrotomatine   α-tomatine

FIGURE 4.8 Tomato saponin structures.

with the desired parameters. However, surface tension of water is not affected by solenopsins extract probably due to its nonpolar behavior. Though, it has the capability to condition the water and reduce the formation of biofilms. So alkaloids might have a different mechanism from the biosurfactant for the suppression of biofilm [41].

4.4 SYNTHESIS OF SURFACTANTS

Surfactants are designed by linking together sterol-based hydrophobic moiety like cholesterol, bile acids (ursodeoxycholic acid) which have tendency to self-assemble with hydrophilic moieties such as polyethylene glycol, amino acid and oligopeptides [42–44]. Due to their unique structure surfactant can bring the two immiscible phases together by lowering the interfacial tension between them [45]. Thus, water immiscible drugs can easily be used for drug delivery due to homogenization of two phases with the help of surfactant (Table 4.2 and Figure 4.9) [46, 47].

4.4.1 RENEWABLE SURFACTANTS VIA HYDROAMINOMETHYLATION OF TERPENES

Another efficient method of preparation of renewable surfactants is via the hydroaminomethylation of terpenes which is a three step catalytic reaction, including hydroformylation followed by condensation of the aldehyde with an amine to an enamine and hydrogenation to form saturated amines (Figure 4.10) [37].

4.4.1.1 Sterol Ethoxylates

Sterol ethoxylates have minor oral toxicity along with eye and skin irritation. Sterol ethoxylates are synthesized by reaction of hydrophobic moiety which is a carbon rich fatty alcohol with a

TABLE 4.2

Examples of Common Steroidal Surfactants

Hydrophobic Moiety	Hydrophilic Moiety	Surfactant
Sorbitan fatty acid esters	Polyoxyethylene	Polysorbate 80
Cholesterol	Glutamic acid	Cholesteryl-glutamic acid
Cholesterol	PHEG	Cholesteryl-PHEG
UDCA	PHEG	UDCA-PHEG
UDCA	PGA	UDCA-PGA

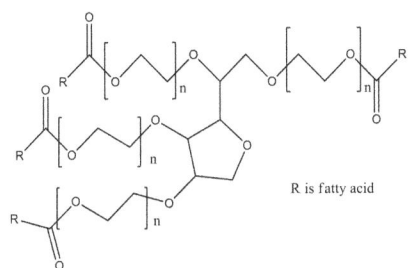

Polysorbate 80

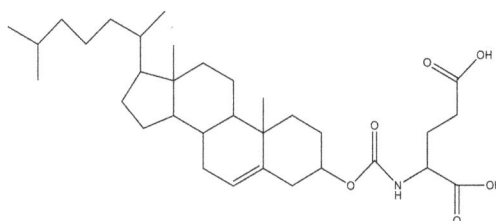

Cholesteryl-glutamic acid

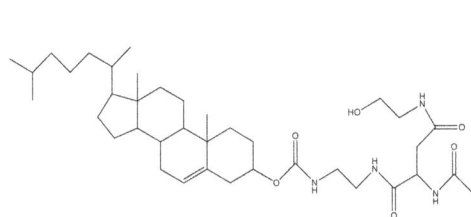

Cholesteryl-PHEG

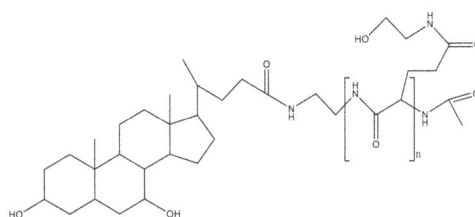

UDCA-PHEG

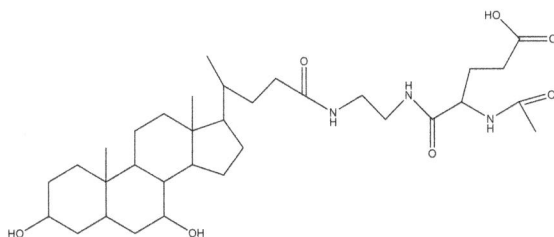

UDCA-PGA

FIGURE 4.9 Examples of common steroidal surfactants.

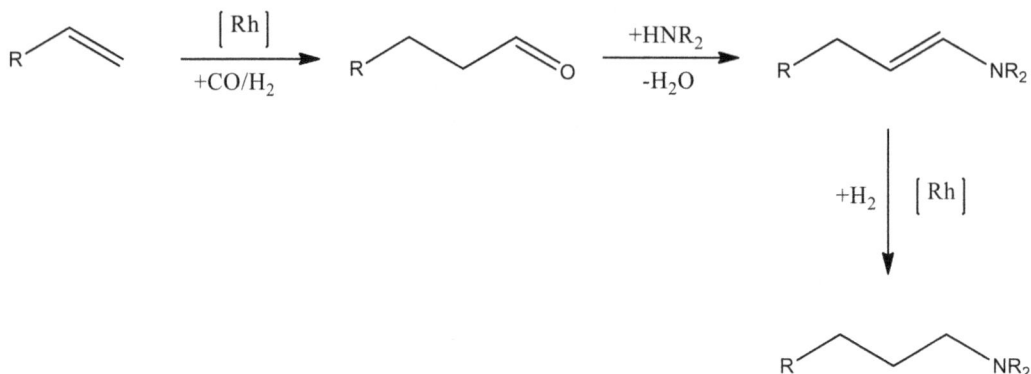

FIGURE 4.10 Hydroaminomethylation of terpenes.

hydrophilic polyoxyethylene chain [48–50]. The natural source of fatty alcohol is oil from palm tree, coconut tree and oil of rapeseed [51, 52]. Sterol ethoxylates derived from coconut oil differ from that of palm oil due to different distribution of carbon atoms, as a result ethoxylates differ in their properties and applications [53]. Secondary alcohols have less reactivity than primary alcohols due to steric hindrance. Most commonly ethoxylates are synthesized by reaction of fatty alcohol with ethylene oxide using a base catalyst, e.g., polysorbate 80. Ethoxylation of sterols leads to polydispersity in the hydrophilic chain of the oxyethylene group, but it can be due to the hydrophobic group as it is obtained from natural sources. Polydispersity due to the hydrophobic group can be minimized by checking its purity. Less polydisperse surfactant is synthesized by reaction of methoxypoly (oxyethylene) methane sulfonate with cholesterol by the use of sodium hydride catalyst.

4.4.1.2 Steryl Glucosides
Steryl glucoside was isolated as phytosteryl 6-O-acyl-D-glucopyranoside from potatoes. It was found mainly in younger tissues like leaves than other parts of the plant. Stigmasterol, beta-sitosterol and campesterol can exist in free form or ester form. Synthesis of sterols in plants occurs in free form followed by steryl glucoside formation.

4.4.1.3 Sterol Peptides
To solution of triethylamine and peptide, cholesteryl formate dissolved in anhydrous DMF added in an inert atmosphere. The crude product was centrifuged, and the supernatant layer was separated from by-products (Figure 4.11) [54].

4.4.1.4 Sterol Cyclodextrin
Cyclodextrins are also important for enhancing bioavailability and pharmacokinetics of drugs upon conjugation with sterols like cholesteryl-beta-cyclodextrin.

4.4.1.5 Sterol-Chitosan Oligosaccharide-Stearic Acid
Amino group of chitosan oligosaccharide and carboxyl group of stearic acid are coupled to synthesize chitosan oligosaccharide copolymer (CSO-SA) [55, 56]. CSO-SA, an anticancer drug form micelle in aqueous medium, can efficiently permeate into cancer cells. Cis-aconitic anhydride dissolved in dioxane is added to doxorubicin and dissolved in pyridine and stirred overnight at 4°C [57, 58]. The reaction mixture is then extracted by using aqueous solution of sodium bicarbonate and chloroform to get CA-DOX. CA-DOX dissolved in DMSO is added to CSO-SA and dissolved in water and then subjected to ultrasonic treatment to obtain DOX-CSO-SA (Figure 4.12) [59, 60].

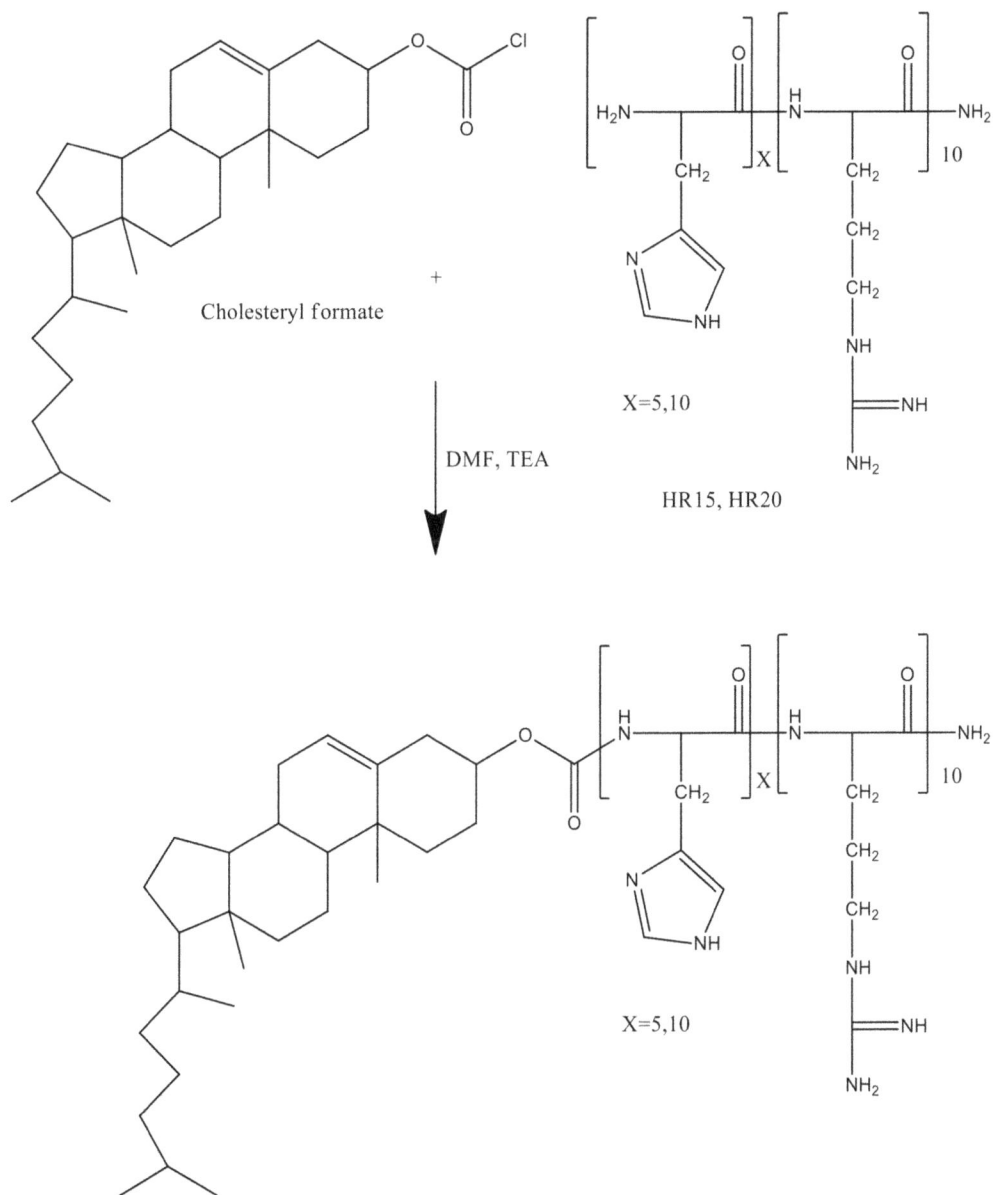

FIGURE 4.11 Sterol peptides.

4.5 PHYSICOCHEMICAL PROPERTIES OF SURFACTANTS

4.5.1 SELF-ASSEMBLING CAPACITY

Surfactants have the peculiar feature of forming a cluster called micelles. Hydrophilic heads are in the exterior, while the hydrophobic tails are cloistered in the interior forming the body of a micelle. CMC is the concentration at which formation of micelles starts. The micelles control the solubility of nonpolar compounds in polar solvents as well as affect the essential property of liquids called viscosity. The corresponding size of the hydrophobic and hydrophilic portions determines the structure of surfactant, thereby varying the shape of micelle from spherical to rod-like or to lamellar.

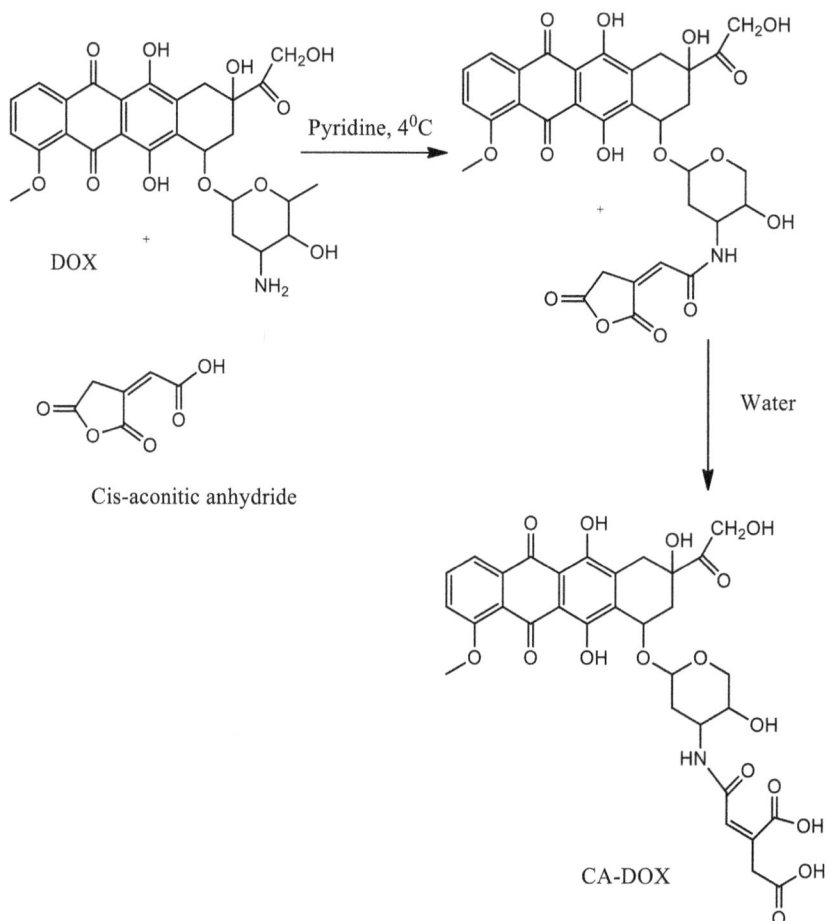

FIGURE 4.12   Sterol-chitosan oligosaccharide-stearic acid.

4.5.2   SOLUBILITY

Due to the nonionizable hydrophilic group such as sugars, glycerol and its derivative, nonionic surfactants do not dissociate in aqueous solution. The hydrophilicity of these surfactants can be increased by the polycondensation of ethylene oxide as polyoxyethylene esters (PEGs) and poloxamers or pluronics [61, 62]. The low water solubility (0.36 mg/ml) of cholesteryl-glutamic acid is the result of low hydrophilic/lipophilic ratio attributed to the small polar head of negatively charged amino acid. Polar heads of UDCA-PHEG, UDCA-PGA and cholesteryl-PHEG are larger than glutamic acid which increase their solubility to higher than 100 mg/ml.

4.5.3   DRUG SOLUBILIZATION CAPACITY

Drug solubility depends upon the molecular structure and self-assembling capacity of surfactants which is an important parameter in drug delivery. Polysorbate 80 dissolves the drug up to 17.3 mg/ml, whereas cholesteryl-PHEG increases solubility up to 2 mg/ml. Cholesterol is more effective in drug solubilization with PHEG-based surfactants than UDCA because of less hydrophobicity of UDCA than cholesterol as the aromatic rings contain two hydroxyl groups [63, 64].

4.5.4 CYTOTOXICITY

Cytotoxicity determines the interactions of surfactant molecules with endothelial cells of blood vessels. Cytotoxicity depends upon the interaction of the polar head of surfactant with the cell wall as well as its tendency to undergo self-assembly. It is concluded from MTT assay that surfactant cytotoxicity follows the given order: UDCA-PHEG < polysorbate 80 < cholesteryl-PHEG < UDCA-PGA.

4.5.5 BIOCOMPATIBILITY

Hemolysis assays check the harm created by surfactants to RBCs. In UDCA-PGA and UDCA-PHEG, hemoglobin is not released which makes it compatible for intravenous injection.

4.5.6 FOAMING ABILITY

The best foaming properties are shown by sterol with 16 oxyethylene units.

4.5.7 MELTING POINT

The melting point of the surfactant decreases with increase in oxyethylene units in the polar head.

4.5.8 CLOUD POINT

Cloud point of surfactant with a longer oxyethylene chain is higher than that of a shorter oxyethylene chain.

4.5.9 STABILITY

It is stable in healthy tissue at pH 7.2 but undergoes degradation in moderately acidic conditions of tumor cells.

4.6 APPLICATIONS OF SURFACTANTS

4.6.1 BIOACTIVITIES OF SURFACTANTS [65]

Saponins are known to have their various bioactivities and help in regulating biochemical reactions of the human body. Some important known activities are cardioprotective activities, antidiabetic activities, etc.

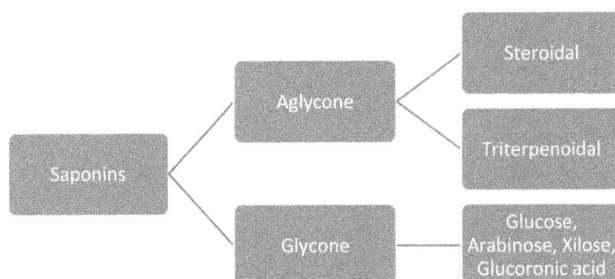

4.6.2 Cardioprotective Activity and Hypocholesterolemic Effect

The cholesterol-lowering activity of saponins has been well documented in animals and humans, different pathways being studied and some elucidated. Saponins are recorded to have an effect in decreasing the absorption of cholesterol from the intestine. Besides, the faecal cholesterol excretion increases noticeably when saponins are included in the daily diet, whereas the conversion of cholesterol into bile acid by the liver is significantly promoted. Some phytosterols such as diosgenin and its derivatives have the capability to reduce the serum cholesterol in the intestinal tract by suppressing the cholesterol absorption, thereby recognized as cardioprotective agents.

4.6.3 Adjuvant Activity

Saponins are used as adjuvants in oral vaccines formulation due to their stimulation of the responses of the immune system against the pathogens. An adjuvant generally supports the immunogenicity of antigen, improving the efficacy of the vaccine, thus minimizing the dose of antigen needed. Extracted saponins from *Quillaja saponaria*, *Ginseng*, *Astragalus* are reviewed to have the ability to stimulate the cell-mediated immune system and the antibody production.

4.6.4 Anticancer Activity

Saponins are regarded to be promising in applying in cancer therapy. Also, the cytotoxicity was evaluated by using triterpene saponin, isolated from the fruit of *Ardisia kivuensis Taton* and from the leaves of *Aralia elata*. These saponins proved to be potential in developing unique cytotoxic drugs to fight against both sensitive and drug-resistant cancers.

4.6.5 Antibacterial and Antifungal Activities

Saponins from citrus peel have been reported to have antibacterial and antifungal properties. Moreover, the peel of citrus fruits is rich in flavonoid, glycosides, sitosterol and volatile oils which can be efficiently used as drugs or as food supplements due to the fact that these compounds can act synergistically in association with antibiotics. Also, steroidal saponins possess antifungal properties against various plant pathogens.

4.6.6 Antidiabetic Activity

Diabetes mellitus, a metabolic disorder indicated by enhanced blood glucose levels, thereby damaging the eyes, kidneys, heart, blood vessels and nerves. Saponins have been shown to decrease diabetes mellitus through a system involving increase of plasma insulin levels as the saponins trigger pancreas to release insulin.

4.7 COMMERCIAL APPLICATIONS

4.7.1 Drug Delivery

Surfactants are used in drug delivery by solubilizing the water immiscible drugs. Surfactants having cholesterol with a long poly ethylene glycol chain stay in the blood for a long time but not much target specific. Glycosides of cholesteryl oligo ethylene glycols are target specific as well as possess long blood circulation time. The drug solubilization is due to association of drug with the polyoxyethylene part [66].

4.7.2 Cosmetic Industry

Nowadays natural substances such as lecithins, phytosterols, are mainly used in the cosmetics industry. Sterol ethoxylates act as good emulsifiers in cosmetic formulations. Phytosterol are used in shampoos to improve the condition of dry hair and reduce the electrostatic charge as well as in moisturizing lotions, facial cleansing lotions, aftershaves and lipsticks. Span 80 is used as an emulsifier with unsaturated fat constituents. Cremophor RH40 is used in scents based on water and alcohol to avoid turbidity.

4.7.3 GENE THERAPY

Gene therapy gives potential treatment for the diseases caused by genetic disorders like cancer. Peptide-based carriers are more biocompatible due to their biodegradability. Histidine-, arginine- and lysine-based peptides are used as gene transfection vectors. Hydrophobic compounds are conjugated with peptides to improve DNA binding. Amphiphilic peptides bonded to cholesterol undergo self-assembly in aqueous medium to form cationic core micelles, which will increase the DNA binding efficiency. These micelles have low cytotoxicity.

4.7.4 FOOD PROCESSING INDUSTRY

Polysorbate 80 is used as an emulsifier in ice-cream to make it smoother and resistant to easy melting. Their utilization in the foodstuff business incorporates fat stabilization in food oils, enhanced solubility in instant drinks and soups and coverings for fruits and vegetables etc.

4.7.5 LAUNDRY

Natural surfactants show excellent emulsion formation ability with vegetable oils. Their outstanding rapport establishes them as the favorable washing detergents.

4.7.6 LOWERING OF CO_2 EMISSIONS

Surfactants are able to cut down the CO_2 emissions. One such example is of the oleochemical surfactants by which 1.5 million tons of CO_2 emissions were escaped in 1998 [67–71].

4.7.7 MISCELLANEOUS

Other promising usages include paper, pulp and metal processing, ceramics, fixing dyes, etc.

Some of the commercial products based on surfactants are listed in Table 4.3.

Polysorbate is used as a solubilizer in mouthwash. In laboratory, polysorbate 80 is used to identify the type of strain of mycobacterium as it causes the solution to change its color and few of the surfactants are also used in cleaning artefacts [72–74].

TABLE 4.3
Commercial Natural Surfactants and Their Uses

Surfactants	Product
Polysorbate 20 (Tween 20)	Targretin soft gelatin capsule
Polysorbate 80 (Tween 80)	Gengraf hard gelatin capsule
Sorbitan monooleate (Span 80)	Gengraf hard gelatin capsule
Polyoxyl-35-castor oil (Cremophor EL)	Gengraf hard gelatin capsule, Ritonavir soft gelatin capsule
Polyoxyl-40-hydrogenated castor oil (Cremophor RH40)	Neoral soft gelatin capsule, Ritonavir oral solution
Polyoxyethylated glycerides (Labrafil M 2125Cs)	Sandimmune soft gelatin capsules
Polyoxyethylated oleic glycerides (Labrafil M 1944Cs)	Sandimmune oral solution
D-Alpha-tocopheryl polyethylene glycol 1000 succinate (TPGS)	Agenerase soft gelatin capsule, Agenerase oral solution

4.8 CONCLUSIONS

Saponins are a phytochemical group present in various plants. Saponin-rich plants earn deep-seated vision as a potential source of natural surfactants as they acquire the capacity to take over harmful synthetic surfactants due to their great surface activity, biological activities and broad distribution in nature. However, natural surfactants face challenges due to low productivity and expensive down-streaming processing. Still, more research in studying this group of secondary metabolites is necessary keeping in view that many of saponins action mechanisms are not well understand. At present, when the concern about consumers' health is increasing, full comprehension is the only key to effectively exploit the potential benefits of saponins. In this sense, future investigations should focus toward optimizing the bioavailability of saponins to enhance their in vivo chemotherapeutic effects.

REFERENCES

1. Garofalakis, G., Murray, B. S., & Sarney, D. B. (2000). Surface activity and critical aggregation concentration of pure sugar esters with different sugar headgroups. *Journal of Colloid and Interface Science*, *229*(2), 391–398.
2. Dave, N., & Joshi, T. (2017). A concise review on surfactants and its significance. *International Journal of Applied Chemistry*, *13*, 663–672. [Online]. Available: www.ripublication.com.
3. Bhadani, A., Kafle, A., Ogura, T., Akamatsu, M., Sakai, K., Sakai, H., & Abe, M. (2020). Current perspective of sustainable surfactants based on renewable building blocks. *Current Opinion in Colloid & Interface Science*, *45*, 124–135. doi: 10.1016/j.cocis.2020.01.002.
4. Gershanik, T., & Benita, S. (2000). Self-dispersing lipid formulations for improving oral absorption of lipophilic drugs. *European Journal of Pharmaceutics and Biopharmaceutics*, *50*(1), 179–188.
5. Pouton, C. W. (2000). Lipid formulations for oral administration of drugs: non-emulsifying, self-emulsifying and 'self-microemulsifying' drug delivery systems. *European Journal of Pharmaceutical Sciences*, *11*, S93–S98. doi: 10.1016/S0928-0987(00)00167-6.
6. Constantinides, P. P. (1995). Lipid microemulsions for improving drug dissolution and oral absorption: physical and biopharmaceutical aspects. *Pharmaceutical Research*, *12*(11), 1561–1572. doi: 10.1023/A:1016268311867.
7. Narang, A. S., Delmarre, D., & Gao, D. (2007). Stable drug encapsulation in micelles and microemulsions. *International Journal of Pharmaceutics*, *345*(1–2), 9–25. doi: 10.1016/j.ijpharm.2007.08.057.
8. Flanagan, J., & Singh, H. (2006). Microemulsions: a potential delivery system for bioactives in food. *Critical Reviews in Food Science and Nutrition*, *46*(3), 221–237. doi: 10.1080/10408690590956710.
9. Wang, T., Chipot, C., Shao, X., & Cai, W. (2011). Structural characterization of micelles formed of cholesteryl-functionalized cyclodextrins. *Langmuir*, *27*(1), 91–97. doi: 10.1021/la103288j.
10. Folmer, B. M. (2003). Sterol surfactants: from synthesis to applications. *Advances in Colloid and Interface Science*, *103*(2), 99–119. doi: 10.1016/S0001-8686(01)00100-2.
11. Sánchez, L., Mitjans, M., Infante, M. R., García, M. T., Manresa, M. A., & Vinardell, M. P. (2007). The biological properties of lysine-derived surfactants. *Amino Acids*, *32*(1), 133–136. doi: 10.1007/s00726-006-0318-x.
12. Cserháti, T., Forgács, E., & Oros, G. (2002). Biological activity and environmental impact of anionic surfactants. *Environment International*, *28*(5), 337–348. doi: 10.1016/j.memsci.2004.04.007.
13. Lee, M. K., Kim, S., Ahn, C. H., & Lee, J. (2010). Hydrophilic and hydrophobic amino acid copolymers for nano-comminution of poorly soluble drugs. *International Journal of Pharmaceutics*, *384*(1–2), 173–180. doi: 10.1016/j.ijpharm.2009.09.041.
14. Al-Baghdadi, O. B., Prater, N. I., Van der Schyf, C. J., & Geldenhuys, W. J. (2012). Inhibition of monoamine oxidase by derivatives of piperine, an alkaloid from the pepper plant *Piper nigrum*, for possible use in Parkinson's disease. *Bioorganic & Medicinal Chemistry Letters*, *22*(23), 7183–7188. doi: 10.1016/j.bmcl.2012.09.056.
15. Lutz, J. F., Pfeifer, S., Chanana, M., Thünemann, A. F., & Bienert, R. (2006). H-bonding-directed self-assembly of synthetic copolymers containing nucleobases: organization and colloidal fusion in a non-competitive solvent. *Langmuir*, *22*(17), 7411–7415. doi: 10.1021/la061382a.
16. Lech-Maranda, E., Korycka, A., & Robak, T. (2006). Pharmacological and clinical studies on purine nucleoside analogs–new anticancer agents. *Mini Reviews in Medicinal Chemistry*, *6*(5), 575–581. doi: 10.2174/138955706776876212.

17. Jin, Y., Xin, R., Ai, P., & Chen, D. (2008). Self-assembled drug delivery systems: 2. Cholesteryl derivatives of antiviral nucleoside analogues: synthesis, properties and the vesicle formation. *International Journal of Pharmaceutics*, *350*(1–2), 330–337. doi: 10.1016/j.ijpharm.2007.08.037.

18. Jin, Y., Xing, L., Tian, Y., Li, M., Gao, C., Du, L., et al. (2009). Self-assembled drug delivery systems. Part 4. In vitro/in vivo studies of the self-assemblies of cholesteryl-phosphonyl zidovudine. *International Journal of Pharmaceutics*, *381*(1), 40–48. doi: 10.1016/j.ijpharm.2009.07.024.

19. Rai, S., & Bhattarai, (2020). A. A review on saponin: plant derived natural surfactants: a review on saponin: plant-derived natural surfactants. Retrieved from: https//encyclopedia.pub/1665, no. i.

20. Holmberg, K. (Ed.). (2003). *Novel surfactants: preparation applications and biodegradability, revised and expanded* (Vol. 114). CRC Press.

21. Panchawat, S., Rathore, K. S., & Sisodia, S. S. (2010). A review on herbal antioxidants. *International Journal of PharmTech Research*, *2*(1), 232–239.

22. Hadi, J. N. and Mohd Hassan, Norazian and Ahmad, Kausar (2011). *Natural surfactants for pharmaceutical emulsion*. In: Current Issues in Pharmacy. IIUM Press, Kuala Lumpur, pp. 178–195.

23. Killeen, G. F., Madigan, C. A., Connolly, C. R., Walsh, G. A., Clark, C., Hynes, M. J., et al. (1998). Antimicrobial saponins of *Yucca schidigera* and the implications of their in vitro properties for their in vivo impact. *Journal of Agricultural and Food Chemistry*, *46*(8), 3178–3186. doi: 10.1021/jf970928j.

24. Francis, G., Kerem, Z., Makkar, H. P., & Becker, K. (2002). The biological action of saponins in animal systems: a review. *British Journal of Nutrition*, *88*(6), 587–605. doi: 10.1079/bjn2002725.

25. Ribeiro, B. D., Alviano, D. S., Barreto, D. W., & Coelho, M. A. Z. (2013). Functional properties of saponins from sisal (*Agave sisalana*) and juá (*Ziziphus joazeiro*): critical micellar concentration, antioxidant and antimicrobial activities. *Colloids and Surfaces A: Physicochemical and Engineering Aspects*, *436*, 736–743. doi: 10.1016/j.colsurfa.2013.08.007.

26. Moghimipour, E., & Handali, S. (2015). Saponin: properties, methods of evaluation and applications. *Annual Research & Review in Biology*, 207–220. doi: 10.9734/arrb/2015/11674.

27. Vincken, J. P., Heng, L., de Groot, A., & Gruppen, H. (2007). Saponins, classification and occurrence in the plant kingdom. *Phytochemistry*, *68*(3), 275–297. doi: 10.1016/j.phytochem.2006.10.008.

28. Phillipson, J. D. (2001). Phytochemistry and medicinal plants. *Phytochemistry*, *56*(3), 237–243. doi: 10.1016/S0031-9422(00)00456-8.

29. Cheok, C. Y., Salman, H. A. K., & Sulaiman, R. (2014). Extraction and quantification of saponins: a review. *Food Research International*, *59*, 16–40. doi: 10.1016/j.foodres.2014.01.057.

30. Sparg, S. G., Light, M. E., & Van Staden, J. (2004). Biological activities and distribution of plant saponins. *Journal of Ethnopharmacology*, *94*(2–3), 219–243. doi: 10.1016/j.jep.2004.05.016.

31. Blunk, D., Bierganns, P., Bongartz, N., Tessendorf, R., & Stubenrauch, C. (2006). New speciality surfactants with natural structural motifs. *New Journal of Chemistry*, *30*(12), 1705–1717. doi: 10.1039/b610045g.

32. Alvès, M. H., Sfeir, H., Tranchant, J. F., Gombart, E., Sagorin, G., Caillol, S., et al. (2014). Terpene and dextran renewable resources for the synthesis of amphiphilic biopolymers. *Biomacromolecules*, *15*(1), 242–251. doi: 10.1021/bm401521f.

33. Kregiel, D., Berlowska, J., Witonska, I., Antolak, H., Proestos, C., Babic, M., et al. (2017). Saponin-based, biological-active surfactants from plants. *Application and Characterization of Surfactants*, *6*(1), 184–205. dx.doi.org/10.5772/68062.

34. di Gregorio, M. C., Travaglini, L., Del Giudice, A., Cautela, J., Pavel, N. V., & Galantini, L. (2018). Bile salts: natural surfactants and precursors of a broad family of complex amphiphiles. *Langmuir*, *35*(21), 6803–6821. doi.org/10.1021/acs.langmuir.8b02657.

35. Barnadas-Rodríguez, R., & Cladera, J. (2015). Steroidal surfactants: detection of premicellar aggregation, secondary aggregation changes in micelles, and hosting of a highly charged negative substance. *Langmuir*, *31*(33), 8980–8988. doi.org/10.1021/acs.langmuir.5b01352.

36. Faßbach, T. A., Gösser, N., Sommer, F. O., Behr, A., Guo, X., Romanski, S., et al. (2017). Palladium-catalyzed hydroamination of farnesene—long chain amines as building blocks for surfactants based on a renewable feedstock. *Applied Catalysis A: General*, *543*, 173–179. doi.org/10.1016/j.apcata.2017.06.014.

37. Faßbach, T. A., Gaide, T., Terhorst, M., Behr, A., & Vorholt, A. J. (2017). Renewable surfactants through the hydroaminomethylation of terpenes. *ChemCatChem*, *9*(8), 1359–1362.

38. Hussain, G., Rasul, A., Anwar, H., Aziz, N., Razzaq, A., et al. (2018). Role of plant derived alkaloids and their mechanism in neurodegenerative disorders. *International Journal of Biological Sciences*, *14*(3), 341. doi: 10.7150/ijbs.23247.

39. Kwon, J., Seo, Y. H., Lee, J. E., Seo, E. K., Li, S., Guo, Y., et al. (2015). Spiroindole alkaloids and spiroditerpenoids from *Aspergillus duricaulis* and their potential neuroprotective effects. *Journal of Natural Products*, *78*(11), 2572–2579. doi: 10.1021/acs.jnatprod.5b00508.
40. Oleszek, M., & Oleszek, W. (2020). Saponins in food. *Handbook of dietary phytochemicals*, 1–40.
41. Carvalho, D. B. D., Fox, E. G. P., Santos, D. G. D., Sousa, J. S. D., Freire, D. M. G., et al. (2019). Fire ant venom alkaloids inhibit biofilm formation. *Toxins*, *11*(7), 420.
42. Matsuoka, K., & Moroi, Y. (2002). Micelle formation of sodium deoxycholate and sodium ursode-oxycholate (Part 1). *Biochimica et Biophysica Acta (BBA) – Molecular and Cell Biology of Lipids*, *1580*(2–3), 189–199. doi: 10.1016/S1388-1981(01)00203-7.
43. Sanchez, L., Mitjans, M., Infante, M. R., & Vinardell, M. P. (2006). Potential irritation of lysine deriva-tive surfactants by hemolysis and HaCaT cell viability. *Toxicology Letters*, *161*(1), 53–60. doi: 10.1016/j.toxlet.2005.07.015.
44. Brito, R. O., Marques, E. F., Gomes, P., Falcao, S., & Söderman, O. (2006). Self-assembly in a catan-ionic mixture with an amino acid-derived surfactant: from mixed micelles to spontaneous vesicles. *The Journal of Physical Chemistry B*, *110*(37), 18158–18165. doi: 10.1021/jp061946j.
45. Francis, M. F., Piredda, M., & Winnik, F. M. (2003). Solubilization of poorly water soluble drugs in micelles of hydrophobically modified hydroxypropylcellulose copolymers. *Journal of Controlled Release*, *93*(1), 59–68. doi: 10.1016/j.jconrel.2003.08.001.
46. Ménard, N., Tsapis, N., Poirier, C., Arnauld, T., Moine, L., Lefoulon, F., et al. (2011). Physicochemical characterization and toxicity evaluation of steroid-based surfactants designed for solubilization of poorly soluble drugs. *European Journal of Pharmaceutical Sciences*, *44*(5), 595–601. doi: 10.1016/j.ejps.2011.10.006.
47. Yang, D. B., Zhu, J. B., Huang, Z. J., Ren, H. X., & Zheng, Z. J. (2008). Synthesis and application of poly (ethylene glycol)–cholesterol (Chol–PEGm) conjugates in physicochemical characterization of nonionic surfactant vesicles. *Colloids and Surfaces B: Biointerfaces*, *63*(2), 192–199. doi: 10.1016/j.colsurfb.2007.11.019.
48. le Maire, M., Champeil, P., & Møller, J. V. (2000). Interaction of membrane proteins and lipids with solubilizing detergents. *Biochimica et Biophysica Acta (BBA) – Biomembranes*, *1508*(1–2), 86–111. doi: 10.1177/2284026518773227.
49. Hill, K., & Rhode, O. (1999). Sugar-based surfactants for consumer products and technical applications. *Lipid/Fett*, *101*(1), 25–33. doi: 10.1002/(sici)1521-4133(19991)101:1<25::aid-lipi25>3.0.co;2-n.
50. Johansson, I., & Svensson, M. (2001). Surfactants based on fatty acids and other natural hydrophobes. *Current Opinion in Colloid & Interface Science*, *6*(2), 178–188. doi: 10.1016/S1359-0294(01)00076-0.
51. Piispanen, P. (2002). *Synthesis and characterization of surfactants based on natural products* (Doctoral dissertation, Kemi). [Online]. Available: www.diva-portal.org/smash/get/diva2:9177/FULLTEXT01.pdf.
52. Garavito, R. M., & Ferguson-Miller, S. (2001). Detergents as tools in membrane biochemistry. *Journal of Biological Chemistry*, *276*(35), 32403–32406. doi: 10.1074/jbc.R100031200.
53. Stubenrauch, C. (2001). Sugar surfactants—aggregation, interfacial, and adsorption phenomena. *Current Opinion in Colloid & Interface Science*, *6*(2), 160–170. doi: 10.1016/S1359-0294(01)00080-2.
54. Guo, X. D., Tandiono, F., Wiradharma, N., Khor, D., Tan, C. G., Khan, M., et al. (2008). Cationic micelles self-assembled from cholesterol-conjugated oligopeptides as an efficient gene delivery vector. *Biomaterials*, *29*(36), 4838–4846. doi: 10.1016/j.biomaterials.2008.07.053.
55. Hu, F. Q., Liu, L. N., Du, Y. Z., & Yuan, H. (2009). Synthesis and antitumor activity of doxorubicin con-jugated stearic acid-g-chitosan oligosaccharide polymeric micelles. *Biomaterials*, *30*(36), 6955–6963. doi: 10.1016/j.biomaterials.2009.09.008.
56. Engels, F. K., Mathot, R. A., & Verweij, J. (2007). Alternative drug formulations of docetaxel: a review. *Anti-Cancer Drugs*, *18*(2), 95–103. doi: 10.1097/CAD.0b013e3280113338.
57. Duncan, R. (2006). Polymer conjugates as anticancer nanomedicines. *Nature Reviews Cancer*, *6*(9), 688–701. doi: 10.1038/nrc1958.
58. Yamaoka, T., Tabata, Y., & Ikada, Y. (1995). Fate of water-soluble polymers administered via different routes. *Journal of Pharmaceutical Sciences*, *84*(3), 349–354. doi: 10.1002/jps.2600840316.
59. Kohori, F., Yokoyama, M., Sakai, K., & Okano, T. (2002). Process design for efficient and controlled drug incorporation into polymeric micelle carrier systems. *Journal of Controlled Release*, *78*(1–3), 155–163. doi: 10.1016/S0168-3659(01)00492-8.
60. Pasut, G., & Veronese, F. M. (2007). Polymer–drug conjugation, recent achievements and general strate-gies. *Progress in Polymer Science*, *32*(8–9), 933–961. doi: 10.1016/j.progpolymsci.2007.05.008.

61. Kirby, J., Nishimoto, M., Chow, R. W., Pasumarthi, V. N., Chan, R., Chan, L. J. G., et al. (2014). Use of nonionic surfactants for improvement of terpene production in *Saccharomyces cerevisiae*. *Applied and Environmental Microbiology*, *80*(21), 6685. doi: 10.1128/AEM.02155-14.

62. Schick, M. J. (Ed.). (1987). *Nonionic surfactants: physical chemistry*. CRC Press.

63. Gao, P., Guyton, M. E., Huang, T., Bauer, J. M., Stefanski, K. J., & Lu, Q. (2004). Enhanced oral bioavailability of a poorly water soluble drug PNU-91325 by supersaturatable formulations. *Drug Development and Industrial Pharmacy*, *30*(2), 221–229. doi: 10.1081/DDC-120028718.

64. Humberstone, A. J., & Charman, W. N. (1997). Lipid-based vehicles for the oral delivery of poorly water soluble drugs. *Advanced Drug Delivery Reviews*, *25*(1), 103–128. doi: 10.1016/S0169-409X(96)00494-2.

65. Nguyen, T. L., Fǎrcaş, A. C., Socaci, S. A., Tofanǎ, M., Diaconeasa, Z. M., et al. (2020). An overview of saponins–A bioactive group. *Bulletin UASVM Food Science and Technology*, *77*, 1. doi: 10.15835/buasvmcn-fst:2019.0036.

66. Gursoy, R. N., & Benita, S. (2004). Self-emulsifying drug delivery systems (SEDDS) for improved oral delivery of lipophilic drugs. *Biomedicine & Pharmacotherapy*, *58*(3), 173–182. doi: 10.1016/j.biopha.2004.02.001.

67. Huang, J., Peng, L., Zeng, G., Li, X., Zhao, Y., Liu, L., et al. (2014). Evaluation of micellar enhanced ultrafiltration for removing methylene blue and cadmium ion simultaneously with mixed surfactants. *Separation and Purification Technology*, *125*, 83–89. doi: 10.1016/j.seppur.2014.01.020.

68. Zahid, A., Lashin, A., Rana, U. A., Al-Arifi, N., Ullah, I., Dionysiou, D. D., et al. (2016). Development of surfactant based electrochemical sensor for the trace level detection of mercury. *Electrochimica Acta*, *190*, 1007–1014. doi: 10.1016/j.electacta.2015.12.164.

69. Ullah, I., Shah, A., Badshah, A., Shah, N. A., & Tabor, R. (2015). Surface, aggregation properties and antimicrobial activity of four novel thiourea-based non-ionic surfactants. *Colloids and Surfaces A: Physicochemical and Engineering Aspects*, *464*, 104–109. doi: 10.1016/j.colsurfa.2014.10.002.

70. Ali, N., Khan, A., Bilal, M., Malik, S., Badshah, S., & Iqbal, H. (2020). Chitosan-based bio-composite modified with thiocarbamate moiety for decontamination of cations from the aqueous media. *Molecules*, *25*(1), 226. doi: 10.3390/molecules25010226.

71. Rasheed, T., Shafi, S., Bilal, M., Hussain, T., Sher, F., & Rizwan, K. (2020). Surfactants-based remediation as an effective approach for removal of environmental pollutants—a review. *Journal of Molecular Liquids*, 113960. doi: 10.1016/j.molliq.2020.113960.

72. Giorgi, R., Baglioni, M., Berti, D., & Baglioni, P. (2010). New methodologies for the conservation of cultural heritage: micellar solutions, microemulsions, and hydroxide nanoparticles. *Accounts of Chemical Research*, *43*(6), 695–704. doi: 10.1021/ar900193h.

73. Qasim, M., Islam, W., Ashraf, H. J., Ali, I., & Wang, L. (2020). Saponins in insect pest control. *Co-Evolution of Secondary Metabolites*, 897–924. doi: 10.1007/978-3-319-76887-8_39-1.

74. Chelazzi, D., Bordes, R., Giorgi, R., Holmberg, K., & Baglioni, P. (2020). The use of surfactants in the cleaning of works of art. *Current Opinion in Colloid & Interface Science*, *45*, 108–123. doi: 10.1016/j.cocis.2019.12.007.

5 Sugar-Based Surfactants

Anindita De, Mridula Guin and Preeti Jain
Department of Chemistry and Biochemistry,
Sharda University, Greater Noida, India

CONTENTS

5.1 INTRODUCTION

Surfactants are amphoteric organic molecules that combine polar and nonpolar moieties and have specific physical and chemical properties. They are widely used in food, medicine, textile, agriculture, adhesives, leather, cleaning and cosmetics industries. They are located at interfaces of oil/water/soil or between phases of different polarities and associate themselves in micelles or liquid crystal aggregates [1]. They easily get adsorbed at the surface of the interface of solution when dissolved in solvents like water and oil, which is due to the availability of hydrophilic and hydrophobic groups in their structure. Their main feature is to adjust the physicochemical properties of the aqueous solution, like interfacial tension or viscosity. This feature makes their utility in various fields like cleaning detergents, food, cosmetics, cleaning soaps, paints, pesticides, inks, lubricants, polymers, textiles, pharmaceuticals, mining and gas recovery [2].

Surfactants are commodities and special chemicals with the following characteristics: (i) all humans use them in daily life, (ii) large tonnage and (iii) they are scrap products in wastewater or the environment. Therefore, they need to be carefully designed to prevent undesirable pollution or toxic effects. Mass production also shows that renewable resources are better than fossil resources and their design and manufacturing should be given priority [3]. Today, as the chemical industry paradigm shifts to more sustainable chemicals and processes, people are interested in the development of newer biosurfactants using clean and safe production methods. The purpose is to encounter the belief of consumers for benign products and the expectations of administrations aiming to implement regulations for restricting risks and hazards in the chemical production industry. In particular, although petroleum-based surfactants still dominate due to their cost-effective characteristics, it is expected that "bio-based surfactants" will be used at an industrial scale, especially in the field of personal care products and detergents [4].

There has been a long history of using biological resources to prepare surfactants for two main reasons: (i) the natural hydrophobicity and hydrophilicity of certain biomolecules (e.g. lipids and carbohydrates) to provide necessary duality [5,6] and (ii) to use low-value crops to add value. The production of sugar-based surfactants is relatively new in comparison to the derivatization of fat/oil. Sugar-based surfactants can also be used in biology, particularly in protein solubilization and membrane research due to their high Critical Micelle concentration(CMC) values. Moreover, they can also be used to reconstruct membrane proteins with lipids in biomimetic systems. Studies have shown that surfactants belonging to the alkyl polyglycoside (APG) group are very efficient, especially in dissolving hydrocarbons and organochlorine compounds in the process of washing away oil from the soil, in the ingredients of pesticides and agrochemicals and also have been proven to be drug carriers (e.g. amphiphilic glucan derivatives). They can be used to replace more toxic and more sensitizing surfactants in protein synthesis (e.g. trehalose fatty acid esters), few of which show antibacterial activity (such as sugar-based Gemini surfactants) or similar chemically pure APGs, due to their non-denaturing properties [1–5].

The hydrophobic part of sugar-based surfactants is generally a long-chain alkyl group and hydrophilic head group is carbohydrate molecules such as glucose, sucrose, etc. [7]. The leading sugar surfactants today are APGs, sorbitol esters and sucrose esters. Regarding the hydrophilic part, although safety and toxicity are considered, there are still a large number of synthetic nonionic and ionic surfactants synthesized at an industrial scale. Nearly half of the surfactants are still made from polyethoxylated units [8]. Therefore, looking for alternatives to the hydrophilic part is an important issue; in this regard, carbohydrates molecules due to the polyhydroxy structure obtained from renewable crops seem to be an ideal choice, although linking sugar molecules to lipid derivatives is somewhat challenging [9]. For all synthesis reactions of amphiphilic compounds, another important challenge is to contain the polar reactant and the oppositely polar catalyst in the same solvent. They tend to be at least partially separated, forming an interface or concentration gradient, which prevents accurate monitoring of stoichiometry [10]. If there are catalysts, one problem is to know their existence in this complicated medium, or whether they are in a phase rich in hydrophilic reactants, or at the interface. In heterogeneous media, the preparation of monosubstituted products tends to stay on the interface and can easily be replaced by di- or polysubstituted derivatives again, which is faster in comparison to the conversion of naked starting sugars [11]. Therefore, the strong influence of solvent and heterogeneity of medium is observed on the resulting DS, and when targeting low-substituted compounds, the additives through the use of solvent mixtures must maintain sufficient homogeneity [12]. Another challenge is the solubility of sugar molecules in fatty acids to obtain products in high yields. The miscibility can be increased by acetylating the sugar molecules before esterification and then performing deacetylation, which greatly avoids the synthesis of diesters and improves selectivity (Figure 5.1).

Amphiphilic products are inherently amphiphilic, which helps to improve the homogeneity of the mixture (autocatalysis) [13].

Among various sugar molecules, only a few meet the criteria of quality, price and availability (Figure 5.2). In addition to synthetic carbohydrate-based surfactants (CBS), there are also naturally occurring surfactants that are called glycolipids in biochemistry and are mainly found in the cell membranes of various living organisms, such as rhamnolipids.

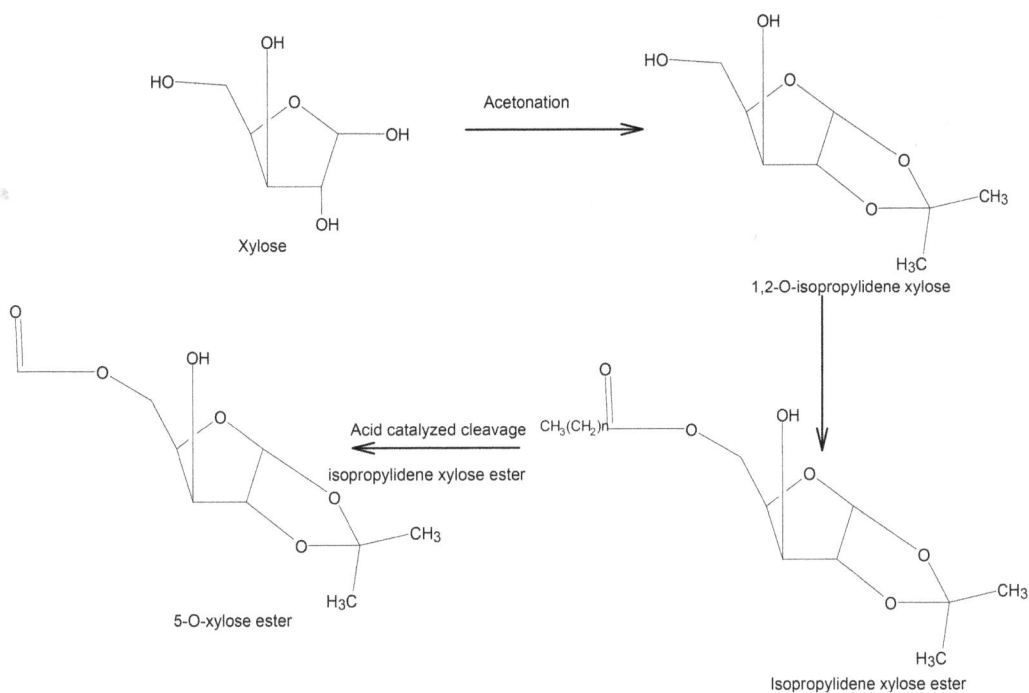

FIGURE 5.1 Preparation of xylose ester by esterification of xylose and fatty acid by acetonation method.

5.2 PRODUCTS AND APPLICATIONS

5.2.1 Sorbitan Esters

Sorbitol or glucitol is another name for the hydrogenated glucose product and is a very easily available polyol. It can be synthesized at a large scale in industry and has been used as a constituent of food, cosmetics and pharmaceutical industries. When it is heated together with fatty acids, esterification occurs. Like other polyols and carbohydrates, esters are formed at the primary and secondary positions (Figure 5.3). It may migrate from one OH to another [14]; therefore, the final

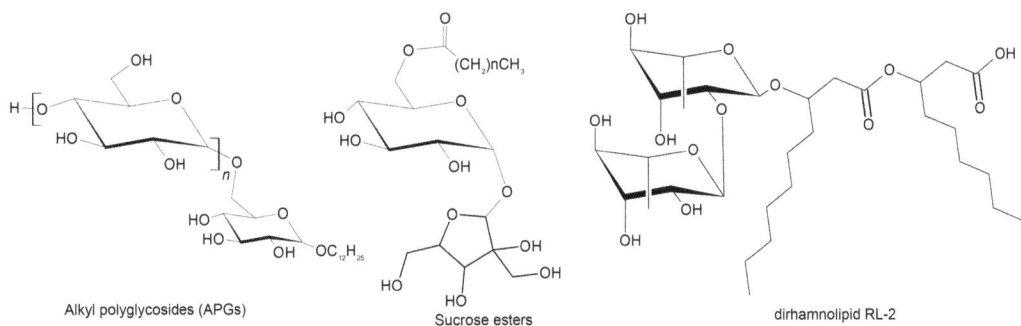

FIGURE 5.2 Typical examples of synthetic (alkyl polyglycosides, sucrose esters) and natural carbohydrate-based surfactants (dirhamnolipid RL2).

product is a thermodynamic blend of various esters in all positions. The esterification of sorbitol may result in different properties of monoesters, diesters or polyesters (Figure 5.3). The acidity of the medium and heating process promotes the subsequent cyclic dehydration of sorbitol into sorbitan and isosorbide, which have only four and two hydroxyl groups, respectively, and the polar products after esterification are much less [15]. As revealed in Figure 5.4, there is a

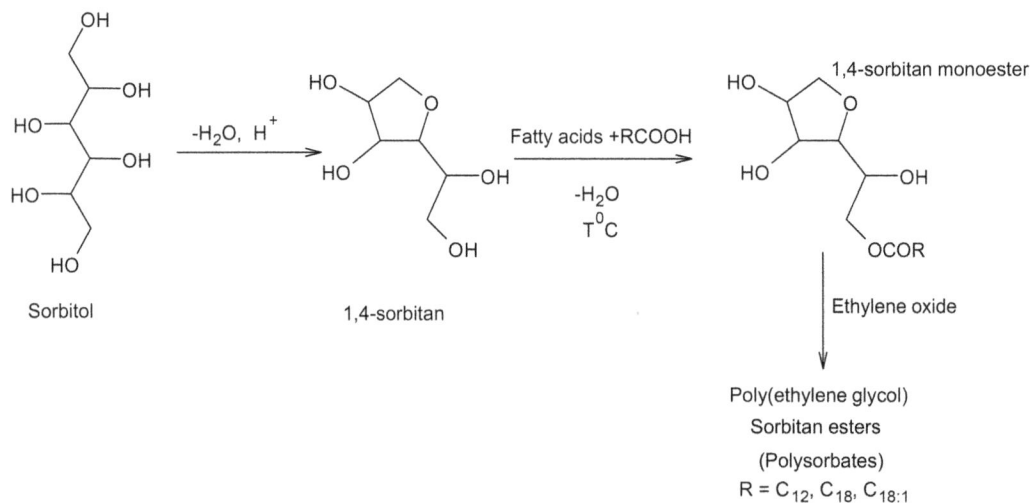

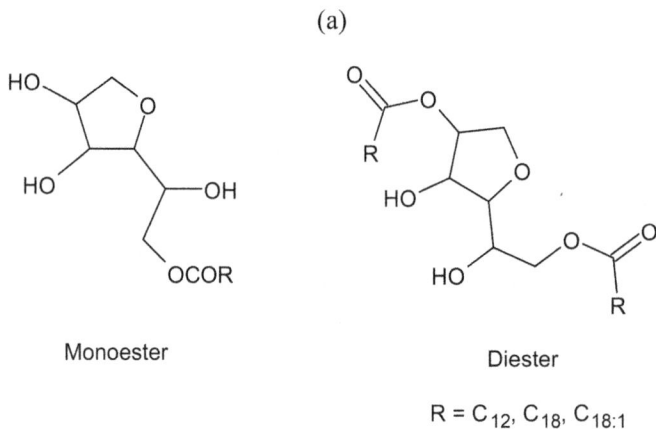

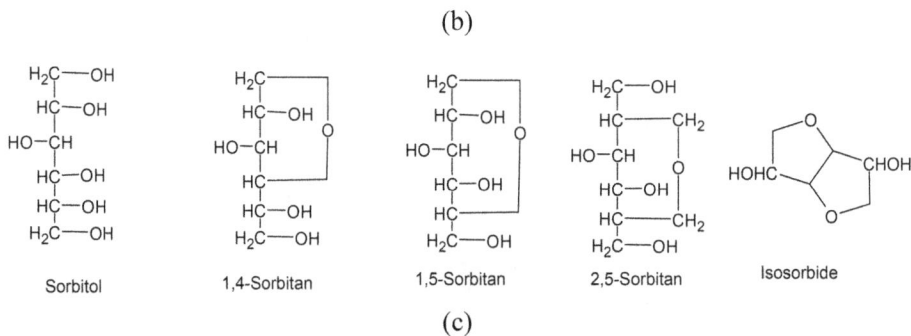

FIGURE 5.3 (a) Intramolecular dehydration of sorbitol and its alkali-catalyzed fatty acid esterification to synthesize sorbitol esters, (b) sorbitol monoester and diester and (c) the name of the by-product compound.

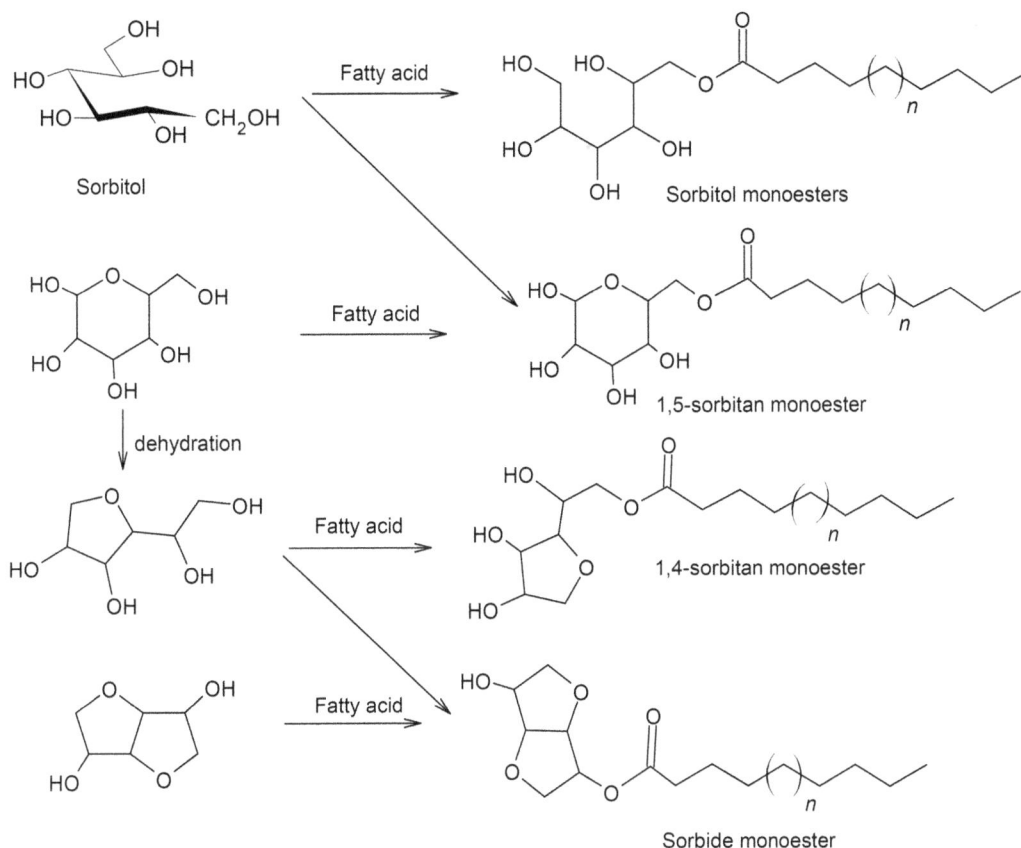

FIGURE 5.4 Depiction of competing esterification and dehydration reaction of sorbitol and fatty acids. These products are mainly used as emulsifiers for medicine, food, cosmetics, pesticides, emulsion polymerization, explosives and other technical applications (Table 5.1).

competition in the esterification and dehydration reaction of sorbitol which results in the mixture of various products. Many enzymatic esterification reactions of sorbitan have also been reported [15]. Therefore, a variety of products with water solubility can be obtained. Sorbitol esters with fatty acids are nonionic and have low HLB lipophilic emulsification potential (Table 5.1). The lipophilicity/hydrophilicity of sorbitol esters is largely affected by the amount of ester formed, chain length and type of fatty acids [15].

TABLE 5.1
Few Products and Physical Properties of Sorbitan Esters

Generic Name	Common Name	Physical Form (25°C)	HLB* (±1)
Sorbitan monolaurate	Span 20	Liquid	8.6
Sorbitan monopalmitate	Span 40	Solid	6.7
Sorbitan monostearate	Span 60	Solid	4.7
Sorbitan monooleate	Span 80	Liquid	4.3
Sorbitan tristearate	Span 65	Solid	2.1
Sorbitan trioleate	Span 85	Liquid	1.8

5.2.2 Sucrose Ester

Sucrose is a nonreducing disaccharide with multiple functions: tri-primary alcohol, five secondary alcohol and two anomeric carbon atoms. The preparation of sucrose esters is very complex because sucrose molecules are very delicate to temperature, and because they have a high functionality of eight hydroxyl groups, it is difficult to achieve selectivity in the esterification reaction [16]. A general chemical method is to directly esterify sucrose with fatty acids (Figure 5.5) by dissolving sucrose in solvents (such as DMF and DMSO) (Figure 5.5), or use methyl esters in the presence of alkaline catalysts (such as K_2CO_3). On the other hand, preparation of sucrose esters by transesterification reaction of fatty acids and methyl esters is shown in Figure 5.6 without using other additive solvents. In a typical esterification reaction, a composite mixture of mono-, di-, tri- and tetra-esters is usually obtained, and the resulting mixture is highly hydrophobic. In addition, the structural differences caused by these selectivity issues will promote major variation in physical and chemical properties, and hence their applications. For example, mono- or disubstituted fatty esters of sucrose will result in the inversion of micelles, and two different monosubstituted micelles with chains located at one or the other of the foundation structure of sugar will also display morphological variations [17].

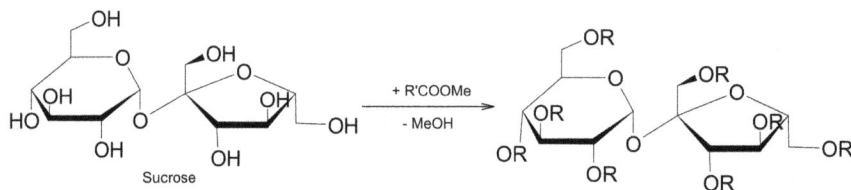

FIGURE 5.5 Typical esterification reaction of sucrose.

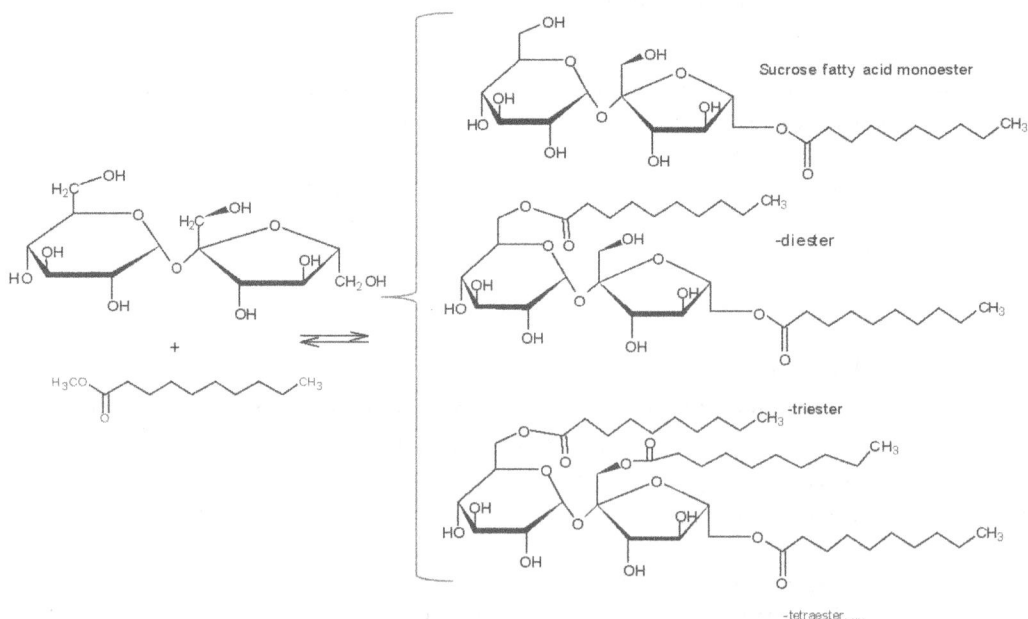

FIGURE 5.6 Preparation of sucrose esters by transesterification reaction of fatty acids and methyl esters.

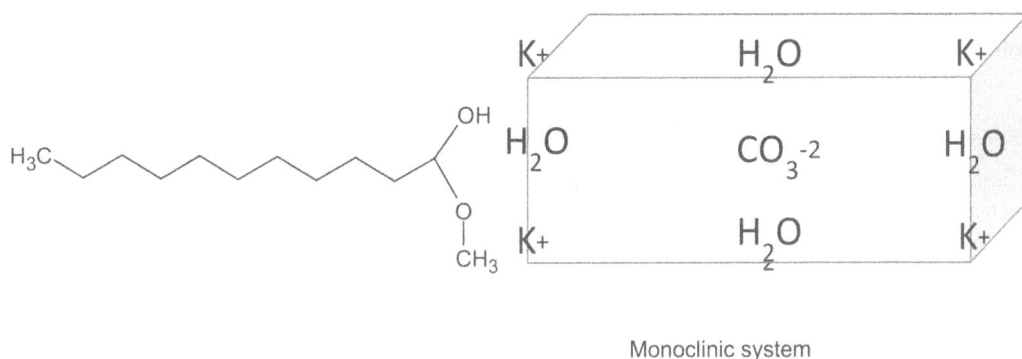

Monoclinic system

FIGURE 5.7 Adsorption of methyl ester ion on the surface of heterogeneous catalyst K_2CO_3.

Researchers have developed several methods to increase the selectivity of the reaction and provide economical purification methods, such as liquid-liquid extraction and crystallization. Scientists have made many improvements in the preparation of these surfactants to increase the selectivity of the product: (i) reaction with acyl chloride, (ii) application of two-phase system using PEG as emulsifier and (iii) use of enzyme catalyst [17,18]. An optimized and solvent-free process is described by Cognis, in which a continuous phase is developed with methyl esters that were adsorbed on the surface of heterogeneous catalyst K_2CO_3 and reacted with sucrose ester that was present beforehand to form sucrose polyester as an intermediate (Figure 5.7) [19]. The standard technique (traditional method, Figure 5.8) is still a combination of transesterification and purification with an improved solvent-free transesterification process.

Sucrose ester products are highly hydrophobic and have a limited range of applications. In the late 1960s, Dainippon Manufacturing was the first company to commercialize these products.

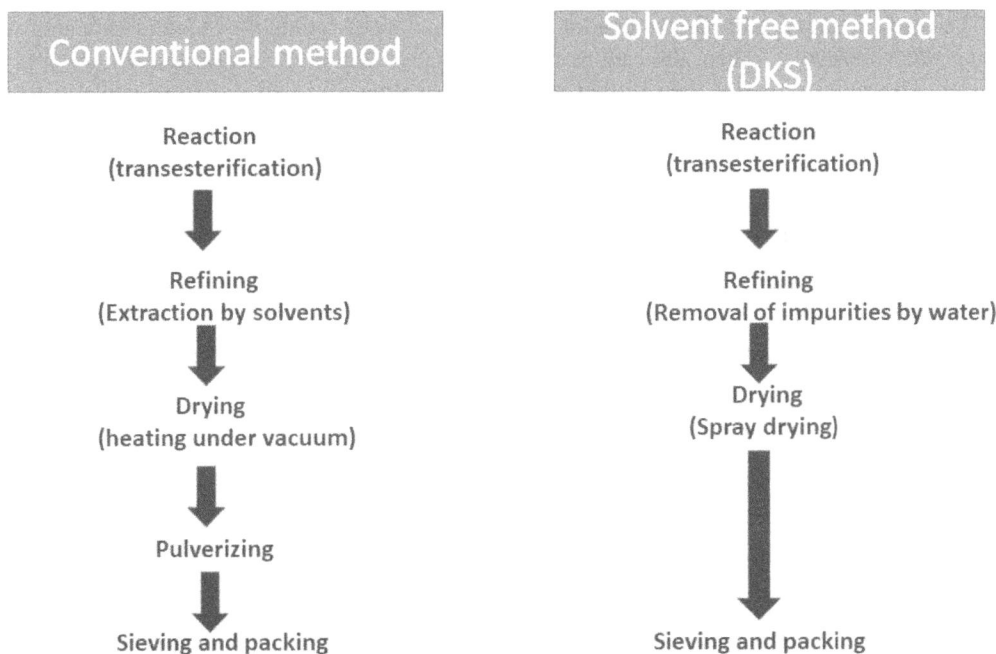

FIGURE 5.8 Process scheme adopted for industrial-scale production of sucrose esters [13].

Two methods are usually on an industrial scale used for synthesis: (i) conventional method and (ii) solvent-free method.

5.2.3 ALKYL POLYGLYCOSIDE (APGs)

These are important classes of glucose-derived nonionic surfactants with interesting properties such as low toxicity, biocompatibility and easy availability from renewable sources. APG was first discovered in the 1890s, but about a century later, Henkel Group began mass production. Today, the global consumption of APG has reached billions of dollars and exhibits a wide range of potential applications, including pharmaceuticals, detergents, food processing, cosmetics industries, as well as catalytic and adsorption applications, nano-biotechnology and environmental remediation.

APG is a polymerized acetal of glucose and fatty alcohol, and their structure is similar to gly-colipid (Figure 5.9). However, since the average degree of polymerization (DP) is very low, they are often referred to as oligomers. The reaction of glucose with alcohols with 8–16 carbon atoms produces a mixture of alkyl mono-, di-, tri- and oligosaccharides, which is a mixture of α and β anomers. The products are characterized by alkyl chain length and DP, which can be well-defined as the number of glucose units present in the oligomer.

5.2.3.1 Synthesis of APGs

Despite these benefits, the cost-effective synthesis of APG surfactants is still a difficult task. One of the earliest reported methods is Fischer glycosylation, in which glucose and fatty alcohol are reacted in the presence of an acid catalyst. Since then, several methods, such as the Koenigs-Knorr method [20], Schmidt method [21], Lewis acid method [22], base-catalyzed alkylation [23] and enzyme-catalyzed synthesis have been proposed [24]. However, in large-scale production, Fisher's method is preferable to other methods because it can better control the degree of product polymerization with no emissions (Figure 5.10).

Fischer proposed two different approaches: direct synthesis and deacetylation process. In the direct method, carbohydrates react directly with excess alcohol to obtain monoglycoside as the main product (Figure 5.11).

The reaction starts with the protonation of the sugar, which produces oxonium cations after losing water molecules. Then, the cations react with fatty alcohol molecules to form alkyl gly-cosides. The activated sugar then further reacts with another alkyl glycoside molecule to form

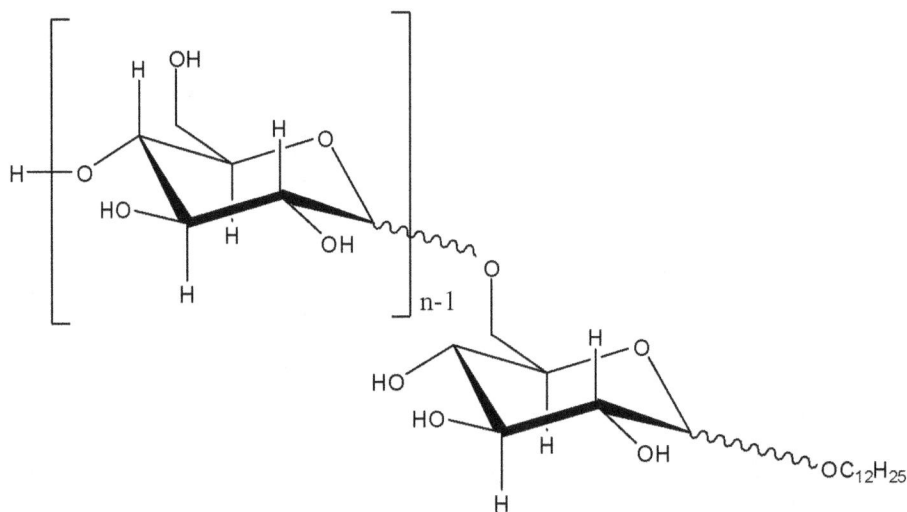

FIGURE 5.9 Molecular structure of APGs.

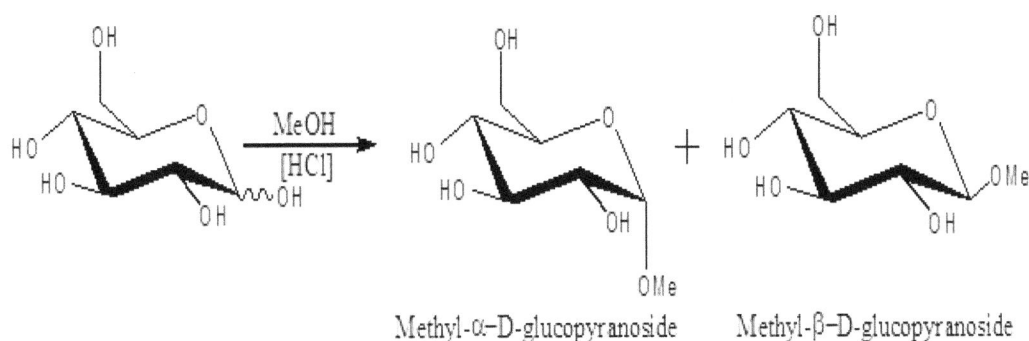

FIGURE 5.10 Synthesis of glucoside according to Fischer.

an oligoglycoside, which can react again with another fatty alcohol molecule or oligomer [25]. The final product obtained is a complex mixture, because each glycoside can assume four different cyclic forms (two pyranosides and two furanosides). In addition, there are several possible ways of covalently linking two sugar molecules to the alcohol moiety. Therefore, a thorough analysis is required to obtain precise composition range and find the polymerization degree range. In direct synthesis, fine powder of monomer and anhydrous sugar is used to avoid unnecessary side reactions.

When the used alcohol contains a long-chain hydrophobic portion, the reaction rate becomes slow due to the low solubility of carbohydrates. What follows is that undesirable side reactions will produce dehydration products, oligomers and dialkyl ethers (due to the self-condensation reaction of

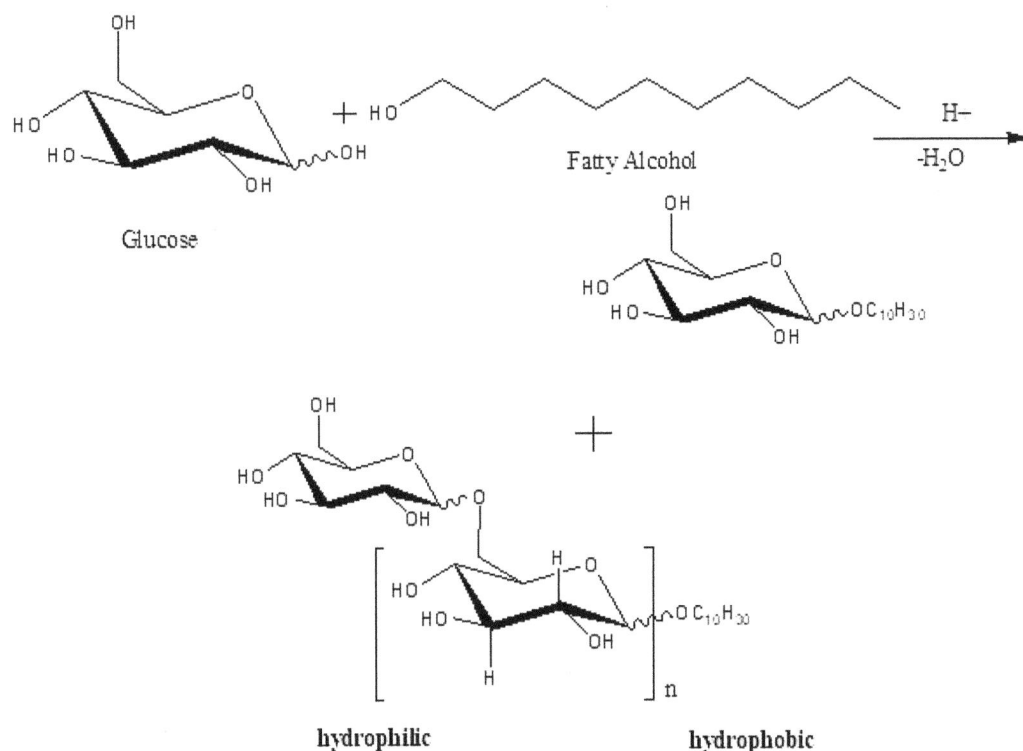

FIGURE 5.11 Direct synthesis of APGs.

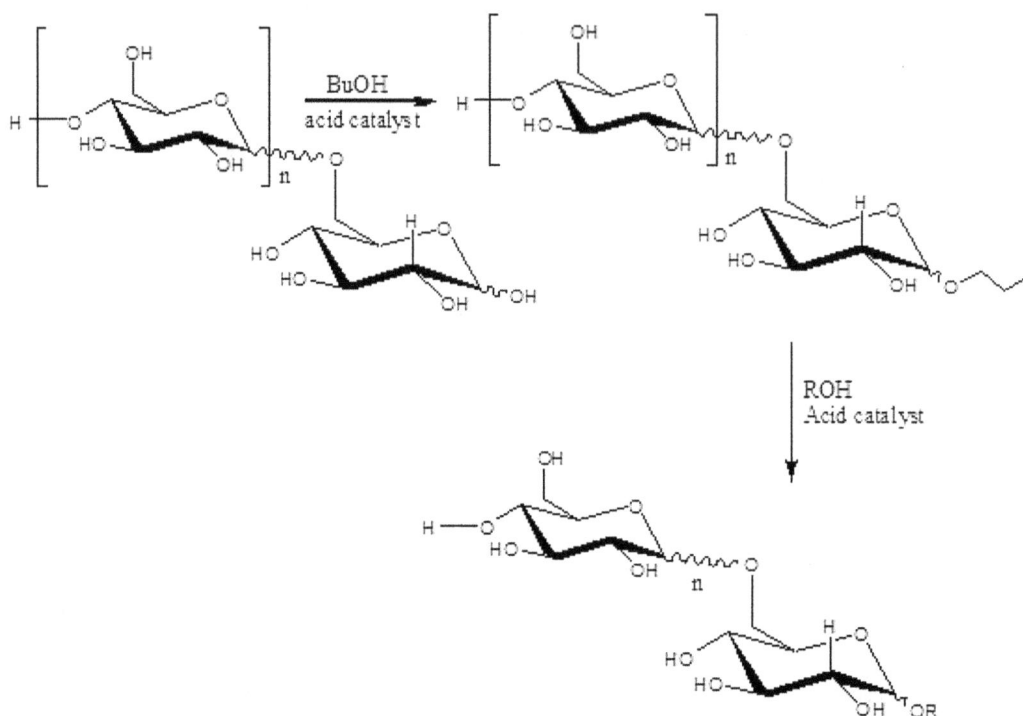

FIGURE 5.12   Synthesis of APGs via butyl glycoside route.

fatty alcohols). This problem can be achieved by using two steps involving butanol in the first stage and forms butyl glucoside intermediate, which undergoes acetyl exchange reaction with long-chain fatty alcohol (Figure 5.12). The subsequent reaction is driven by the easy evaporation of butanol. In addition to glucose, dextran is also used in the synthesis process, which depolymerizes in situ. In the presence of additives such as 2,5-furandicarboxylic acid (FDCA) [26], the glycosylation of decanol is promoted, because less by-products are produced, and the final-colored product is less, which is very preferred in cosmetic formulations.

Transacetalization is an alternative process where oligo- and polyglucose like starch are used along with alcohol like butanol and propylene glycol and an acid catalyst (Figure 5.13). The oligoglycoside intermediate obtained in the first step is treated with a long-chain alcohol in the presence of acid catalyst. If the molar ratio of sugar to alcohol remains the same, the oligomer distribution of the product in the direct and transacetalization process is also the same [27]. The temperature required for this process is very high (>140°C), and the final product needs to be refined. Therefore, compared with direct synthesis, this method incurs greater production costs.

5.2.3.1.1   Raw Material and Catalyst for Synthesis

The hydrophobic part of APG is provided by alcohol, which can be natural or synthetic alcohol, or a mixture of fatty alcohols. The alkyl chain lengths of the fatty alcohols used vary from one source to another. For example, coconut/palm kernel oil provides C_{12-14} chain length, while rapeseed oil produces C_{16-18} chain length. The hydrophilic part of the surfactant is produced by monomers or polymerized carbohydrates. Agricultural products such as barley, wheat, corn and potatoes are usually suitable choices [28]. A few sources of carbohydrate are described in Figure 5.14.

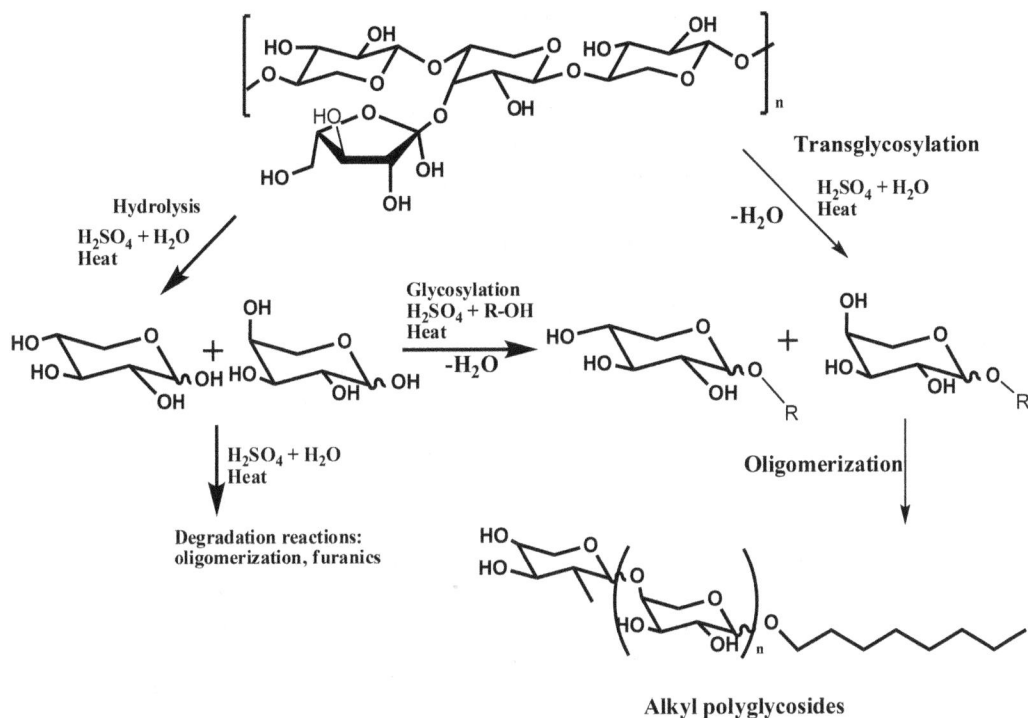

FIGURE 5.13 Transglycosylation of hemicelluloses in fatty alcohol medium [27].

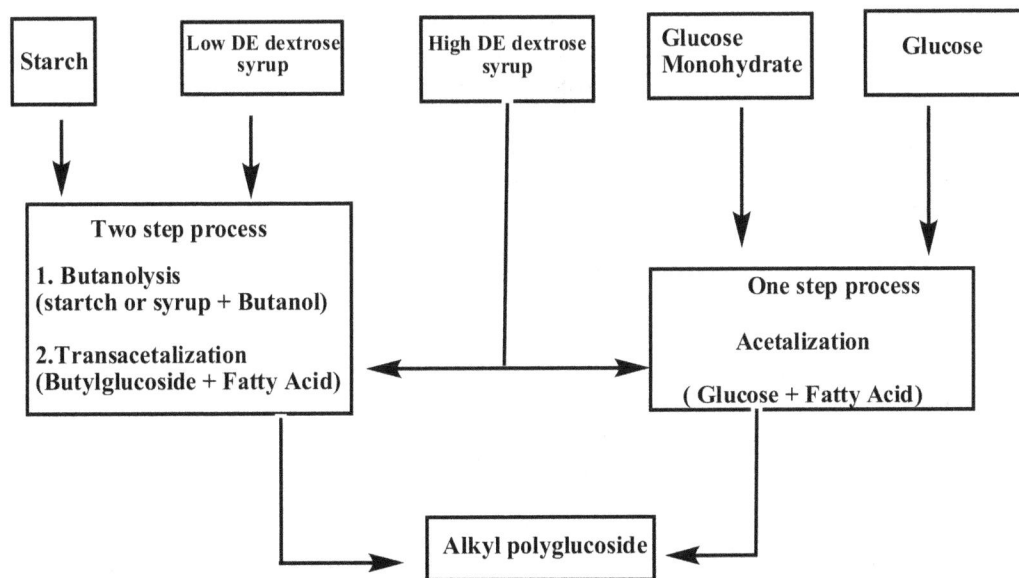

FIGURE 5.14 Carbohydrate source for industrial-scale alkyl polyglucoside synthesis [28].

Various research groups have studied the transglycosylation reaction of hemicellulose in the presence of fatty alcohols to produce a mixture of alkyl polypentoside [25]. The arabinose present in cereal products contributes to the production of alkyl polypentoside, while wood-derived carbohydrates are less reactive due to their low solubility in water. The use of highly polar solvents (e.g. DMSO) can increase solubility, thereby increasing product yield [25].

Cellulose is one of the most common and naturally available biopolymers, and many efforts have been made to synthesize APG from cellulose. This method essentially combines the depolymerization (acid-catalyzed hydrolysis) of polysaccharides into monosaccharide units and reglycosylate them with external fatty alcohols. However, high temperatures required for hydrolysis and glycosylation are incompatible with the stability of APG. Some modifications have been made to make the method more advantageous, for example catalytic ball milling with perfluorosulfonic acid polymers can produce APG in high yields. In this method, cellulose and Aquivion PFSA are depolymerized in a ball mill for 3 hours and then reacted with n-dodecanol. In the process, the total output has reached 70% [25].

Several research groups are actively working in this field and have proposed effective catalytic systems for this purpose. Villandier and his colleagues used Amberlyst-15 Dry to react cellulose with APGs with chain lengths up to C_{12} [29]. Deng et al. used heteropoly acid H3PW12O40 as a catalyst to perform deglycosylation reaction in methanol to obtain methyl glycosides and then use longer alcohol for glycosylation transfer in the presence of Amberlyst-15 Dry [30]. According to reports, perfluorosulfonic acid Aquivion PW98 can obtain good decyl glucoside yields from cellulose and decanol through a mechanically catalyzed depolymerization step [31]. Agricultural products such as wheat straw containing hemicellulose as the main component can also be directly converted. Conversion requires some pretreatment steps, where all the components of the starting material, including lignin derivatives and hydroxymethylfurfural (HMF), except for the formation of alkyl pentosides [32]. It was also found that ionic liquids can directly convert xylan into alkyl xyloside [33].

The use of catalysts can promote the formation of glycosidic bonds such as Acids H_2SO_4, HCl, H_3PO_4 or BF_3, sulfonic acids or their salts, such as sulfonic acid resins, alkyl sulfates, alkylbenzene sulfonates, alkyl sulfonates and sulfosuccinic acid. The catalyst is neutralized at the end of the reaction, with a base such as NaOH [34]. One of the disadvantages of all methods known to date is the formation of polydextrose as a by-product, which can increase the viscosity of the mixture. As a result, it becomes difficult to separate APG and recover unreacted starting materials. This problem can be solved by using a high alcohol/glucose ratio, which creates a safety hazard and requires expansion of production scale. Using catalysts can minimize these problems.

In the presence of the H_2SO_4 catalyst, the alcohol: glucose ratio is 2:1, and 20% of polydextrose by-product is obtained. In the presence of p-toluenesulfonic acid (PTSA), the yield of by-products was 11%. The percentage can be further reduced to 9.2% with an arylsulfonic acid catalyst [35].

The synthetic method inspired by biotechnology is another convenient way to synthesize APG. Commonly used enzymes are glycosidases (enzymes that cut glycosidic bonds) and glycosyltransferases (enzymes that form glycosidic bonds). It has been observed that in the presence of large amounts of alcohol and small amounts of water, glycosidase activity gets reversed, leading to glycoside synthesis [36]. Purified enzymes and even crude plant extracts containing enzymes have been used to synthesize hydrophobic APG. However, the efficiency of this method decreases as the alcohol chain length increases. The availability of enzymes and the possibility of using simpler carbohydrate donors are areas of concern. However, with the latest advances in enzyme engineering and the expectation of directed mutagenesis technology (providing customized enzymes), researchers will be able to address these shortcomings, and future processes will include more and more biocatalysts in the APG production process (Table 5.2).

TABLE 5.2

Summary of Glycosidases Utilized for Biotechnological Accesses Towards APGs [37]

Substrate	Enzyme	Alcohol	Product
Glucose	Almond β-glucosidase	6-Hydroxyhexanol	6-Hydroxyhexyl β-D glucopyranoside
Glucose	Almond β-glucosidase	Octanol	Octyl β-D glucopyranoside
Cellobiose	Candida molischiana 35M5N β-glucosidase	Decanol	Decyl β-D glucopyranoside
Lactose	CloneZyme Gly-001-02 from Diversa	Heptanol	Heptyl β-D glucopyranoside
Methyl Glucoside	Cell bound β-glucosidase of *Pichia etchellsii*	Dodecanol	Dodecyl β-D glucopyranoside
Birchwood Xylan	Xylanase from *Thermobacillus xylanilyticus*	Octanol	Octyl β-D-xylotrioside

5.2.3.2 Physicochemical Properties of APGs and their Application

The physicochemical properties mainly depend on the molecular composition. Commercial APG is a mixture of glycosides of different chain lengths, different types and DP. The DP can be adjusted by changing the molecular ratio of sugar to fatty alcohol and the type of acid catalyst used. The lower DP value destroys the hydrophilic/lipophilic balance, thus affecting the relative properties of surfactants.

Commercial APG is a mixture of glycosides of different chain lengths, different types and DP. The DP can be adjusted by changing the molecular ratio of sugar to fatty alcohol and the type of acid catalyst used. The lower DP value destroys the hydrophilic/lipophilic balance, thus affecting the relative properties and performances of surfactants. In addition, DP and saccharide type will also affect the CMC value. The molecular structure also affects properties such as foam production properties. For example, the glucose-based APG produced the largest foam yield and the highest stability in terms of C_{12}/C_{12} chain length. In the case of xylose-based surfactants, the same results were also observed for C_{10} tail xyloside. They are used in the household detergent and cosmetic industries that require rich and stable foam. Xylose-based APG is used in hand soaps and dishwashing liquids because they generate a lot of foam.

Wetting power is another useful characteristic often considered in the laundry and textile cleaning industries. In these areas, APG of D-xylose shows better performance. Therefore, they have used them in crop production, and a lower dosage will ensure higher herbicide efficiency (Tables 5.3 and 5.4) [25].

TABLE 5.3

Surface and Foam Characteristics of Decyl Glucoside Compositions Obtained from Various Lignocellulosic Materials [13]

Origin	CMC (mg/L)	γCMC(mN/m)
Poplar	433	27
Wheat straw	483	28
Wheat bran	493	28
Xylan	230	27
D-Glucose	994	26
D-Xylose	301	28

TABLE 5.4

Use of APGs in Detergent Products (Based on the Alkyl Chain Length)

Type of Product	Alkyl Chain Length
Manual dishwashing detergents	$(C_{12/14})$
Cleaners	$(C_{12/14}; C_{8/10})$
All-purpose cleaners	$(C_{12/14}; C_{8/10})$
Bathroom cleaners	$(C_{8/10})$
Liquid toilet cleaners	$(C_{8/10})$
Window cleaners	$(C_{8/10})$
Floor care formulations	$(C_{12/14})$
Liquid laundry detergents	$(C_{12/14})$
Powder/extrudate detergents	$(C_{12/14})$

APG is mainly used in the formulation of cosmetics and personal care products as well as hard surfaces and washing powders. They are nontoxic and nonirritating to the skin. Therefore, it is a very effective nonionic surfactant and can be used in handwashing detergents where chemicals come into contact with the skin. APG with a benzene ring in the aglycone can also be used in the field of gelling agents. When the benzene ring is a galloyl residue, the compound has both surfactant and antioxidant properties. Several well-known multinational organizations, such as Henkle, BASF, Noble, Cognis, etc., have participated in the commercial production of APG.

5.2.4 Fatty Acid Glucamides

Fatty acid glucamide is an amide of glucose and fatty acid derivatives. Their synthesis involves the reaction of methylamine with glucose to form the corresponding N-methylglucamine, and then the intermediate is converted into fatty acid amides by the reaction of fatty acid methyl esters (Figure 5.15). Contrary to the structure of APG, fatty acid glucamide contains only one glucose unit in the fatty acid chain. Consequently, fatty acid glucamide has lower solubility in water and tends to crystallize better [39].

$C_{12/14}$- and $C_{16/18}$-based fatty acid glucamides are extensively used in powdered and liquid detergents. One common example is the compound MEGA, abbreviated form of N-alkanoyl-N-methylglucamine (Figure 5.16).

Like APG, glucamides also show synergy with other kinds of surfactants, and because of their polyol structure, they possess low irritation. The disadvantage of glucamides is their affinity for calcium ions. They must always be formulated with chelating agents to avoid precipitation. N-methyl lauroyl glucosamine is used in dish soap, with an annual output of approximately 5,000 tons. However, if N-methylglucamine is present as a synthetic residue or metabolite during the biodegradation process, N-methyl lauroyl glucamine may form N-nitrosamine, which is suspected of being carcinogenic [40].

Derivatives of glucamide such as glucosamine oxide and betin and anionic glucamide and difunctional glucamide are used more frequently.

To further improve the surfactant properties, the APGs are derivatized by suitable chemical transformations. Simple nucleophilic substitution reactions are used to synthesize many APG derivatives. APG with alkyl chains C_8–C_{16} is used for derivatization reaction, the average DP is 1.1–1.5, and three series of derivatives can be generated. The presence of a large number of hydroxyl groups makes them surpass functionalized molecules. The derivatization

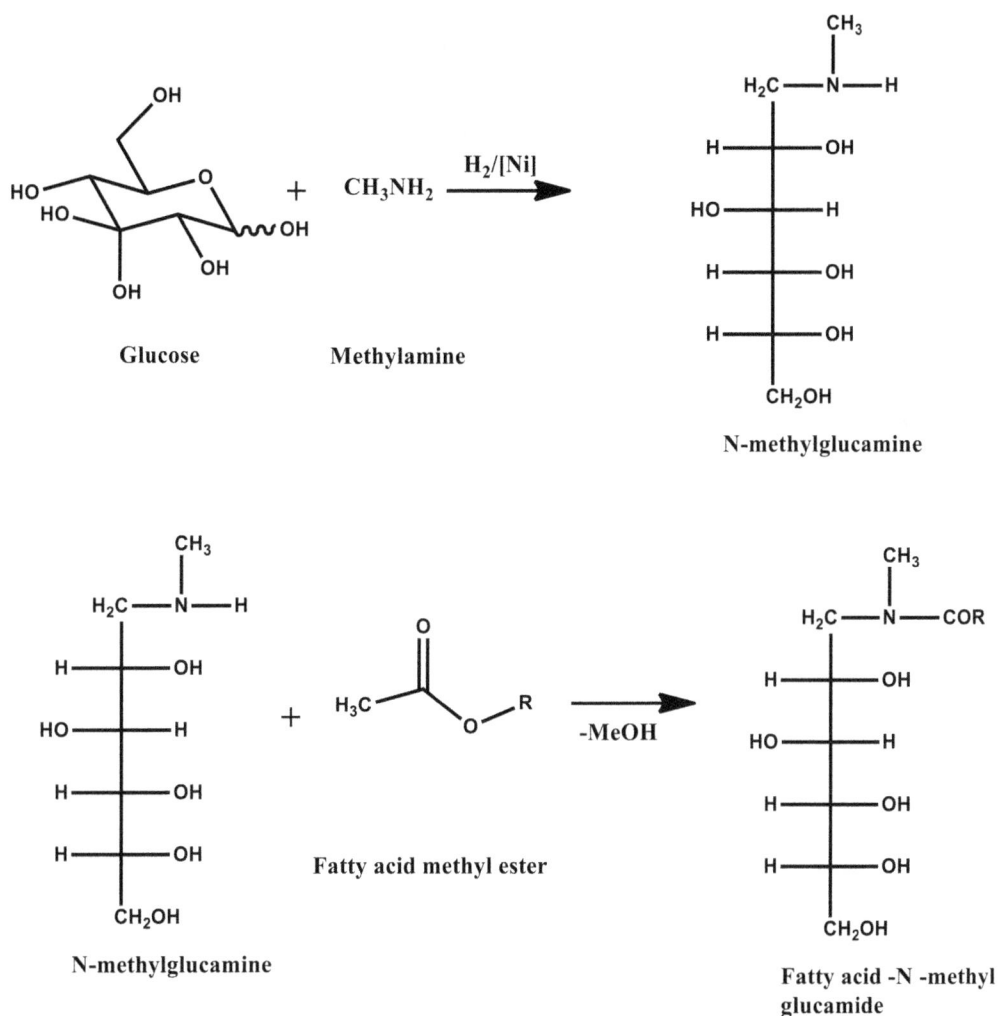

FIGURE 5.15 Synthesis of the fatty acid glucamide is in two steps through the reductive alkylation reaction of glucose and methylamine using Raney nickel as the hydrogenation catalyst, to obtain *N*-methylglucosamine, which is then combined with fatty acid methyl ester in the second step.

reaction mainly involves the chemical conversion of the hydroxyl group at the C_6 position of glucose. Avoid using a large DP value, as it will lead to complex mixtures and analysis. The derivatization reaction is carried out by using APG with a DP value of about 1.1 to achieve better and simplified production [28]. Some derivatives available in the market are given in Figure 5.17.

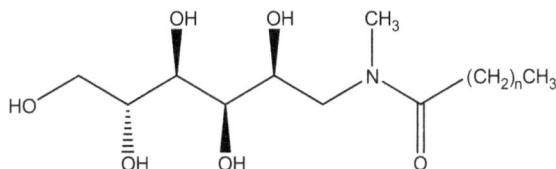

FIGURE 5.16 *N*-alkanoyl-*N*-methylglucamide (MEGA).

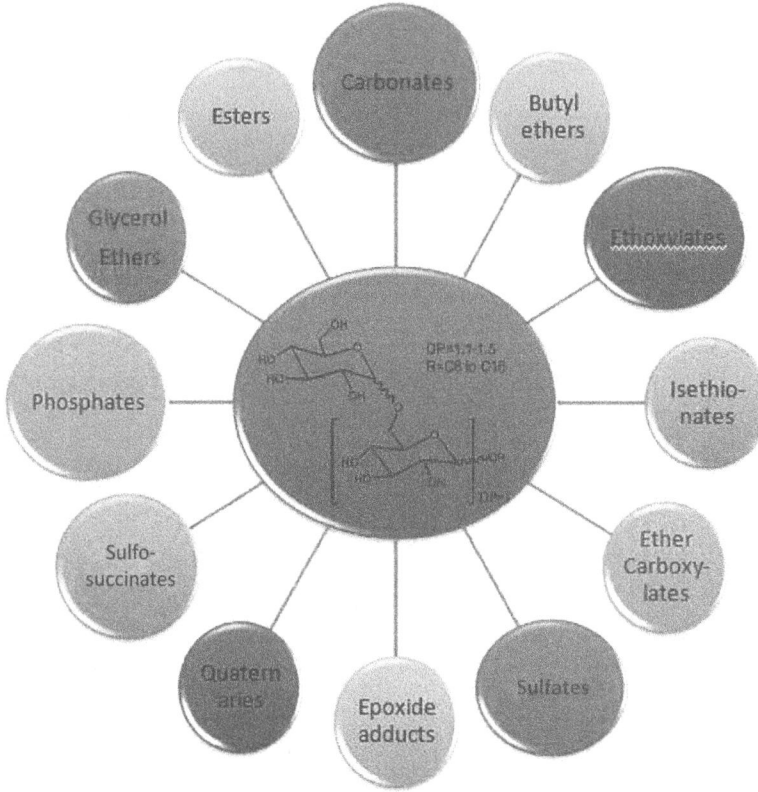

FIGURE 5.17 Alkyl polyglycoside derivatives.

5.2.4.1 Methyl Glucoside Ester

It is synthesized by the base-catalyzed esterification reaction of methyl glucoside with methyl ester of stearic or oleic acid (Figure 5.18). In the next step, the ethoxylation reaction produces polyethylene glycol methyl glucoside esters. These derivatives are lipophilic in nature and are almost insoluble in water. They behave as excellent emulsifying agents and have found many applications in the cosmetic industry. By varying the degree of substitution and hydrocarbon chain length, the rheological properties can be adjusted to enhance the emulsification [41].

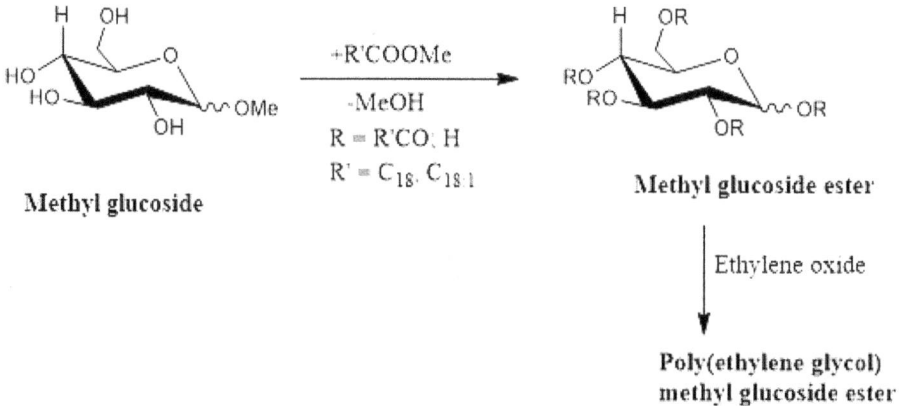

FIGURE 5.18 Synthesis of methyl glucoside ester.

5.2.4.2 Anionic Derivatives of Alkyl Polyglycosides

Three most important nonionic APG esters are citrates, tartrates and sulfosuccinates (Figure 5.19). These are prepared by the esterification of APG (C_{12}/C_{14}) with tartaric acid, citric acid and maleic anhydride followed by sulfonation of succinate [42].

One important anionic surfactant is APG carboxylate which is extensively used in daily products such as shampoos, body washes, face washes, etc. They show superior foaming properties compared to nonionic counterparts. The industrially cost-effective synthesis method of this surfactant utilizes the reaction of sodium monochloroacetate with an aqueous solution of APG (Figure 5.20).

5.2.4.3 Anionic Phosphorous Derivative

The latest work by Seweryn et al. [43] established the possibility of using anionic phosphorus derivatives of APGs in natural raw materials to produce eco-friendly cosmetics. The products formulated with anionic phosphorus derivatives with different alkyl chain lengths (C_4; C_8; C_{10}; C_{12}) instead of sodium lauryl sulfate can provide safe and gentle shower gels and shampoos. The products are found to contain remarkably lower zein number and pH change of bovine serum albumin solution indicating less irritant effect on skin.

5.2.5 HEMICELLULOSE-BASED SURFACTANT

A large number of polypentosides are synthesized from hemicellulose whereby hydrolysis of hemicellulose produces a mixture of glycosides with main components xylose and arabinose. Desired alkyl polypentosides are prepared by reacting these pentoses with fatty alcohols. Their properties

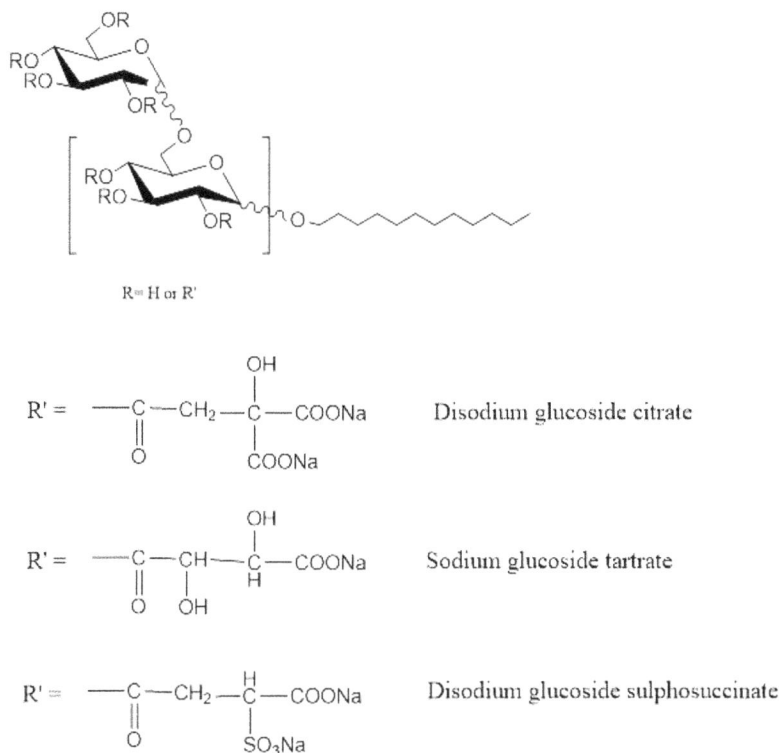

FIGURE 5.19 Some anionic alkyl polyglycoside derivatives.

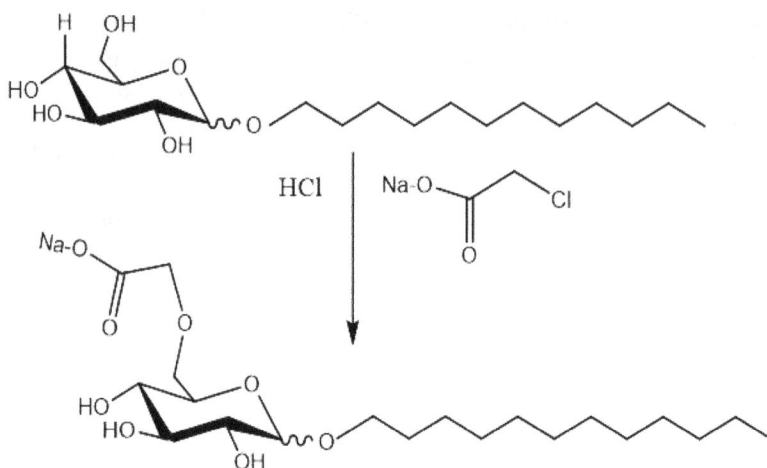

FIGURE 5.20 Industrial synthesis of alkyl polyglycoside carboxylate.

are similar to that of APG. Wheat bran hemicellulose has been successfully converted into pentose-based surfactants in a single-step procedure (Figure 5.21).

5.2.6 INSULIN DERIVATIVES

Insulin is a polydisperse polysaccharide, primarily composed of β(2-1)-linked fructose molecule. The glucose unit is at the reducing end of the structure. Insulin derivatives are polymeric sugar-based surfactants. They are used in paints, surface coatings, cosmetics and agrochemicals as emulsifiers. Insulin is extracted from chicory roots and the corresponding surfactant derivative is branded in the name of Inutec in the market.

5.2.7 NONIONIC DERIVATIVES OF ALKYL POLYGLYCOSIDES

By looking into the wide range of application of APG, people are inspired to derivatize APGs for developing surfactants with special property. Their foaming and wetting properties are substantially improved by chemical transformation during derivatization. Various nonionic derivatives such as esters to ethoxylates and ionic derivatives of sulphates and phosphates have been prepared by Rhode et al. [44].

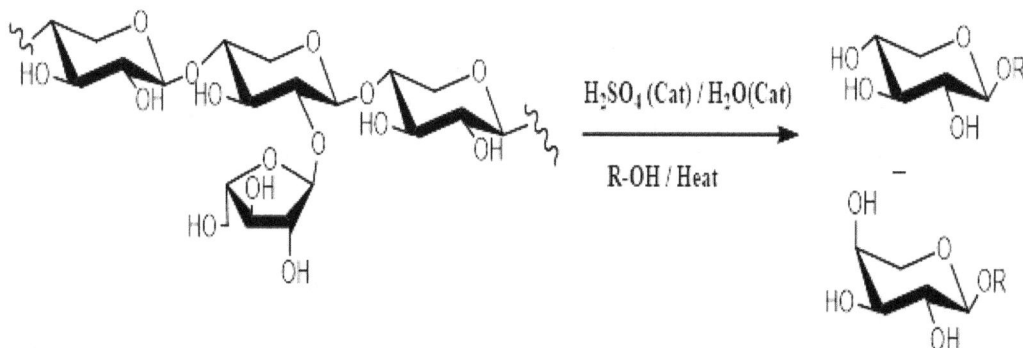

FIGURE 5.21 Synthesis of alkyl pentosides from hemicellulose.

They synthesized three series of derivatives using APGs with alkyl chains (R = C_8–C_{16}) having carbon atoms of 8, 10, 12, 14 and 16 and an average DP of 1.1–1.5. Surfactant properties are modified by substituting hydrophilic as well as hydrophobic groups in preparing APG derivatives of glycerol ethers, butyl ethers and carbonates. Mostly the derivatization is performed on the free –OH group at C_6 position by selective reaction using protective groups. The presence of a large number of hydroxyl groups makes the chemical transformation of APG more complicated and produces a mixture of products. The separation and characterization of the products is a tedious and costly process. Thus, APGs with a DP of 1.1 known as monoglycosides are used for derivatization.

5.2.8 SYNTHESIS OF GLYCEROL ETHER DERIVATIVES OF APG

Three different methods are reported for the preparation of APG glycerol ethers (Figure 5.22). In the first method, APG reacts with glycerol by nucleophilic substitution method in basic conditions where a mixture of mono-, di- and triglycerol ethers is obtained as products. The second etherification reaction involves a basic catalyst for the ring opening of an epoxide. The third method utilizes glycerol carbonates to synthesize APG glycerol ethers by eliminating CO_2 gas.

5.2.9 SYNTHESIS OF APG CARBONATES

The transesterification reaction of alkyl monoglycosides with diethyl carbonate produces the carbonate derivative of APG (Figure 5.23). In addition to monocarbonate, a mixture of higher order carbonates in certain ratios is also obtained by modification of reaction time and distillate volume.

5.2.10 SYNTHESIS OF APG BUTYL ETHERS

To modify the cleansing and foaming property, butyl ether derivatives of APG are synthesized. Industrially butyl ether derivatives are prepared using a water-free process using the reaction of

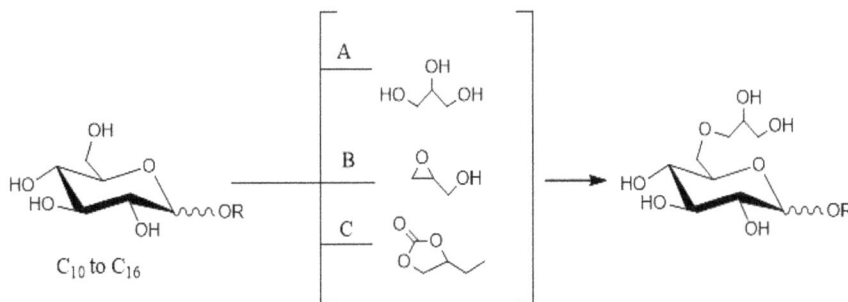

FIGURE 5.22 Synthesis of glycerol ether derivatives.

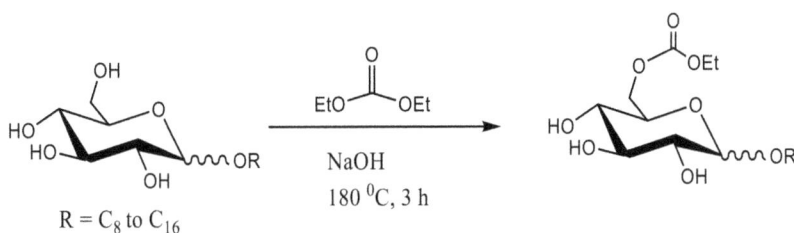

FIGURE 5.23 Synthesis of carbonate derivatives of APG.

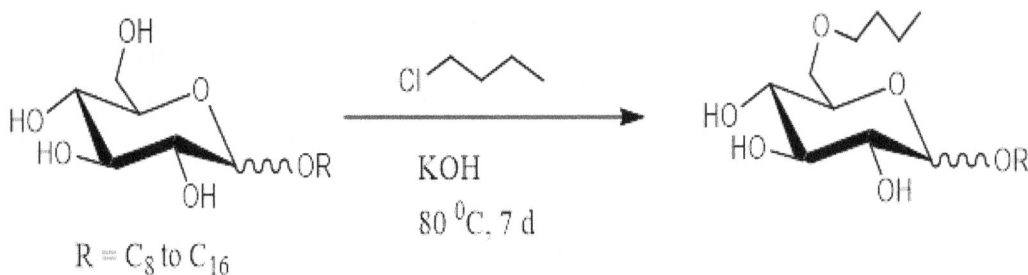

FIGURE 5.24 Synthesis of alkyl polyglycoside butyl ethers.

APG with butyl chloride in the presence of potassium hydroxide as catalyst (Figure 5.24). The average degree of esterification is 1–3 butyl groups per glycoside unit. C_{12} APG is the best for this reaction, whereas APG with $n = 8$ or 16 produces the least amount of product.

5.3 CONCLUSIONS

This chapter presented a general view of preparation and general applications of sugar-based surfactants in comparison to the existing synthetic surfactants. The chapter mainly focuses on three categories of sugar molecules, namely APG, sucrose esters and sorbitan esters and their derivatives to produce newer biodegradable surfactants with potential applications. The approaches adopted to synthesize these compounds at an industrial scale have also been reviewed. The limitations in their production at an industrial scale and possible solutions investigated by researchers in the last few decades are analyzed and discussed. APGs and their derivatives are the most successful sugar-based surfactants nowadays due to their outstanding performance, multifunctionality, competitive price and high product safety.

ACKNOWLEDGEMENT

The authors gratefully acknowledge the management of Sharda University as it provided all kinds of support while writing this chapter.

REFERENCES

1. Rosen, M. J., & Kunjappu, J. T. (2012). *Surfactants and interfacial phenomena*. Wiley, Hoboken, NJ.
2. Zoller U. (2008). *Handbook of detergents, part E: applications*. Taylor & Francis, California.
3. Kjellin, M., & Johansson, I. (eds). (2010). *Surfactants from renewable resources*. Wiley, Stockholm.
4. Holmberg, K. (2001). Natural surfactants. *Current Opinion in Colloid & Interface Science, 6*, 148–159
5. Hubbard, A. (2004). *Novel surfactants: preparation, applications, and biodegradability*. Marcel Dekker, New York, NY.
6. Jessop, P. G., Ahmadpour, F., Buczynski, M. A., Burns, T. J., Green, N. B. II, Korwin, R., ... & Wolf, M. H. (2015). Opportunities for greener alternatives in chemical formulations. *Green Chemistry, 17*, 2664–2678.
7. Foley, P., Kermanshahi-pour, A., Beach, E. S., & Zimmerman, J. B. (2012). Derivation and synthesis of renewable surfactants. *Chemical Society Reviews, 41*, 1499–1518.
8. Razafindralambo, H., Blecker, C., Paquot, M. (2012). Carbohydrate-based surfactants: structure-activity relationships. In: *Advances in chemical engineering*. InTech, Rijeka
9. Almeida, M. V. D., & Hyaric, M. L. (2005). Carbohydrate derived surfactants. *Mini-Reviews in Organic Chemistry, 2*, 283–297
10. Nakamura, S. (1999). Application of sucrose fatty acid esters as food emulsifiers. In: Karsa, D. R. (ed) *Industrial applications of surfactants IV*, 1st edn. The Royal Society of Chemistry, Manchester, pp 73–87

11. Hayes, D. G., Kitamoto, D., Solaiman, D., Ashby, R. (eds). (2009). *Biobased surfactants and detergents: synthesis, properties, and applications*. AOCS Press, Urbana
12. Queneau, Y., Chambert, S., Besset, C., Cheaib, R. (2008) Recent progress in the synthesis of carbohydrate-based amphiphilic materials: the examples of sucrose and isomaltulose. *Carbohydrate Research*, *343*, 1999–2009
13. Hill, K., & Rhode, O. (1999). Sugar-based surfactants for consumer products and technical applications. *Lipid/Fett*, *101*(1), 25–33.
14. Gozlan, C., Deruer, E., Duclos, M. C., Molinier, V., Aubry, J. M., Redl, A., … & Lemaire, M. (2016). Preparation of amphiphilic sorbitanmonoethers through hydrogenolysis of sorbitan acetals and evaluation as bio-based surfactants. *Green Chemistry*, *18*(7), 1994–2004.
15. Cottrell, T., & Van Peij, J. (2004). Sorbitan esters and polysorbates. *Emulsifiers in Food Technology*, *CH12*, 162–185.
16. Plat, T., & Linhardt, R. J. (2001). Syntheses and applications of sucrose-based esters. *Journal of Surfactants and Detergents*, *4*(4), 415–421.
17. Garti, N., Clement, V., Leser, M., Aserin, A., & Fanun, M. (1999). Sucrose ester microemulsions. *Journal of Molecular Liquids*, *80*(2-3), 253–296.
18. Parker, K. J., James, K., & Hurford, J. (1977). Sucrose ester surfactants—a solventless process and the products thereof.
19. Hill, K. (2010). Surfactants based on carbohydrates and proteins for consumer products and technical applications. *Surfactants from Renewable Resources*, *65*, 1–104.
20. Böcker, T., & Thiem, J. (1989). Synthese und eigenschaften von kohlenhydrattensiden. *Tenside, Surfactants, Detergents*, *26*(5), 318–324.
21. Focher, B., Savelli, G., & Torri, G. (1990). Neutral and ionic alkylglucopyranosides. Synthesis, characterization and properties. *Chemistry and Physics of Lipids*, *53*(2–3), 141–155.
22. Vill, V., Böcker, T., Thiem, J., & Fischer, F. (1989). Studies on liquid-crystalline glycosides. *Liquid Crystals*, *6*(3), 349–356.
23. Vulfson, E. N., Patel, R., & Law, B. A. (1990). Alkyl-β-glucoside synthesis in a water-organic two-phase system. *Biotechnology Letters*, *12*(6), 397–402.
24. Chahid, Z., Montet, D., Pina, M., &Graille, J. (1992). Effect of water activity on enzymatic synthesis of alkylglycosides. *Biotechnology Letters*, *14*(4), 281–284.
25. Estrine, B., Marinkovic, S., & Jérome, F. (2019). Synthesis of alkyl polyglycosides from glucose and xylose for biobased surfactants: synthesis, properties, and applications. *Biobased surfactants* (pp. 365–385). AOCS Press, Elsevier, Cambridge, MA.
26. van Es, D. S., Marinkovic, S., Oduber, X., &Estrine, B. (2013). Use of furandicarboxylic acid and its decyl ester as additives in the Fischer's glycosylation of decanol by D-glucose: physicochemical properties of the surfactant compositions obtained. *Journal of Surfactants and Detergents*, *16*(2), 147–154.
27. Ludot, C., Estrine, B., Hoffmann, N., Le Bras, J., Marinkovic, S., &Muzart, J. (2014). Manufacture of decyl pentosides surfactants by wood hemicelluloses transglycosidation: a potential pretreatment process for wood biomass valorization. *Industrial Crops and Products*, *58*, 335–339.
28. Geetha, D., & Tyagi, R. (2012). Alkyl poly glucosides (APGs) surfactants and their properties: a review. *Tenside surfactants detergents*, *49*(5), 417–427.
29. Villandier, N., &Corma, A. (2010). One pot catalytic conversion of cellulose into biodegradable surfactants. *Chemical Communications*, *46*(24), 4408–4410.
30. Deng, W., Liu, M., Zhang, Q., Tan, X., & Wang, Y. (2010). Acid-catalysed direct transformation of cellulose into methyl glucosides in methanol at moderate temperatures. *Chemical Communications*, *46*(15), 2668–2670.
31. Karam, A., De Oliveira Vigier, K., Marinkovic, S., Estrine, B., Oldani, C., & Jérôme, F. (2017). Conversion of cellulose into amphiphilic Alkyl glycosides catalyzed by aquivion, a perfluorosulfonic acid polymer. *ChemSusChem*, *10*(18), 3604–3610.
32. Marinkovic, S., Bras, J. L., Nardello-Rataj, V., Agach, M., & Estrine, B. (2012). Acidic pretreatment of wheat straw in decanol for the production of surfactant, lignin and glucose. *International Journal of Molecular Sciences*, *13*(1), 348–357.
33. Sekine, M., Kimura, T., Katayama, Y., Takahashi, D., & Toshima, K. (2013). The direct and one-pot transformation of xylan into the biodegradable surfactants, alkyl xylosides, is aided by an ionic liquid. *RSC Advances*, *3*(43), 19756–19759.
34. El-Sukkary, M. M. A., Syed, N. A., Aiad, I., & El-Azab, W. I. M. (2008). Synthesis and characterization of some alkyl polyglycosides surfactants. *Journal of Surfactants and Detergents*, *11*(2), 129–137.

35. Borsotti, G., & Pellizzon, T. (1996). *U.S. Patent No. 5,527,892*. Washington, DC: U.S. Patent and Trademark Office.

36. Turner, P., Svensson, D., Adlercreutz, P., & Karlsson, E. N. (2007). A novel variant of Thermotoganeapolitana β-glucosidase B is an efficient catalyst for the synthesis of alkyl glucosides by transglycosylation. *Journal of Biotechnology, 130*(1), 67–74.

37. Wang, L., & Queneau, Y. (2018). Carbohydrate-Based Amphiphiles: Resource for Bio-based Surfactants.

38. Lukic, M., Pantelic, I., & Savic, S. (2016). An overview of novel surfactants for formulation of cosmetics with certain emphasis on acidic active substances. *Tenside Surfactants Detergents, 53*(1), 7–19.

39. Fréville, V., Van Hecke, E., Ernenwein, C., Salsac, A. V., & Pezron, I. (2014). Effect of surfactants on the deformation and detachment of oil droplets in a model laminar flow cell. *Oil & Gas Science and Technology–Revue d'IFP Energies nouvelles, 69*(3), 435–444.

40. Brancq, M. (2001). Composition herbicide complement du glyphosate et au moins un alkyl polyxyloside. EP1303190A1.

41. Ruiz, C. C. (Ed.). (2008). *Sugar-based surfactants: fundamentals and applications* (Vol. 143). CRC Press, United States.

42. Seweryn, A., & Bujak, T. (2018). Application of anionic phosphorus derivatives of alkyl polyglucosides for the production of sustainable and mild body wash cosmetics. *ACS Sustainable Chemistry & Engineering, 6*(12), 17294–17301.

43. Rhode, O., Weuthen, M., & Nickel, D. (1996). New nonionic derivatives of alkyl polyglycosides— synthesis and properties. *Alkyl Polyglycosides: Technology, Properties and Applications*, 139–149. ISBN:9783527614691.

6 Flavonoid-Based Surfactants

Ritika Kubba, Deepali Ahluwalia, Anil Kumar and Sudhir G. Warkar
Department of Applied Chemistry, Delhi Technological University, Delhi, India

CONTENTS

6.1   INTRODUCTION

'Surfactants' – the diminution of three words Surface-Active-Agents – basically are amphiphilic molecules that lower the interfacial tension or to be more specific the surface tension, mainly for the liquid-liquid systems, liquid-solid systems and the aqueous media (air-water). For instance, in the aqueous media, these molecules of amphiphilic nature are absorbed in the air-water interface. Particular alignment has been attained at the interface, where the hydrophilic part being water-loving is in the water and the hydrophobic part being water-fearing is in the air (see Figure 6.1). Such an alignment results in the reduction in interfacial tensions or surface tensions.

Due to these properties, surfactants are one of the most illustrative chemical materials. A wide variety of applications in agriculture areas, detergent industry and cosmetic, etc. are witnessed [1, 2]. Scheme 6.1 represents various fields in which surfactants have made their contributions.

Surfactants definitely are a boon to mankind and proved out to be very crucial in chemical industries. However, the majority of surfactants are synthetic and due to their wayward and tenacious nature, they bring out environmental and toxicological problems [3]. So, it won't be mindful to continue using such toxic surfactants to ease our lives today and in the longer run creating risks to human existence. Most of the surfactants are made from nonrenewable petroleum feedstocks. Apart from their toxicity, it takes millions of years for petroleum to be formed from the plant and

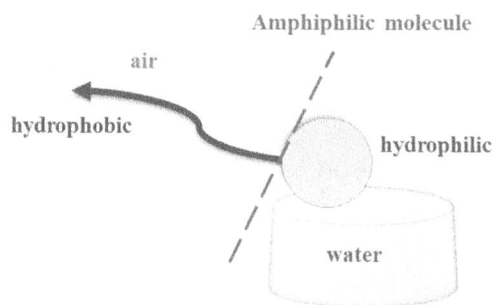

FIGURE 6.1   Structure of surfactant.

animal remains. Certainly, petroleum is finite and nonbiodegradable, if we would continue using it with the same pace, the day is not far when we will regard petroleum as dinosaurs that are extinct today. Saving petroleum for the future generation can be regarded as one of the huge benefits of conserving petroleum.

Sodium lauryl sulfate (SLS)-based detergents are the most common nowadays. It has been noticed that in the field of personal care products, shampoos mostly contain SLS-based detergents [4] and commonly used conditioning agents are cationic polymers and surfactants. These surfactants may result in nitrosamine development, which are said to be perilous to health. Formaldehyde-based chemicals are also been used as preservatives in personal care products, which may affect the skin adversely. We are pretty well aware that the consumption of surfactants is quite high and is expected to rise in the coming years [5] (Scheme 6.2).

It is crucial to switch to natural surfactants to satisfy the modern consumer's needs while keeping in mind the principles of green chemistry. Surfactant production using renewable sources can accord to the reduction of carbon dioxide emissions (CO_2) and hence could attenuate the greenhouse effect.

Natural surfactants are considered to be a very important earth-friendly alternative to the commercially available synthetic surfactants because of their reasonably nontoxic and biodegradable nature [6, 7]. Progressive escalation has been experienced in the interest of such surfactants [8, 9] possibly because of their diversity, eco-friendly approach, large-scale production possibilities, selectivity and their wide variety of applications in environmental rampart [10, 11].

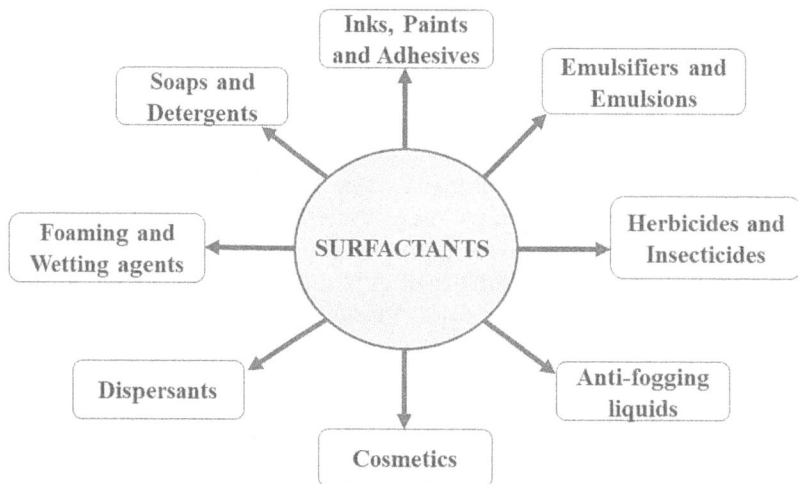

SCHEME 6.1   Various industries making use of surfactants.

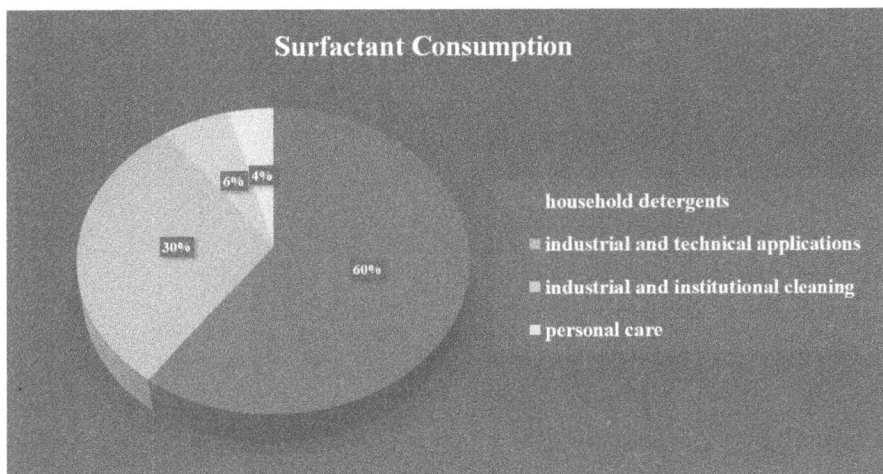

SCHEME 6.2 Consumption of surfactants in different sectors.

Natural surfactants can be of plant or animal origin. These surfactants are obtained from a variety of natural sources and natural products. These natural sources can be plant oils, fats, proteins, steroids, terpenes, alkaloids and moreover mono, di and polysaccharides too proved out to be an important source. Surfactants based on these natural sources have been covered nicely in the previous chapters.

In this chapter, we would focus on the surfactants based on the natural antioxidants called 'flavonoids.' Where on one hand flavonoid extracts of various plants are responsible for the surfactant activity, on the other hand, flavonoids being antioxidant, antimicrobial and anti-inflammatory tend to provide additional features to the surfactants that can be robust and green at the same time. As we enhance in this chapter, we will understand what actually flavonoids are and where these species are found. Then we shall deal with the extraction procedures of these natural antioxidants from a variety of plants. This will be followed by a detailed discussion of flavonoid-based surfactants, their applications and how they will provide additional features to the surfactants.

6.2 FLAVONOIDS

6.2.1 GENERAL

Since always plants have been playing a major role in human health by meeting all the desired expectations. From being an effective tool in maintaining and improving human health to being such a valuable component which could rarely have any side effects on human health, medicinal plants have gained a lot of attention amongst researchers and biologists.

Flavonoids belongs to a significant class of natural products, possessing a polyphenolic structure and are found mainly in food products like fruits, vegetables and beverages [12]. Extracts of various vegetables, fruits, roots, stems, bark, flowers and wine, etc. have been quite beneficial for human health and for treating various diseases long before scientists isolated flavonoids as the effectual species. The color that leaves, fruits and flowers of variety of plants have is mainly due to the presence of flavonoids in them [13, 14]. They also play an important role in the plant's growth and defense against infections [15]. Flavonoids at the same time have been equally instrumental for human health (Scheme 6.3).

Antioxidants are the species that inhibit oxidation. Oxidation can produce free radicals which can lead to chain reactions in the human body leading to damage of cells. Flavonoids exhibit an important scavenging action and can scavenge free radicals and in turn act as a natural antioxidant [15].

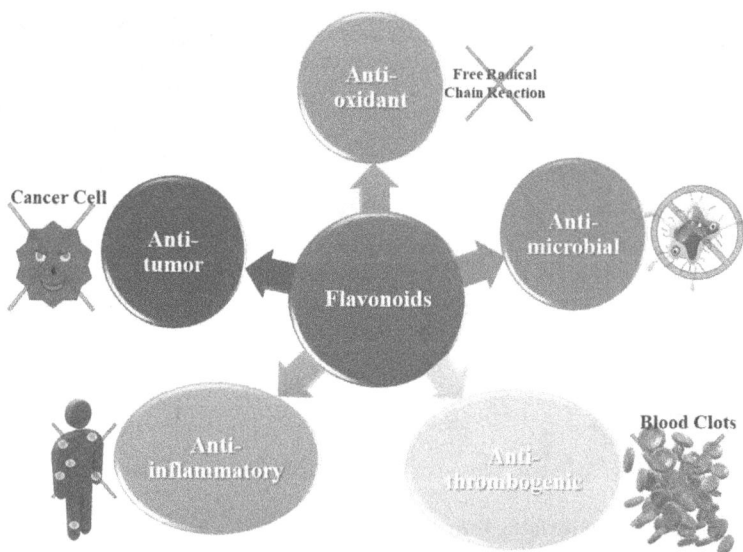

SCHEME 6.3   Benefits of flavonoids for human health.

They contain highly reactive hydroxyl groups which basically stabilize the oxygen free radicals by oxidizing themselves into less-reactive and hence more stable radicals. Almost all flavonoids show antioxidant effects.

Antimicrobial activity of flavonoids has also been identified. Wang et al. in 1998 have discussed antiviral effects of flavonoids in viral infection therapies [16]. Antiviral activities of hesperetin, quercetin and naringin have been studied where these flavonoids have shown anti-infective effects on various viruses (DNA and RNA viruses) [17]. These flavonoids have also been able to show antibacterial activities [18]. Lotus leaves contain quercetin, which can be easily extracted and may prove out to be an exceptional antibacterial agent [19].

Poor immune system and autoimmune diseases can have adverse effects on human health. A variety of inflammatory processes takes place in our body which can affect our immune functions badly. In 2005, Comalada et al. highlighted the anti-inflammatory functions of flavonoids like quercetin which can improve the immune system by inhibiting the initial process of inflammation [20]. Flavonoids extracted from the plant *Spilanthes paniculata* have been recognized as a promising anti-inflammatory natural product [21].

Flavonoids being natural antioxidants may prove out to be an important chemopreventive agent [22]. Reports have been there where flavonoid intake has resulted in a substantial fall in the risks of lung cancer and thus prove out to be an important antitumor agent [23]. Melanoma tumor cell growth and metastatic potential can be inhibited by using quercetin and apigenin [24]. They have also been seen to diminish the risks of angiogenesis in the human body [25].

Risks of acute platelet aggregation can also be reduced by using flavonoids, such as quercetin, kaempferol and myricetin [26]. Alcaraz et al. speculated flavonoids as a capable antithrombogenic agent because of their cyclooxygenase and lipoxygenase pathways action inhibition [27].

6.2.2   CLASSIFICATION OF FLAVONOIDS

Flavonoids can be classified into varied subgroups. The division of flavonoids into various subgroups has been done on the basis of the position of attachment of the C ring and B ring. When the B ring attaches to the C ring via its carbon-3, then flavonoid class is called isoflavonoid.

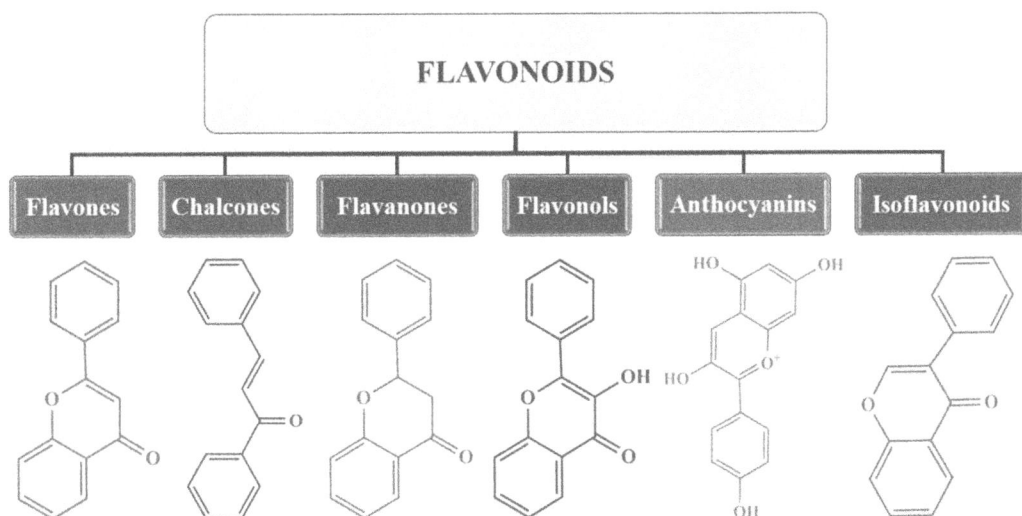

SCHEME 6.4 Classification of flavonoids.

Whereas when linkage is at carbon-4, then it is called neoflavonoids. Flavonoids in which attachment of B ring is found at carbon-2 of the C ring can be further classified on the basis of structural characteristics of C ring: flavones, flavanones, flavonols, anthocyanins and chalcones (Scheme 6.4).

Flavones are found in a variety of citrus fruits and medicinal plants, namely chamomile, mint, ginkgo biloba, etc. They can be further classified into apigenin, rhoifolin, tangeritin and baicalein [28]. When it comes to the discussion of their structure, flavones have unsaturation in the form of double bond between carbon-2 and carbon-3 and a ketonic group at carbon position 4 of the C ring. Many flavones were found to have an additional hydroxyl group on carbon on the position 5 and 7 of ring A, some have it on the carbon-3' and carbon-4' of the ring B. Flavanones are another class of flavonoids, which have nearly same structure as that of flavones, the only difference being the saturated ring C. In other words, unlike flavones, flavonoids have single bond between carbon-2 and carbon-3 of the ring C. The presence of these flavonoids is the major factor behind the bitterness of peels and juices of citrus fruits [29].

Flavonols on the other hand are also quite comparative to flavones. The 'ol' suffix in its name suggests the presence of a hydroxyl group. The only difference in the structure of flavonols and flavones is the presence of hydroxyl group on carbon-3 of the C ring. Tea and red wine are a rich source of flavonoids. Apart from this, berries, onions, tomatoes, grapes are also rich in this class of flavonoids [29].

The subclass of flavonoids responsible for the color in flowers and fruits and even plants is anthocyanins. Their coloration is mainly dependent on the pH and the presence of methyl or acyl group at the hydroxyl group on the A ring and B ring [29]. In the basic structure of flavonoids, (Figure 6.2) if there is an absence of C ring resulting in an open-chain flavonoid, a new class of flavonoids comes into view, namely chalcones. Chalcones can be further classified into phloretin, phloridzin, chalconaringenin and arbutin. Fruits like pears and berries are a rich source of these open-chain flavonoids.

Isoflavonoids are a typical class of flavonoids as compared with the ones discussed previously. When observed closely, it has been noticed that isoflavonoids are different from flavones only on the basis of the carbon which connects the B ring with the C ring. In flavones, the connection is made at carbon-2 of C ring whereas in isoflavonoids B ring is attached at carbon-3. Apart from medicinal plants, microbes are also sources of isoflavonoids [30]. Various subclasses and sources of flavonoids are well tabulated in Table 6.1.

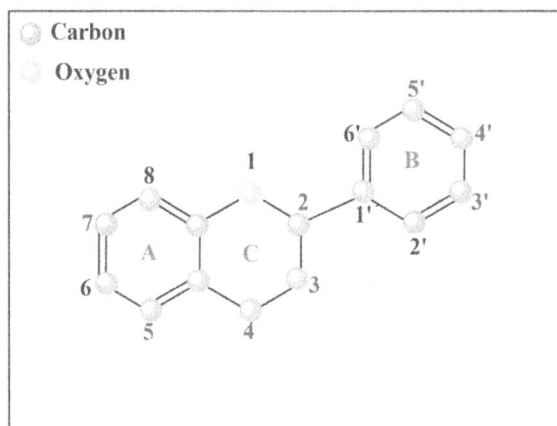

FIGURE 6.2   Basic structure of flavonoids.

6.2.3   Plants Rich in Flavonoids

Medicinal plants are a major source of flavonoids. Plant parts, including leaves, roots, stem, bark, fruits and flowers, all are rich in flavonoid content, leaves being the richest. Nowadays, herbal medicines have gained a lot of importance because of the presence of natural antioxidants viz. flavonoids in them. Plant-based research received an added impulse after flavonoids were found to be effective in palliating a lot of life-threatening diseases. Flavonoids exhibit antioxidant, antitumor, anti-inflammatory, anti-microbial and antithrombogenic activities. These features have inspired many researchers to include the benefits of these naturally occurring products in the surfactant industry. Herein we plan to discuss five such plants which are rich in flavonoids, namely Reetha, Amla, Neem, *Aloe vera* and Shikakai and are capable of showing surfactant property. In Scheme 6.5, we plan to depict important plants that are rich in flavonoids. Structures of some important flavonoids are shown in Figure 6.3.

6.3   EXTRACTION

6.3.1   Neem

Azadirachta indica A. Juss, commonly known as neem, belongs to the Meliaceae family, originated in Indian subcontinent over 4000 years ago (Table 6.2) [31]. The latinized word for neem, *A. indica* in Persian language refers to *Azadi* = free, *rachta* = tree and *Indica* is the Latin name for India.

TABLE 6.1
Flavonoids: Subclasses and Sources

Flavonoid	Subclasses	Sources
Flavones	Apigenin, rhoifolin, baicalein, tangeretin	Fruits and medicinal plants
Flavonols	Quercetin, morin, rutin, kaempferol, myricetin	Fruits, medicinal plants, vegetables
Flavanones	Naringin, hesperitin, eriodictyol, naringenin, hesperidin	Fruits and medicinal plants
Chalcones	Phloretin, phlioridzin, arbutin, chalconaringenin	Fruits, medicinal plants, vegetables
Anthocyanins	Cyanidin, peonidin, malvidin, delphinidin	Fruits, medicinal plants, vegetables, nuts and dry fruits
Isoflavonoids	Genistin, daidzin, glycitein, genistein, daidzein	Legumes, medicinal plants

---

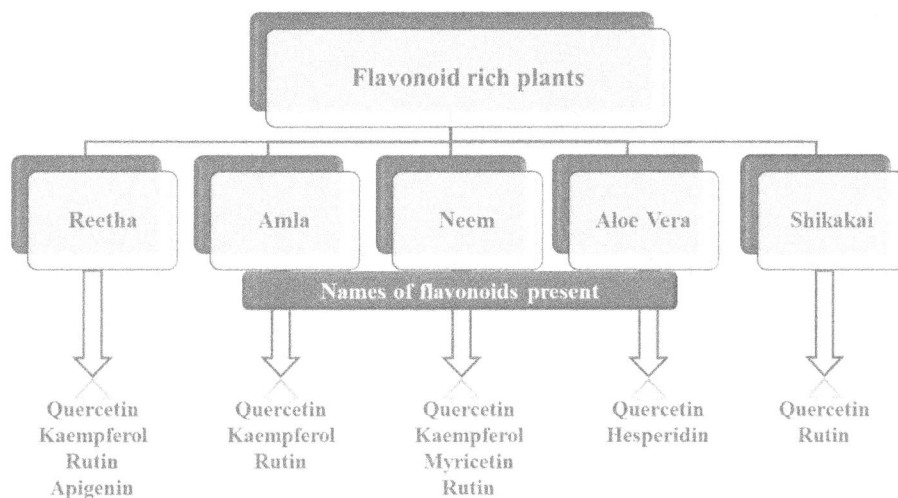

SCHEME 6.5 Plants rich in flavonoids.

The name literally means 'The free tree of India,' free here corresponds to free from insects and diseases. Apart from this, the neem tree has been given several other names owing to its tremendous benefits, such as 'The Divine Tree,' 'Nature's Drugstore,' 'Heal All,' 'Village Dispensary,' 'Indian Lilac' and many more [32]. In the Eastern part of Africa, it is called Mwarobaini (Swahili) which means 'The tree of the 40,' as it is said to cure over 40 different diseases (Table 6.3) [32]. Researchers reported that there exist more than 140 active substances in a neem tree. From truck to seed to twig to leaf, every single component of this species has been found to be of extensive benefits to mankind. To study the composition and extract medicine out of the neem tree, one has to prepare the extract of plant. We, thus, report some of the methods that are extensively used to prepare neem leaf extract in Schemes 6.6–6.9 [33,34].

6.3.2 ALOE VERA

Aloe vera belonging to *Asphodelaceae* (Table 6.2) has been known for its wide applications in pharmaceutical and cosmetic industries [36]. The yellow juice obtained from the pericyclic cells under the plant's skin is mostly utilized by pharmaceutical industries. Latex portion containing aloe-emodin is an anthraquinone which is a gastrointestinal irritant and works as cathartic [37, 38]. The inner central zone of the leaf that has thin-walled mucilaginous cells consists of aloe gel. This gel is used by cosmetic industries as it contains a variety of organic compounds that can be employed for moisturizing, healing and emolliating skin [39–41]. Table 6.4 represents some of the pharmacological properties of *Aloe vera* and diseases that can be treated by it [42]. *Aloe vera* acts as a curative precursor for these diverse diseases owing to its composition, in Scheme 6.10, a summary of the major components present in *Aloe vera* is presented [42, 43]. To study the composition further, one has to prepare an extract of the aloe plant. In Scheme 6.11, one such method of extraction of *Aloe vera* has been shown [44, 45].

6.3.3 AMLA

Phyllanthus emblica L., commonly called Amla or Indian Gooseberry, belongs to *Phyllanthaceae* family (Table 6.2) [46, 47]. The fruit is native to the tropical region of Southeast Asia. It is rich in polyphenols and hydrolysable tannin-derived compounds that act as antioxidants (Scheme 6.12). The extract from the fruit has been used for treating symptoms ranging from constipation to treatment

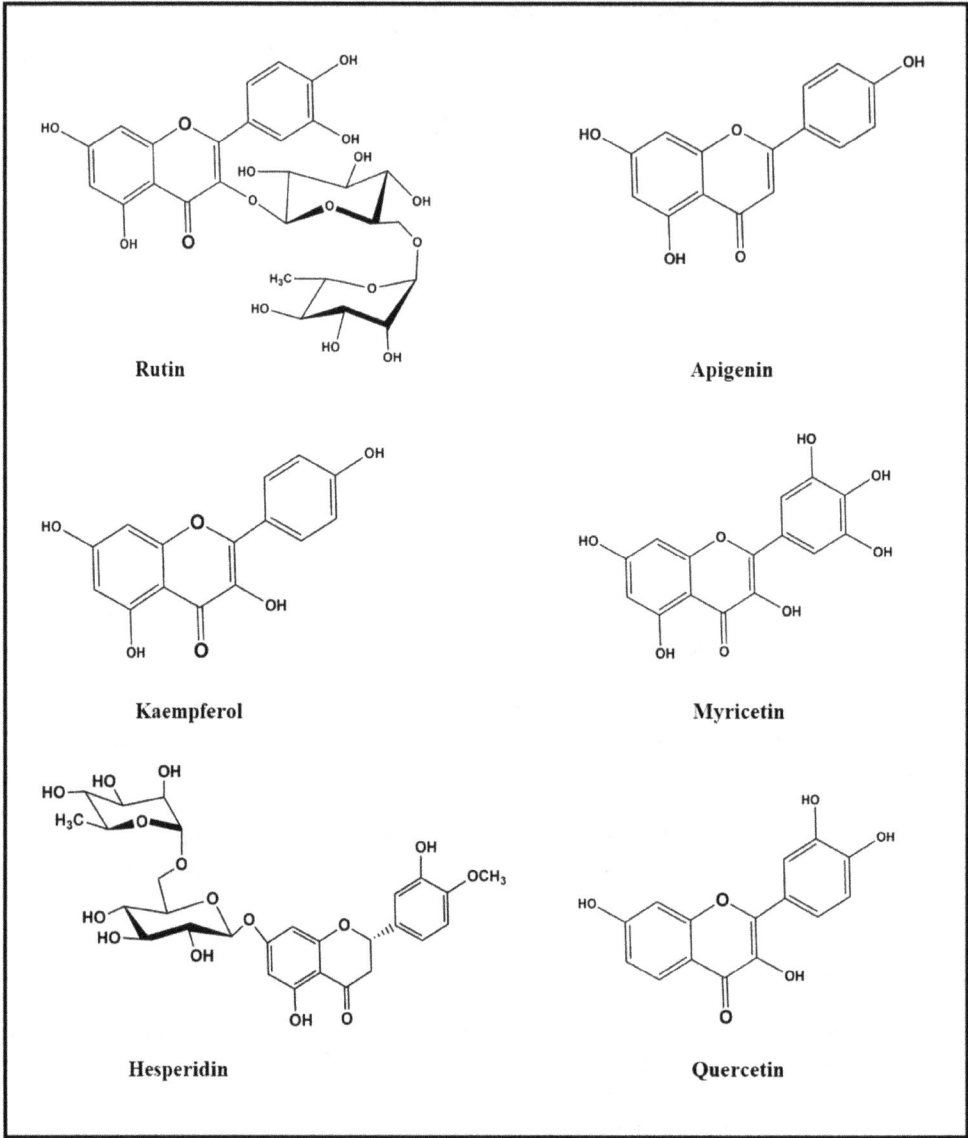

FIGURE 6.3   Structures of some important flavonoids.

TABLE 6.2
Taxonomical Classification

Taxon	Neem	*Aloe vera*	Amla	Reetha	Shikakai
Order	Rutales	Asparagales	Malpighiales	Sapindales	Fabales
Class	Dicotyledon	Monocotyledon	Dicotyledon	Dicotyledon	Magnoliosida
Family	Meliaceae	Asphodelaceae	Phyllanthaceae	Sapindaceae	Fabaceae
Subfamily	Maliadeac	Asphodeloideae	Phyllanthoideae	Sapindoideae	Mimosoideae
Tribe	Melieae	Aloeae	Phyllantheae	Andropogoneae	Acacieae
Genus	Azadirachta	Aloe	*Phyllanthus* L.	*Sapindus* L.	Acacia
Species	Indica	*A. vera*	*Phyllanthus emblica* L.	*Sapindus mukorossi* Gaertn	*A. concinna*

TABLE 6.3

Neem Tree as Nature's Drugstore [35]

Components	Diseases Treated
Seed	Leprosy, intestinal worms, cancer
Flower	Phlegm, bile suppression
Fruit	Eye problem, phlegm, diabetes, intestinal worms, piles, urinary disorders, wounds, leprosy
Leaf	Skin ulcer, eye problem, leprosy, cancer
Gum	Wounds, ulcers, scabies, skin diseases
Oil	Intestinal worms, leprosy
Twig	Asthma, piles, phantom tumor, cough, obstinate urinary disorder, diabetes, intestinal worms
Bark	Cures fever, analgesic

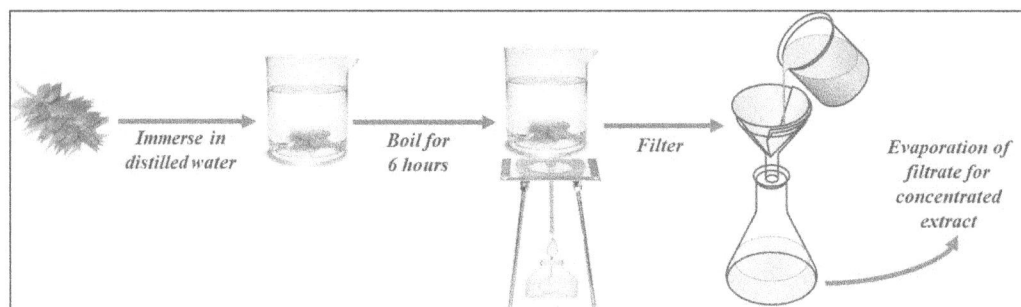

SCHEME 6.6   Decoction method of extraction of *Azadirachta indica* [33].

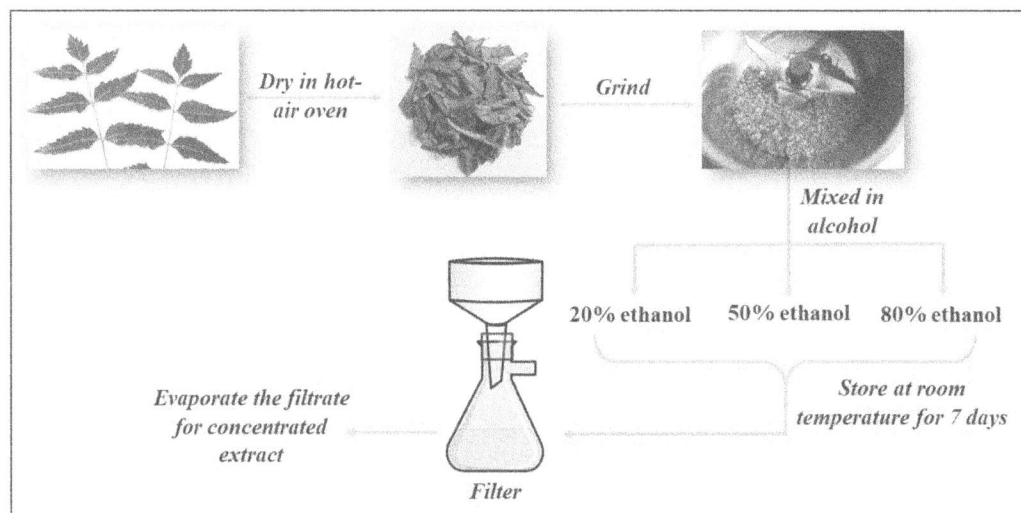

SCHEME 6.7   Maceration method of extraction of *Azadirachta indica* [34].

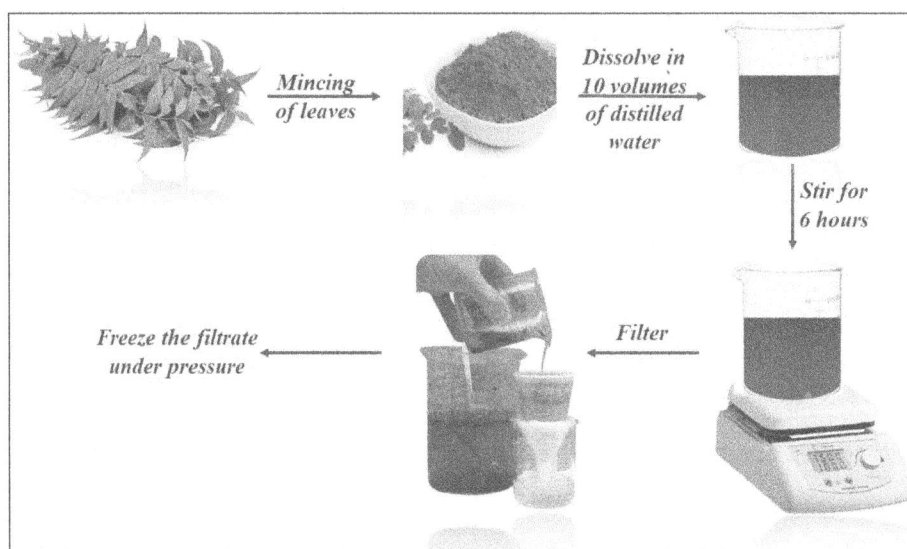

SCHEME 6.8   Freeze-drying method of extraction of *Azadirachta indica* [33].

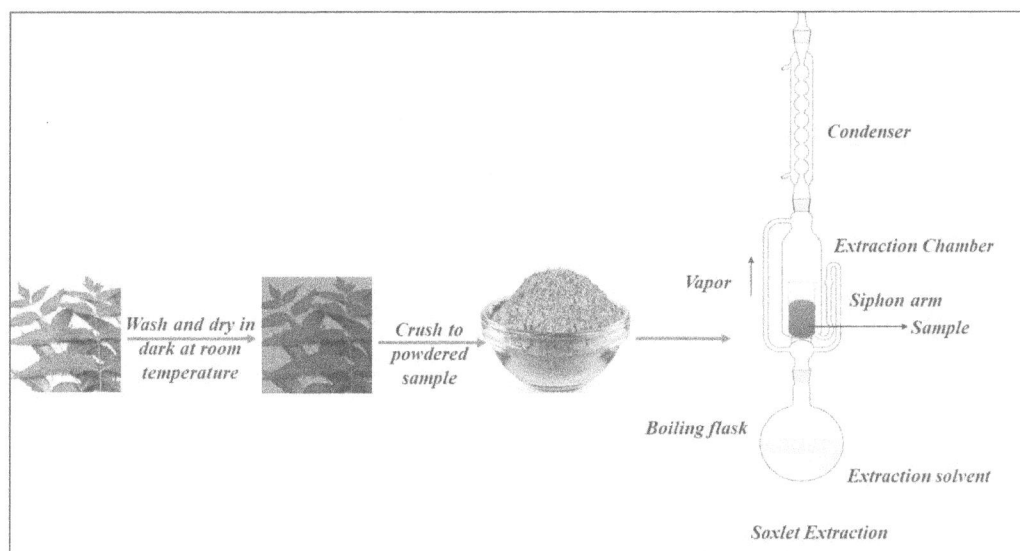

Wash and dry in dark at room temperature

Crush to powdered sample

Vapor

Condenser

Extraction Chamber

Siphon arm

Sample

Boiling flask

Extraction solvent

Soxlet Extraction

SCHEME 6.9 Soxhlet method of extraction of *Azadirachta indica* (solvent = alcohol or water) [34].

of tumor (Table 6.5). It carries small molecules with both antitumor activity and cancer-preventive properties. Amla is widely used in hair products these days [48–50]. It also carries quercetin (a flavonoid, discussed in the next section) that shows antitumor and antiproliferative properties for numerous cancer cell lines from multiple tissue types, leukemia, xenograft transgenic murine model of breast cancer and phase I clinical trial. In Scheme 6.13, we report one such method for preparation of amla extract [51].

TABLE 6.4
Pharmacological Properties and Diseases Treated by Aloe Vera Plant [42]

Pharmacological Properties	Diseases Treated
Adaptogen	Arthritis
No side effect	Asthma
Breaks down and digests dead tissue	Candida
Provides essential nutrients	Lupus erythematosus
Reduces swelling	Digestive and bowel disorders (atonic, irritable bowel syndrome, constipation,
Anti-inflammatory	ulcerative colitis, Crohn's disease)
Moisturizes	
Penetrates tissue	Chronic fatigue syndrome
Relieves itching	Dermatosis (eczema, acne, psoriasis, burns, athlete's foot, frostbite, cold
Antifungal	sores)
Anesthetizes-relieves pain	Sports injuries
Antimicrobial- prevents infection	Ulcers (internal and external)
Stimulates cell growth	
Cleanses and detoxifies	

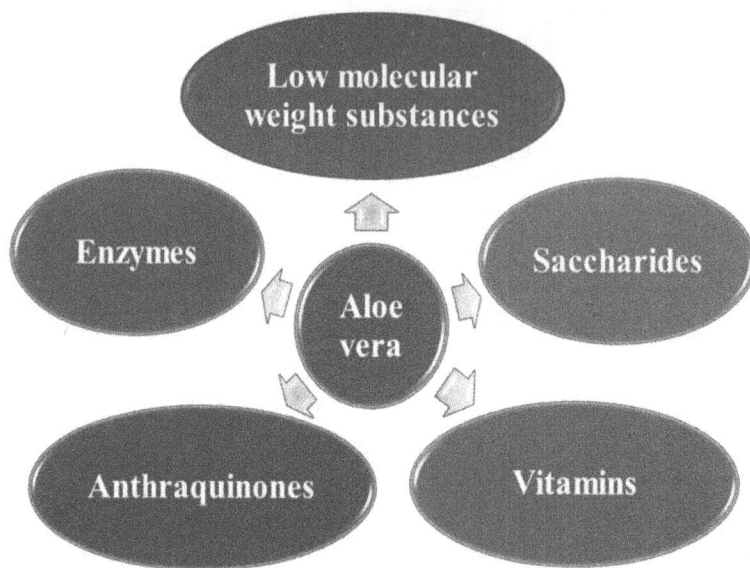

Anthraquinones	**Vitamins**	**Enzymes**
• aloe-emodin	• B1	• amylase
• aloin	• B6	• catalase
• aloetic acid	• B2	• carboxypeptidase
• anthranol	• C	• lipase
• barboloin	• choline	• cyclooxydase
• emodin	• β-carotene	• oxidase
• isobarbaloin	• α-tocopherol	
• ester of cinnamic	• folic acid	
acid		

Saccharides	**Low molecular weight substances**
• cellulose	• cholesterol

SCHEME 6.10   (a) Major components of *Aloe vera* and (b) constituents present in various components of *Aloe vera*. [43]. *(Continued)*

- glucose
- mannose
- aldopentose
- acemannan
- glucomannan
- acetylated glucomannan
- galactogalacturan
- glucogalatomannan
- galactoglucoarabinomannan

- arachidonic acid
- lectin-like substances
- gibberellin
- lignin
- β-sitosterol
- salicylic acid
- steroids
- uric acid
- triglycerides

SCHEME 6.10 *(Continued)*

6.3.4 REETHA

Sapindaceae is an important family of Plantae kingdom that consists of about 2000 species (Table 6.2) [5]. *Sapindus mukorossi* is commonly known as soapberry, soapnut, washnut, aritha, reetha, doadni and dodan. The deciduous tree is mainly found in high ranges of Shivaliks, Indo-Gangetic plains and Sub-Himalayan tracks in heights from 200 to 1500 m [58, 59]. It is a popular ingredient in ayurvedic products such as shampoos, cleansers, for treating psoriasis, eczema, for removing freckles and for eliminating lice from scalp. The fruit contains 10.1% saponins in the pericarps and constitutes 56.5% of the drupe known for inhibition of tumor cell growth [60]. The pericarps are also employed as a protection gear of pests and microorganisms. Other major components of the fruit are depicted in Scheme 6.14, along with the properties exhibited by reetha [61, 62]. To make effective use of this fruit in a variety of bailiwick, beneficial components are to be extracted from the fruit. One such method for extraction of *S mukorossi* is discussed in Scheme 6.15 [63,64].

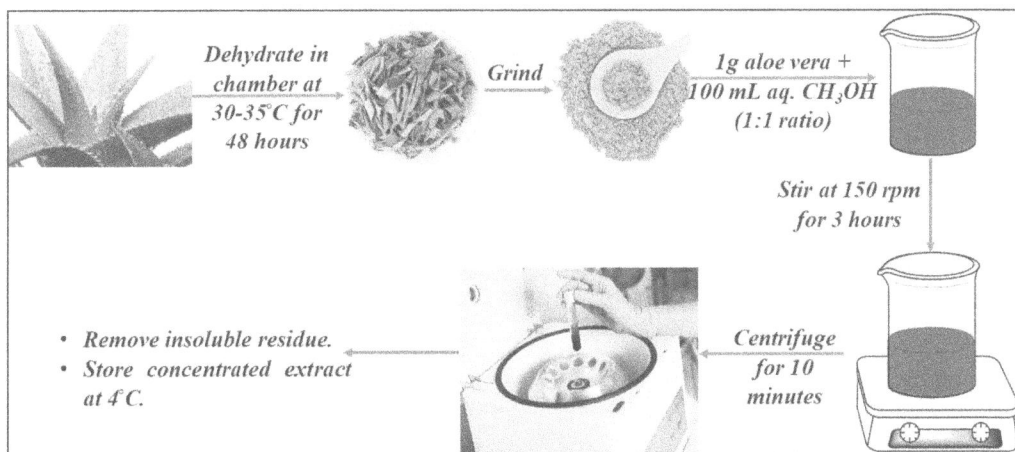

Dehydrate in chamber at 30-35°C for 48 hours

Grind

1g aloe vera + 100 mL aq. CH₃OH (1:1 ratio)

Stir at 150 rpm for 3 hours

- *Remove insoluble residue.*
- *Store concentrated extract at 4°C.*

Centrifuge for 10 minutes

SCHEME 6.11 Procedure for extraction of *Aloe vera* [45].

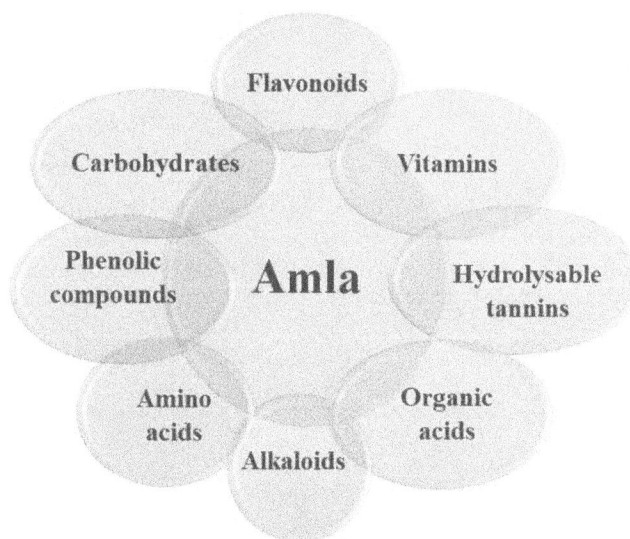

Hydrolysable Tannins

- Punigluconin
- Emblicanin A and B
- Chebulagic acid (Benzopyran tannin)
- Chebulinic acid (Ellagitannin)
- Corilagin (Ellagitannin)
- Geranin (Dehydroellagitannin)
- Ellagotannin

Amino Acids

- Proline
- Glutamic acid
- Alanine
- Aspartic acid
- Lysine
- Cystine

Phenolic compounds

- Methyl gallate

Alkaloids

- Phyllantine

SCHEME 6.12 (a) Major components in *Phyllanthus emblica* L. [52] and (b) constituents present in various components of *Phyllanthus emblica* L. *(Continued)*

> Gallic acid   > Phyllembein

> Ellagic acid   > Phylllantidine

> Trigallalyl glucose

Carbohydrate   **Vitamin**

> Pectin   > Ascorbic acid

Flavonoid   **Organic acid**

> Quercetin   > Citric acid

> Kaemferol

SCHEME 6.12 *(Continued)*

6.3.5 SHIKAKAI

The fruit *Acacia concinna*, i.e. commonly known as Shikakai in India, belongs to the Fabaceae family (Table 6.2) [65]. It is found in tropical rainforests of Southeast Asia, including India, Myanmar and Thailand. It is also known as *Sampoi* in Thailand. The fruit has several other names like 'Hair-cleansing Goddess,' 'Fruit for Hair' and many more, based on its unique properties [66, 67]. It is also used as an important herb in Ayurveda. All the benefits of *A. concinna* are derived from its pods. For years, the pods of the fruit are employed as laxatives, expectorants, skin-tonic, mucoactive agents and an emetic [68]. Shikakai contains a variety of saponins, flavonoids and monoterpenoids, exhibiting important pharmacological properties, as discussed in Scheme 6.16 [69]. To adequately

TABLE 6.5
Traditional and Medicinal Uses of *Phyllanthus emblica* L. (Amla) [60,61]

Vermifuge	Irritation in the bladder	Febrifuge
Diabetes	Painful respiration	Diarrhea
Hemorrhage	Dysentery	Anemia
Jaundice	Retention of urine	Dyspepsia
Scabies	Stops nausea and vomiting	Eye-disorder
Scurvy	For burning in vagina	Aging
Bleeding of nose	Regulating blood sugar	Anti-inflammatory
Cleanses mouth	Increases hemoglobin	
Enhances cell survival	Increases RBC count	

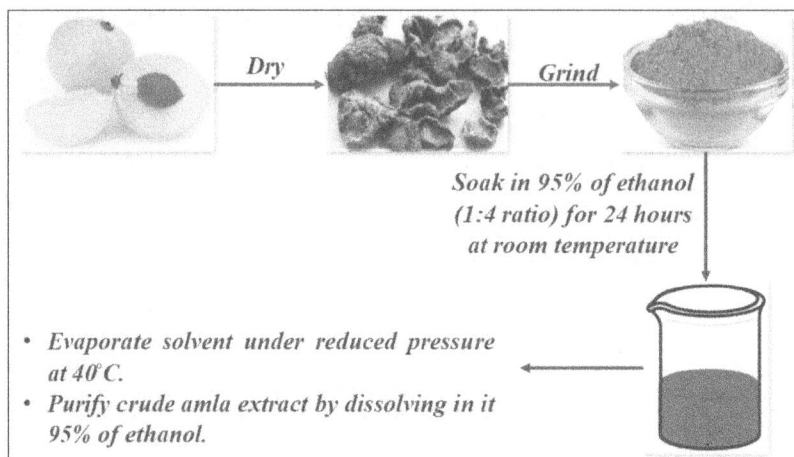

SCHEME 6.13   Extraction of *Phyllanthus emblica* L. (Amla fruit) [55–57].

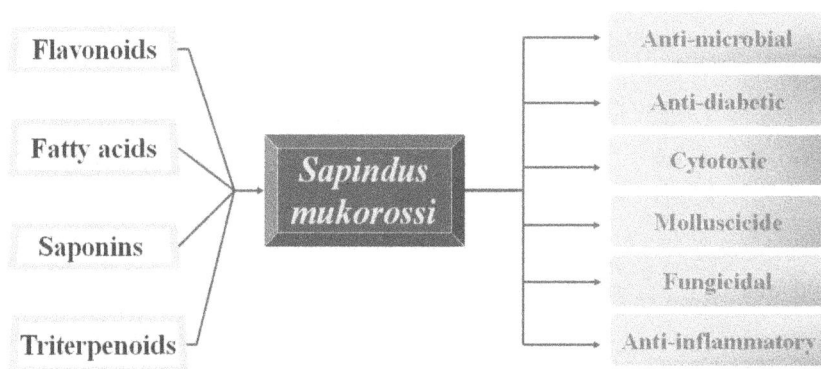

SCHEME 6.14   Major components of *Sapindus mukorossi* (Reetha) fruit and its functions [61, 62].

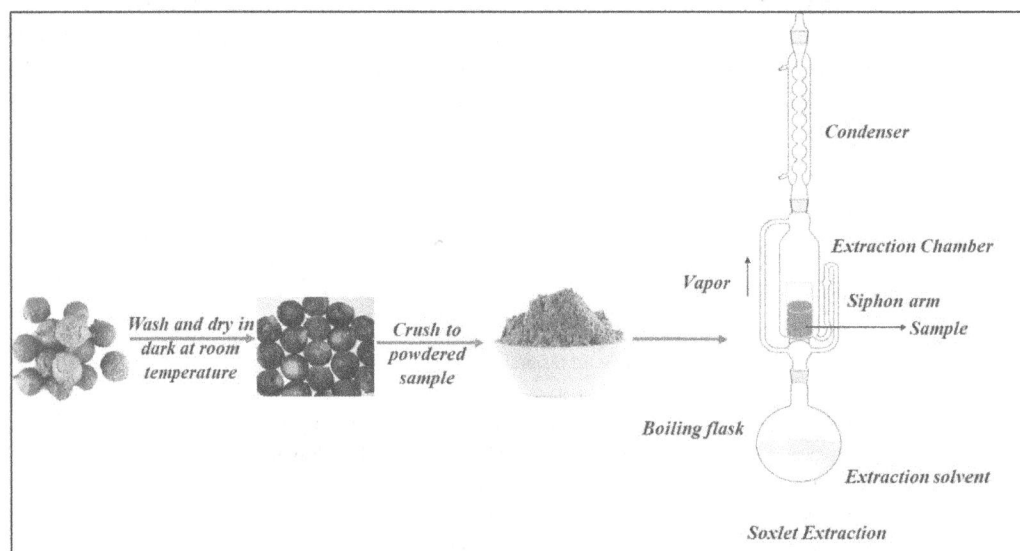

SCHEME 6.15   Soxhlet extraction of *Sapindus mukorossi* (Reetha fruit) [63, 64].

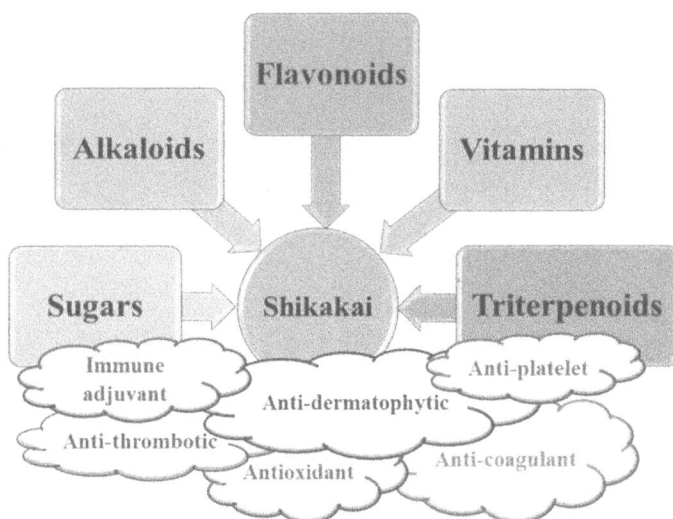

SCHEME 6.16 Major components of *Acacia concinna* (shikakai) and its properties.

utilize the fruit, there is a need for extraction of shikakai. One such method of extraction has been shown in Scheme 6.17 [70].

6.4 BIOCHEMICAL SCREENING TESTS FOR FLAVONOID

By now, the reader is well versed with the plants and fruits, the applications as surfactant of which will be studied in the next two topics. One thing common in all the above-discussed species is the presence of flavonoids (Scheme 6.5). The qualitative analysis for the presence of flavonoids in the plant extract can be done via two simple methods: sodium hydroxide test (Scheme 6.18), wherein the colorless solution indicates the presence of flavonoids in the extract, and Shinoda test, the formation of pink, red or reddish-brown solution upon addition of magnesium turnings and dilute hydrochloric acid to ethanolic extract of the plant indicates the presence of flavonoids (Scheme 6.19) [44, 70].

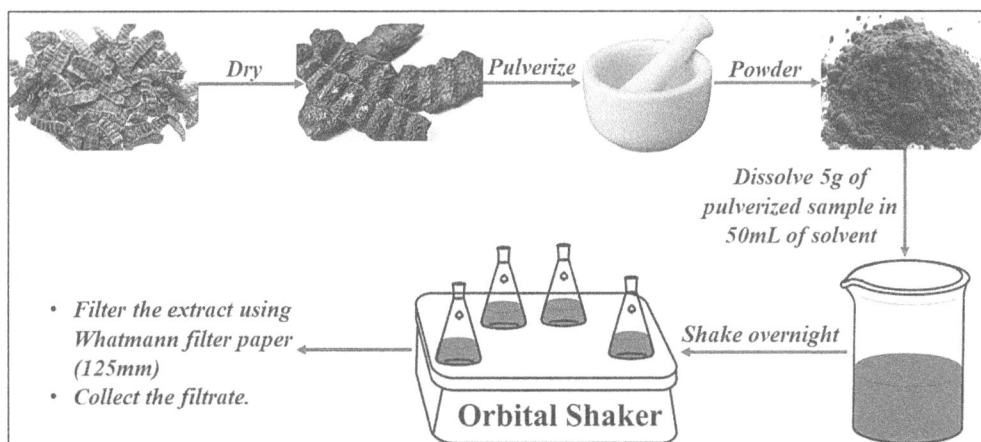

SCHEME 6.17 Extraction of *Acacia concinna* (Shikakai fruit).

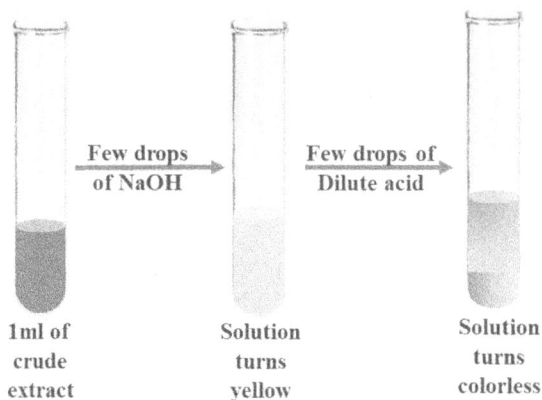

SCHEME 6.18   Hydroxide test for qualitatively analyzing the presence of flavonoid.

6.5 FLAVONOID-BASED SURFACTANTS

We live in a world where the environment is not as clean as it used to be in the past. There are various impurities, dust, dirt, microbes and pollution all around us. These impurities in the long run affect human health adversely resulting in various disorders and diseases like eye irritation, respiratory problems, dermatitis, etc. Thus, maintaining hygiene and cleanliness is as important as is food for human existence.

Be it the human body, skin, hair or even the clothes that are worn, all needs to be tidy and clean to maintain the basic hygiene. Apart from this, the kitchenware, floor and from anything to everything, we come in contact with must be clean and free from any sort of dirt or microbe growth. Household and personal care industry are the most prosperous consumers of detergents.

Synthetic surfactants are widely available and are highly used nowadays. SLS-based surfactants are extensively used in industry. SLS being inexpensive and because of its effective foaming capabilities are an important part of many personal care products. But it is a non-natural material and can prove to be quite harsh for the human body.

Synthetic surfactants also faced difficulties in working with hard water. Even the builders used to soften the water were highly discouraged, the reason being the use of phosphates. Phosphates can cause eutrophication of water which can badly affect the environment [71]. Consumers as well as retailers all are progressively insisting on the natural alternatives.

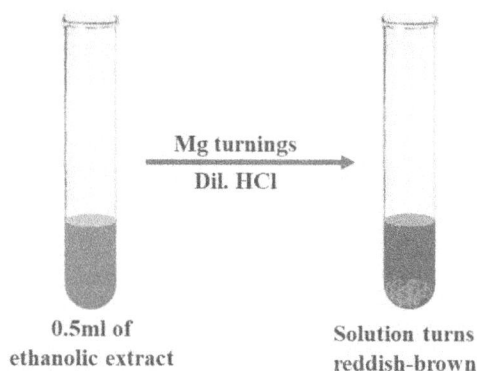

SCHEME 6.19   Shinoda's test for determining the presence of flavonoids.

Natural alternatives would mean those surfactants which are made up from natural ingredients that are of plant or animal origin. Natural surfactants are proving out to be highly commendable as these compounds can work in extreme conditions like salinity, pH or temperature [72]. Food, pharmaceuticals and cosmetic industries are also choosing these natural surfactants over the harsh chemical-based synthetic surfactants. These natural compounds are highly acceptable in the detergent industry because of its wide number of advantages [72]. Some of their advantages over their chemical equivalents are been highlighted in Scheme 6.20.

Since ancient times a wide variety of herbal plants are there but they are gaining a lot of attention now. Utilization of plants is the most important requirement for the production of nature-based surfactants. Plants can't be used as such directly and it won't be fair to expect it to show surfactant activity in that form without extracting out the contents responsible for such properties. Five such generally used plants and their extraction procedures have been discussed in the previous topic.

Phytochemicals are chemicals found in plants [73]. 'Phyto' is a Greek word that means 'plants,' these plant chemicals play an important role in the plant growth and its protection against predators [73, 74]. Phytochemicals can be classified into carotenoids and polyphenols, which mainly contain phenolic acids, flavonoids and lignans. These plant chemicals can be utilized to discover surfactant activities. Plants containing high flavonoid content proved out to be a good fit as a natural surfactant. Phytochemical screening is an effective tool in estimating the phytochemical contents in the plant and thus deciding which plant can show such surfactant features.

Human skin and hair are the most exposed part of the body and thus are highly prone to dirt and infections. Plant-based face washes, soaps and shampoos are catching a lot of attraction these days. In the coming portion, we shall discuss herbal shampoos made from *Aloe vera*, amla, neem, reetha and shikakai. *Aloe vera* gel face washes are also highly recommended nowadays because of being natural; these are very compatible with human skin and at the same time do not create any environmental risks. Synthetic cleansers for fabrics and textiles get trapped in the inter-fiber spaces in fabrics and make them rough, grey and dull [32]. Reetha- and shikakai-based detergents are natural and much milder thus maintaining the natural color, shine and texture of the fabric.

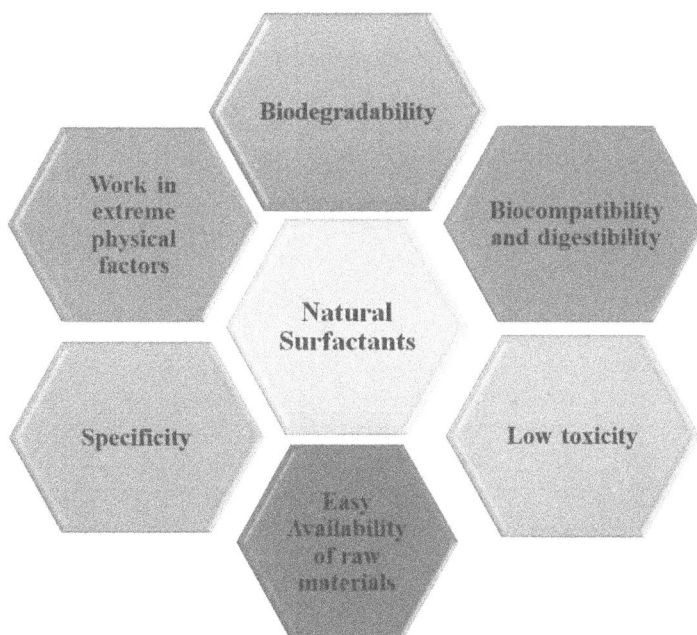

SCHEME 6.20 Advantages of natural surfactants.

Along with fabric detergents, we plan to talk about natural pesticides as well in this chapter. Synthetic and harsh chemical-based pesticides tend to impact the ecosystem and nature negatively. Pesticide market is quite vast and the use of toxic compounds may contaminate the water, vegetation, crops and soil. These drawbacks motivated researchers to discover some natural resorts to synthetic pesticides. Neem oil-based pesticides are one great flavonoid-based material which proved out to be instrumental in solving these issues. Scheme 6.21 depicts various flavonoid-based surfactants.

6.5.1 HERBAL SHAMPOO

When it comes to personal grooming, hair is the most attractive part. Every individual desires good, long and shiny hair. Shampoos play a crucial role in keeping the hair and scalp free from dirt and grease [75]. They prove out to be an acknowledged cosmetic product which is more of a necessity in the present scenario when we all are surrounded by dust and pollution all the time. Cleansing the hair and scalp is not the only motive of including shampoos in our daily life. They tend to be an amazing hair treatment resulting into glossy, non-frizzy, infection free and manageable healthy hair [76, 77].

While discussing the importance of shampoos in our day-to-day life, the foremost question that arises in our mind is 'what is a good shampoo?' This question has been answered wrongly many times. Consumers generally go with the conviction that a good shampoo is the one which produces more foam [77] and they completely forget about the efficacy and safety of the product. To match up consumer's notions, developers included excessive amounts of harsh detergents in shampoos.

Sodium lauryl ether sulfate (SLES) is one of those surfactants that are chiefly added to shampoos. SLES is manufactured from SLS, ethoxylation procedure of SLS is carried out using ethylene oxide. Residuary ethylene oxide may be leftover and might act as an adulterant in the surfactant. Such toxic impurities prove out to be dreadful and result in dull, dry and rough hair by taking away 80% of the oil from the hair. Conditioning agents added in shampoos nowadays produce high amounts of nitrosamine which can result into a health hazard. Some components of synthetic shampoos like formaldehyde-based preservatives can cause eye and skin irritation [78]. Features of a good shampoo have been shown in Scheme 6.22.

SCHEME 6.21   Different flavonoid-based surfactants.

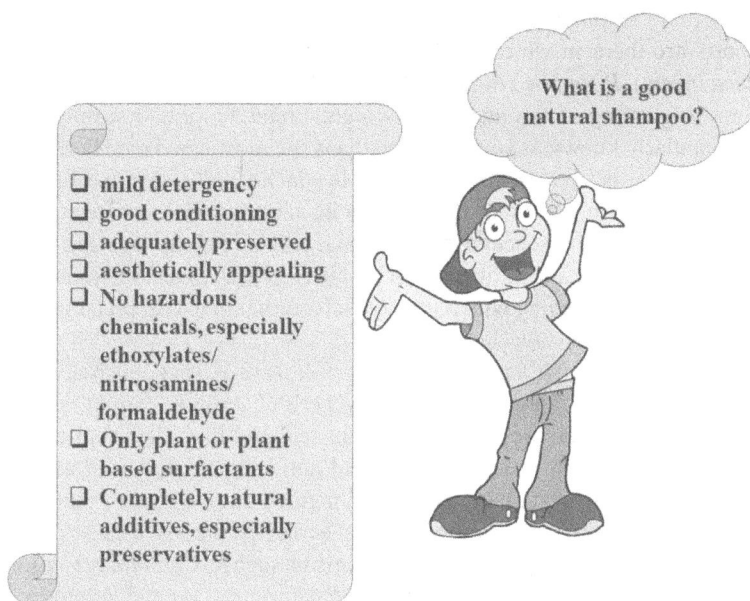

SCHEME 6.22 Features of a good shampoo.

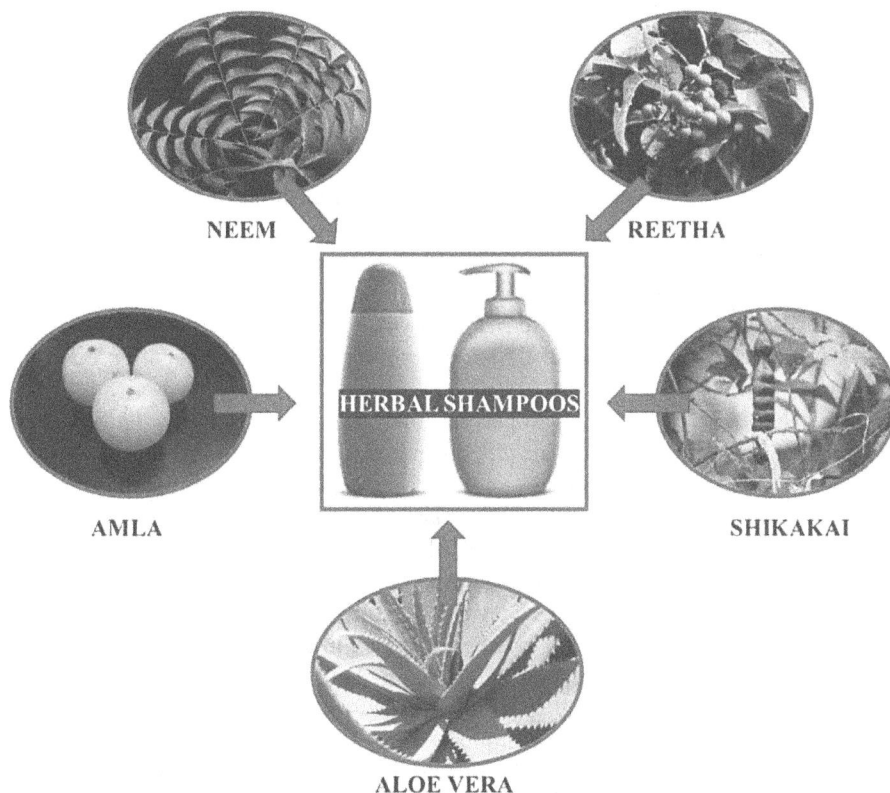

SCHEME 6.23 Herbally formulated shampoos.

There are a plethora of plants that can be advantageous for hair and can be formulated into shampoos. Reports are there in which aqueous extracts of *A. concinna* (shikakai), *S. mukorossi* (reetha), *A. indica* (neem), *P. emblica* (amla), *Aloe vera*, Tulsi, lemongrass, etc. established key part in dandruff removal, glossing of hair and acted as a great remedy for hair scalp problems [79, 80].

S. mukorossi popularly known as soapnut or reetha is the most convinced herb which was found to be quite impressive in the treatment of migraine headache, chloremia and brain disorders like epilepsy. One most important feature of this plant is its antimicrobial features that can be helpful for septic systems. Due to the presence of flavonoid and phenolic components in *S. mukorossi*, it is capable of showing antioxidant and antihepatotoxic activities [81]. Since old times *A. concinna* also known as shikakai is also used along with *S. mukorossi* (reetha) in herbal preparations to contribute in additional cleaning of hair.

Being highly soothing, healing and rejuvenating, *Aloe vera* is very much accepted in the cosmetic industry nowadays [81]. Another rejuvenating herb *P. emblica* (amla) is observed to be a promising antioxidant due to the presence of ascorbic acid and high flavonoid content [82]. Good amount of flavonoid in this herb makes it a potential anti-inflammatory and antigenotoxic agent [82]. *P. emblica* is also an important hair growth and pigmentation tonic in traditional recipes [83].

A. indica (neem) can be a very important additive in herbal shampoos because of its anti-dandruff activity. Neem is mainly effective against fungal organisms such as *Candida albicans* and strains of Gram-negative and Gram-positive organisms such as *Escherichia coli* and *Staphylococcus*, respectively. Herbal anti-dandruff shampoos can also be prepared using *A. indica* (neem) and they prove out to be harmless and more economical and potent than synthetic anti-dandruff shampoos [79].

The main object of this topic of this chapter is to discuss the various herbally formulated shampoos (Scheme 6.23) for hair growth and hair strengthening without any damage. Reported herbal shampoos were prepared by collecting herbal extracts of *S. mukorossi*, *A. concinna*, *Aloe vera* and *P. emblica* [84]. Conditioner plays an important role in formulating a good shampoo. While formulating a natural shampoo, it is very crucial to include a natural conditioner only, to maintain its herbal nature. Cocamidopropyl betaine (CAPB) is a widely used conditioner [85] which has been derived from coconut oil.

Extracts of Reetha, Shikakai, Amla and *Aloe vera* were subjected to phytochemical screening for total flavonoid, phenolic and saponin content.

Polyphenolic contents were determined by the methods discussed in Schemes 6.24–6.26. These flavonoid, saponin and phenolic compounds are capable of showing antimicrobial, anti-dandruff and anti-inflammatory characteristics. Saponin and flavonoid contents are the major factors that render these extracts with surfactant and cleansing action [79]. Results of phytochemical screening (Table 6.6) showed that these herbs are rich in phytochemicals. Shikakai shows higher content of saponins. Amla has higher content of flavonoids while phenolic contents are present in higher amounts in *Aloe vera*.

Flavonoids exhibit a very important property, namely antioxidant activity. Amla was observed to have higher concentrations of flavonoids [79] compared to other herbs and thus if we plan to formulate shampoos with high antioxidant potent, Amla can be a great choice. If we discuss phenolic contents, then it was found to be higher in *Aloe vera* when compared to others, whereas saponin content was highest for Shikakai. Higher amount of saponin makes Shikakai the best fit for cleaning and foaming abilities in herbal shampoos devoid of damaging the hair [79]. At the same time, Amla being the richest in flavonoids can supply extra features to the herbal formulations. Saponins, flavonoids and phenolics when combined together can result in an appreciable herbal formula which can be as natural as desired and render absolutely healthy hair.

While preparing a recipe for natural shampoo, it is very important to keep two crucial aspects in mind, i.e. cleansing and detergency capabilities. Phenolics and flavonoids are the hydrophobic molecules and they display great grease and oil encapsulation properties. By strongly encapsulating the grease and oil, these phytochemicals exhibit tremendous cleansing activity. Amla formulations showed cleaning ability of nearly 90% as compared to others. This result can be correlated to the

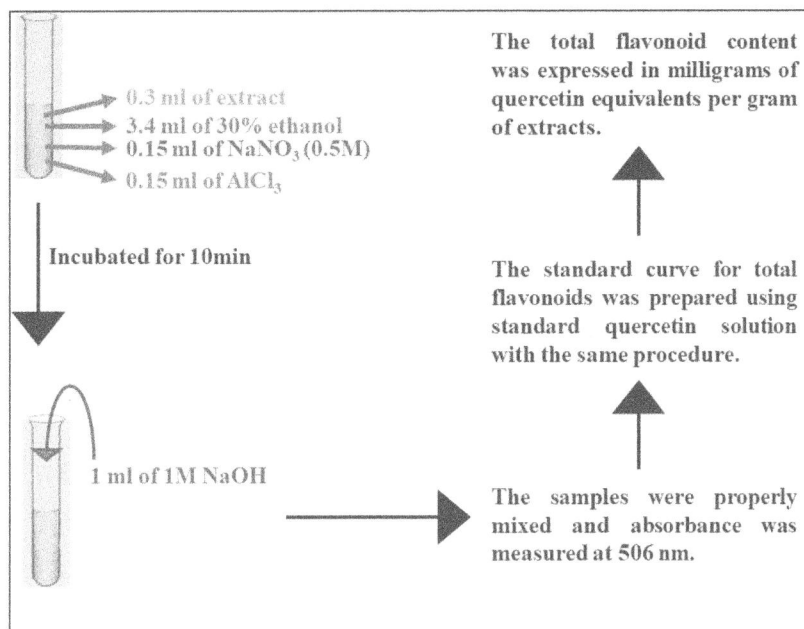

SCHEME 6.24 Method for phytochemical estimation of flavonoid content.

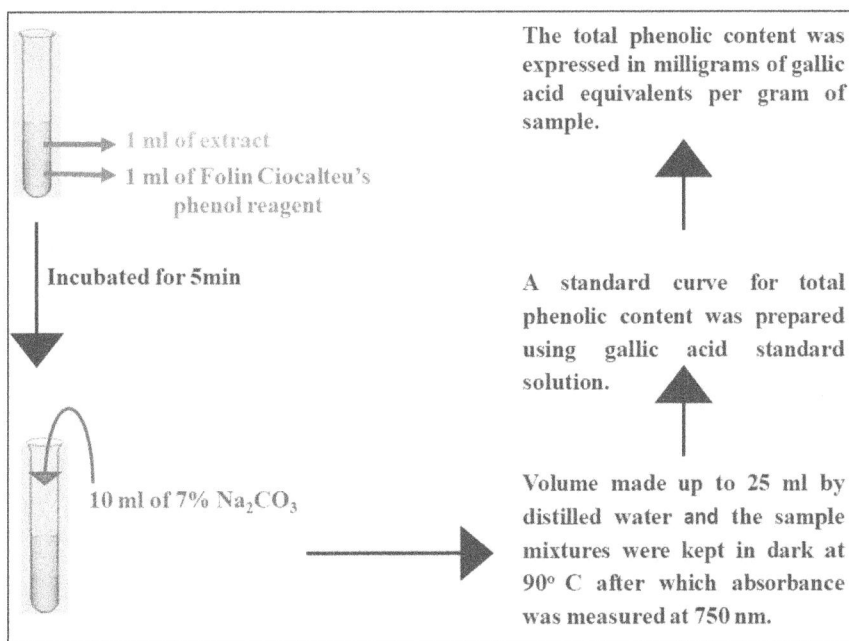

SCHEME 6.25 Method for phytochemical estimation of phenolic content [84].

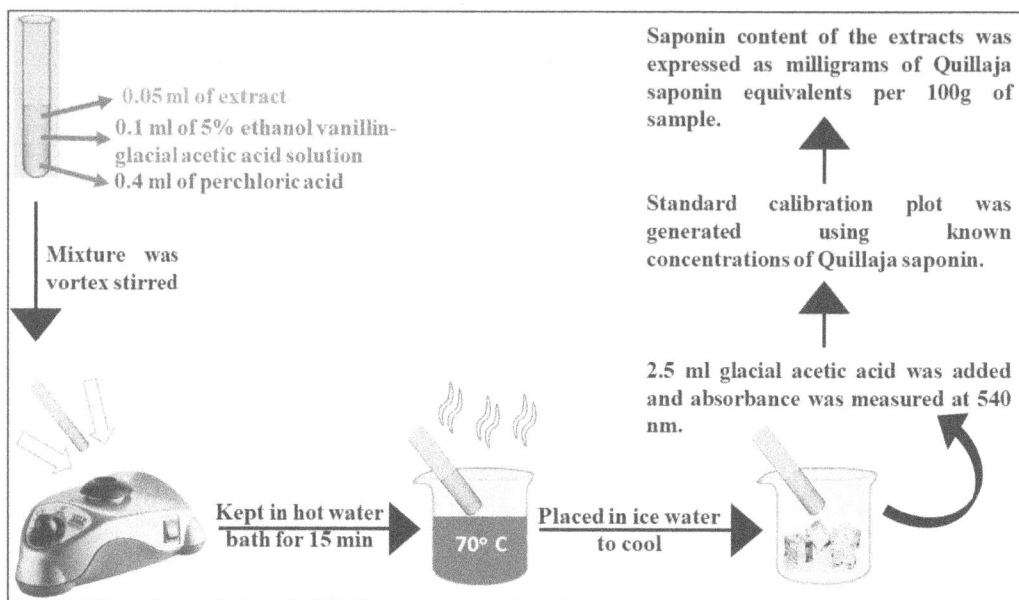

SCHEME 6.26 Method for phytochemical estimation of saponin content [79].

higher flavonoid content present in Amla. Detergency ability on the other hand may be a resultant of higher surfactant content. Saponins present in these herbs further enhance the cleaning ability by reducing the surface tension and proved to be an active surfactant molecule. Shikakai formulations showed significant detergency ability of nearly 95% because of being abundant in saponins [80].

6.5.2 Natural Face Wash

Skin is the largest organ of the human body. As compared to the whole body, skin is the organ that comes into contact with the rest of the world. Body fluids are held in the skin to prevent dehydration and keep us away from harmful microbes otherwise we would get infections. It is the organ that makes us feel the pain, heat and cold because of being full of nerve endings [86]. If humans could not feel these things, then they can get hurt in a bad way and would not even know it. Skin holds everything together and contains lots of tissues, without the skin human bones, muscles and organs would have no support.

Also, skin being the largest organ provides a barrier between the outside world and immensely coordinated systems within the body. It also helps in maintaining the correct body temperature. Whenever the body gets hot, capillaries (the blood vessels near the surface of the skin) enlarge to let the warm blood cool down. Vitamin D production is also an important function of skin.

TABLE 6.6
Phytochemical Content Estimation of Amla, Aloe vera, Shikakai and Reetha

S. No.	Phytochemical Content Estimation	Amla	Aloe vera	Shikakai	Reetha
1.	Total flavonoid content (quercetin equivalents mg/g)	3.33 ± 0.04	2.19 ± 0.06	2.06 ± 0.06	1.82 ± 0.09
2.	Total phenolic content (gallic acid equivalents mg/g)	1.92 ± 0.02	3.71 ± 0.07	1.96 ± 0.04	1.91 ± 0.02
3.	Total saponins content (*Quillaja saponaria* equivalents mg/g)	1.64 ± 0.02	1.56 ± 0.05	2.25 ± 0.03	1.92 ± 0.02

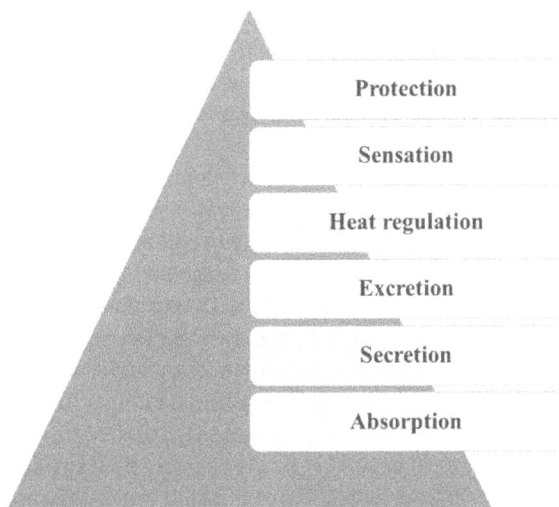

SCHEME 6.27   Functions of the skin.

When the sun shines on it, skin makes vitamin D which is essential for healthy bones. Skin also provides our body with immune defense from a lot of infectious diseases. Functions of the skin are featured in Scheme 6.27.

Along with being a vital and largest body organ, skin is the most complicated organ too. Where, on one hand, skin has several roles to play in life and health maintenance, it also has lots of potential issues. At present, humans face many skin disorders and diseases. Skin is unique in its own way but when it comes to the health of the skin then it demands a lot of concern and attention. There is huge competition to have healthier, younger, clearer and fresher skin. And this focus on skin health further enhances an individual's mental health, confidence and self-esteem because 'when you look good, you feel good.'

To retain healthy skin, a balanced and nutritious diet is a must. Apart from the healthy diet, there are several other factors too, namely hormonal changes that can affect the skin. During puberty, both girls and boys face many changes in the body, acne vulgaris is one of those [87]. Acne vulgaris is a common skin disorder. Pilosebaceous gland functioning undergoes changes during the time of pubertal transition. Diverse processes in the pilosebaceous unit result in overgrowth of bacteria and inflammation which further results into acne. Acne on the face makes it look less pleasant and it becomes a mental concern for many young teenagers. To get rid of the problem of acne, there are many synthetic drugs [88, 89], but the question arises 'are they actually solving the skin issues or creating new skin issues?'.

Face washes or face cleansers are the products which are capable of cleaning the face without drying it out. It should be good for all the skin types. Removing dirt, oil and providing moisture to the dry skin are some of the very important requirements from the face wash. It should wash off all the grim, makeup and oil from the face. Synthetic harsh face washes might remove these oil-soluble impurities from the face but they can't be regarded as 100% effective. Ordinary soaps can't be used on the delicate facial skin as they tend to take away all the moisture and make the face dry and patchy. So, it becomes very crucial to look for some milder and natural alternatives. A natural mild cleanser that can keep the skin fresh, moisturized, clean and germ free is something that facial skin needs.

India's herbal drug industry is the oldest medical care system in the world. Ayurveda and Unani have given the ancient herbal healing ways which makes use of natural products and herbs to confront health problems. This class of medicines may seem new to the Western medical professionals but actually all the prescribed medications do contain plant extracts. In the present scenario, countries across the world are looking forward to the Indian herbal drug industries. The demand for natural- and plant-based surfactants has seen rapid growth in the past few years. Natural products contain biologically active ingredients which can act as a very important natural surfactant and can render younger and energetic-looking skin. Natural face washes are capable of resulting in acne-free skin.

Aloe vera is one very beneficial herbal plant which belongs to Africa by origin. This fleshy plant has many spines and long, narrow edges whereas the leaf flesh is slimy. *Aloe vera* is often used to cleanse the face. In fact, all parts of *Aloe vera* are very useful and can be used for acne-prone skin. Acne therapy can be carried out by reduction of soap usage and introducing *Aloe vera* as a natural surfactant. It helps to exfoliate the dead cells in the skin and prevents gathering of acne-causing bacteria [90]. These natural alternatives proved out to be safer than antibiotics. Basically, due to the long-term usage of antibiotics, acne-causing microorganisms tend to develop resistance to that drug. *Aloe vera* consists of a good amount of flavonoid. Flavonoid being a highly active biological agent makes *Aloe vera* a good fit for natural face wash. Liquid soaps tend to be more practical for the face.

R. Walandini et al. gave a very efficient *Aloe vera*-based facial cleansing gel. This can be formulated by mixing 1 g of *Aloe vera* extract with 200 g of distilled water. Mixture was stirred vigorously to achieve homogeneity. Measurement of pH of the solution was done, it was followed by addition of 0.5%, 0.7%, 2% carbomer 940. Mixture was again stirred; pH was measured and triethanolamine was added so that the solution reaches pH 5. 5 drops of whey kefir were added followed by moringa seed oil and virgin coconut oil. The resultant product was then mixed evenly [90]. Scheme 6.28 shows the formulation of *Aloe vera* extract-based facial cleansing gel. This *Aloe vera*-based cleanser proves to be safe for sensitive skin and is totally eco-friendly. Chances of environmental pollution are negligible and in fact it saves clean water because to rinse, it does not require much amount of water.

6.5.3 NATURAL FABRIC DETERGENTS

Along with the consumption in the personal care sector, surfactants do have substantial consumption in the fabric detergent industry. A wide variety of synthetic detergents are in the market nowadays and they appear to be effective but if seen from the deeper outlook they are neither good for the cloth nor for the environment. These harsh chemicals just make the cloth look dull and gray. Because of being made from reactive and moreover in other words 'over active' ingredients these surfactants while removing the dirt from the surface of the cloth actually takes away the fiber color

SCHEME 6.28 Formulation of Aloe vera extract-based facial cleansing gel.

and affects the quality of the fiber badly. We tend to think we are just adding chemicals from the detergent but we forget that the cloth, the fiber and the dye on it all are chemical substances too and we can't always predict what sort of reactions good or bad might take place when detergents come in contact with the fiber. While choosing a surfactant for fabric cleaning, various factors must be kept in mind as shown in Scheme 6.29.

Deterioration in the fiber quality is one problem whereas environmental pollution is the other. We live in the environment and make use of its resources, so the duty of conserving and protecting it is also ours. Switching to organic and natural surfactants is one very crucial solution that can be thought of. These surfactants are easily available, cheap and eco-friendly [91]. In this portion, we will highlight how Reetha and Shikakai can be a great fit in the fabric cleaning industry too.

Muntaha and Khan discussed how Reetha and Shikakai can be useful for cleaning historic textiles [92]. Ancient Indian textiles have been documented in old sculptures and literature. Evidence of cotton textiles were contrived in the Indus Valley Civilization. A very crucial role is being played by textiles in the social and economic sector of the society. Textiles render a platform for cultural and artistic enhancement. There are various varieties of designs, materials, shapes and colors when it comes to textiles. As the technology is getting advanced, different techniques are being adapted and new materials are getting manufactured. These textiles are beautiful and are our country's pride, preserving them becomes our sole responsibility.

Preservation and conservation of these ancient textiles is in itself a very big challenge. Ancient textiles are of organic nature and very frangible and thus more prone to degradation than inorganic substances. Thus, careful and skilled deliberations about materials are required for textile conservation. Art conservation is not as easy as it looks, for that one has to involve experts from different fields. Fine artists, chemists, carpenters, architects and many more collectively work together to bring about the desired result. Conservators have to work really hard, because they can only have limited tests and trials before selecting the product to be conserved. In conservation of textiles, international organizations and institutions recommend the use of synthetic surfactants. Attaining those conservation-grade products is a costly affair. Due to the environmental concerns, organic detergents are preferred all over the world. Some important features of natural fabric detergent are shown in Scheme 6.30.

Reetha (soapnut) and Shikakai have been extensively used as cleaning silver and gold jewelry, also as shampoos and detergents in India since centuries [69]. These natural products are cheap, easily available and easy to use. They are 100% natural and totally human and environment-friendly materials. Reetha and Shikakai also show tremendous cleansing properties. These age-old natural surfactants are very much accepted as a great organic substitute to synthetic detergents.

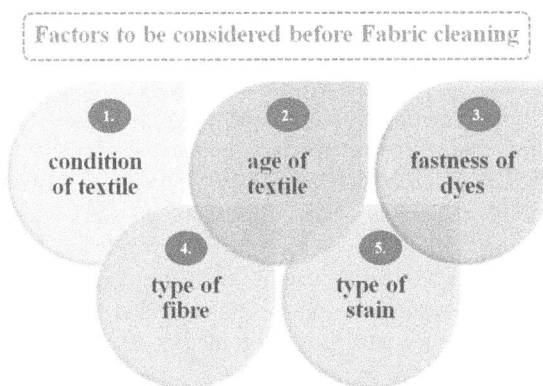

SCHEME 6.29 Various factors to be considered while choosing a surfactant for fabric cleaning.

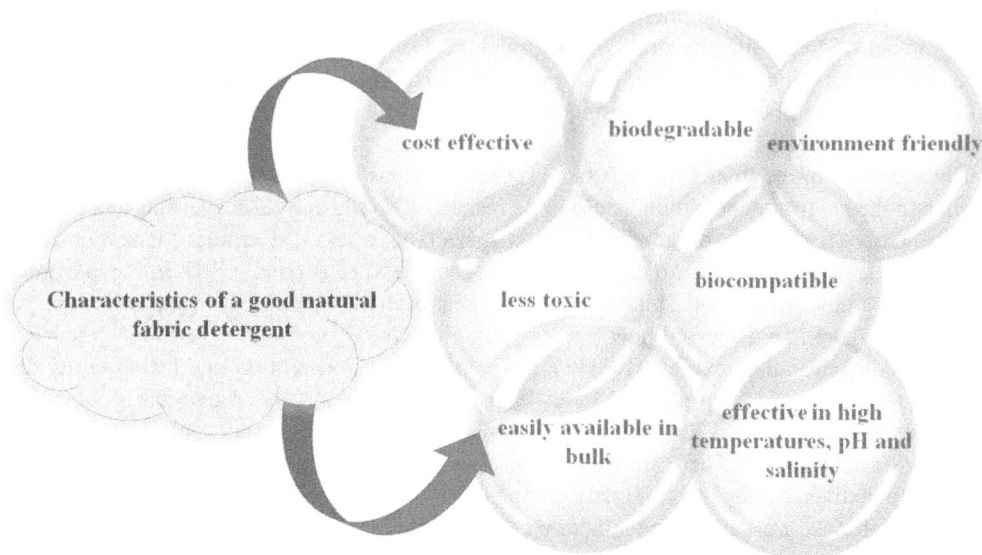

SCHEME 6.30 Important features of natural fabric detergent.

Earlier in this topic, we have discussed the three active phytochemicals present in Reetha and Shikakai. Both these plants are rich in flavonoids. Phenolic and saponin content is also quite high. Saponins are the soapy natured materials that along with flavonoids make Reetha and Shikakai a wonderful fabric detergent. Foaming, cleansing and detergency action of these natural surfactants are also quite commendable which is making it an acceptable choice.

Dirt, dust and stains on the fabric are mostly oily and of greasy nature. Hydrophobic flavonoids and phenolics present in Reetha and Shikakai act as a great oil trapping agent, whereas sugar chains present in saponins of Reetha and Shikakai are hydrophilic in nature and result in excellent foaming, detergency and emulsification properties. One more important property that these surfactants hold is their acidic nature [78]. Synthetic detergents are highly alkaline and damage the protein fiber, whereas Reetha and Shikakai are acidic in nature and thus are suitable for washing protein fiber.

6.5.4 Natural Pesticides

Extensive use of pesticides is there in the modern agriculture businesses. Consumption of food is growing at global scale and thus usage of pesticides is supposed to increase in the near future. Where on one hand technological advancements are important in increasing the business profit, on the other hand, involvement of new pesticides are also crucial so that farmers can manage and grow bigger areas of crops with reduced human labor requirement [93]. Perfect and ideal pesticide will be the one which is selective against certain pets and should be nontoxic to the nontarget organisms. But it becomes a difficult task to have such a pesticide since most of them are toxic to nontarget organisms and humans. Self-poisoning is a major issue that is caused by pesticides in the developing nations. Hazards caused by synthetic pesticides are depicted in Scheme 6.31.

Certain safe and effective pesticides can also be used. But if not used judiciously, they can be harmful to the environment and result into contamination of soil, surface and can pollute groundwater and can kill wildlife. We understand the importance of pesticides but natural alternatives are something this world needs. Neem (*A. indica*) is a very important antimicrobial medicinal plant. The beauty of natural products is their biologically active ingredients. Phytochemicals mainly flavonoids present in neem highlight its antimicrobial and antioxidant properties. Neem oil can be

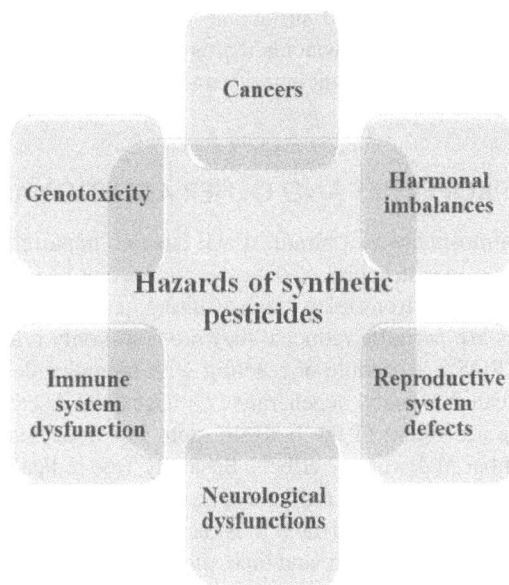

SCHEME 6.31   Hazards of synthetic pesticides.

obtained from various parts of neem plant and can be a tremendous natural pesticide. Scheme 6.32 discusses some important features of neem oil.

Being nontoxic to humans, animals, birds and earthworms, neem oil can be a great inclusion in the pesticides to make it as natural as possible. It proves out to be a safe and powerful pest fighter. It is effective and a cheaper alternative at the same time. Neem oil is difficult to spray that's why their emulsions are usually developed by adding various surfactants such as AOT (aerosol-OT or sodium bis(2-ethylhexyl) sulfosuccinate), SDS (sodium dodecyl sulfate), CTAB (cetyltrimethylammonium bromide) and castor oil-based surfactants. Neem oil-based surfactants can also be developed by sulfonating and can be used for the dispersion of neem oil in water. Thus, this way Neem oil is a chief derivative that can discover its application as a pest control agent [9].

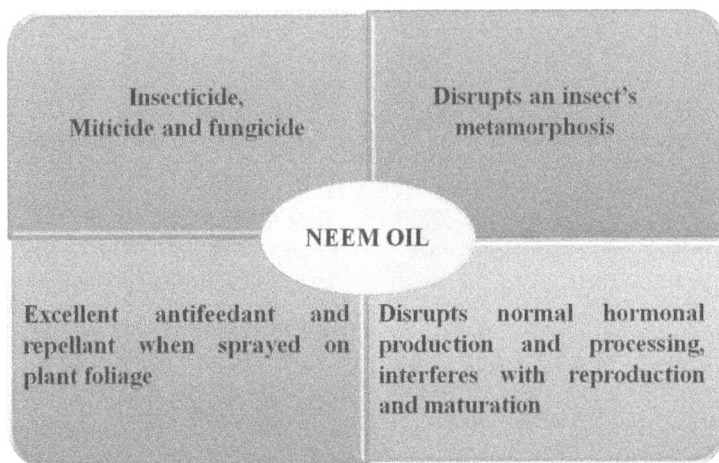

SCHEME 6.32   Some important features of neem oil.

Various applications of flavonoid-based surfactants have been discussed successfully in this topic. One very important property of flavonoids that is to be discussed is their antioxidant property. Next section highlights the antioxidant activity and other additional features that flavonoids are capable of adding.

6.6 ANTIOXIDANT PROPERTIES AND OTHER ADDITIONAL FEATURES

Oxygen is present in the atmosphere as a biradical and has two unpaired electrons with the same spin quantum number. Electronic structure of oxygen is important because its chemical reactivity depends upon it. Absorption of electrons takes place resulting in the formation of various intermediates. When oxygen species are partially reduced, they are commonly referred to as ROS (reactive oxygen species) [94, 95]. ROS are capable of reacting with biomolecules and cellular structures. Such reactions can be the cause of many degenerative damages and diseases [96].

Normally, free radicals are found in the body in quite low concentration. But even that low amount can result in a number of disruptive effects. Basically, free radicals undergo chain reactions and these chain reactions cause changes in millions of molecules and damage enzymes, RNA, DNA in cell membranes and could not get inactivated thus results in production of lipoxygenation products. Free radicals can be of different nature and their site of formation is also different for different cases. Various sites of formation for free radicals can be mitochondria, peroxisome, endoplasmic reticulum, cell membranes, cytosolic membranes, etc. These sites of formation are dependent upon the part of the cell with which the free radicals react. Free radicals of oxygen origin start more reactions in the cell due to the existence of molecular oxygen in aerobic organisms and its easy acceptance of electrons. Processes of autoxidation of hydroquinone and catecholamine and respiration are the responsible reactions for the formation of free radicals [96].

Whatever reactive oxygen metabolites are formed must be eliminated by the antioxidant action. Any disturbance in the balance of oxygen species results in oxidative stress. In this state, accumulation and formation of reactive metabolites are increased and this results in cell destruction which may further cause damage to the genetic information. Some onset diseases because of ROS are shown in Scheme 6.33.

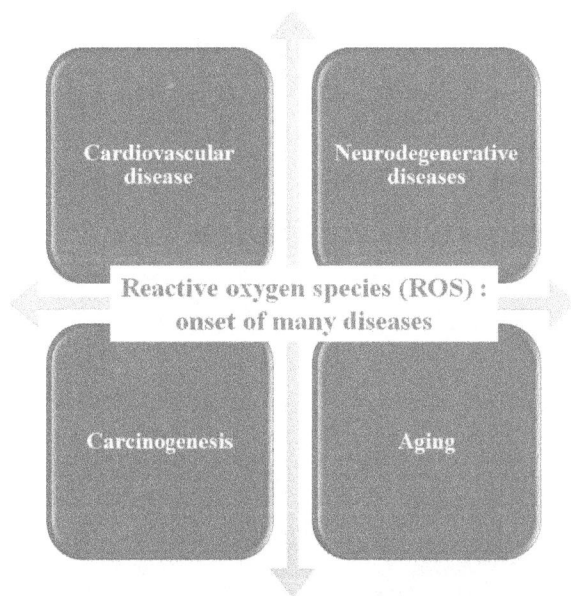

SCHEME 6.33 Major diseases caused by the formation of ROS (reactive oxygen species).

6.6.1 ANTIOXIDANT PROTECTION SYSTEMS

At the membrane level, the process of oxidative alterations of carbohydrates, proteins, lipids and DNA is a universal means of damage to the cell. During evolutionary development, a protective system has been established. Uncontrolled creation of free radicals takes place in the cells. These protective systems are developed to reduce the number of free radicals in the cell. Along with the free radicals, they also limit the precursors that might lead to free radical formation. Scheme 6.34 shows the stepwise antioxidant protection of organisms.

In many in vitro systems, high antioxidant activity of flavonoids has been discussed [97]. In addition to the in vivo antioxidant activity, isorhamnetin, rhamnetin and quercetin can reduce the amount of liver and serum cholesterol [98]. Xanthene oxidase (XOD) activity was also found to be an important feature of flavonoids and thus flavonoids are capable of capturing superoxide radicals. Because of holding such a functionality, flavonoids can result in the reduction of the amount of uric acid and superoxide anion of radicals in tissues and thus helps in the treatment of gout and ischemia [99].

Glycoside-gallate esters of two flavonols may show inhibiting action on HIV-1 (human immunodeficiency virus-1) integrase. This HIV-1 integrase enzyme incorporates the DNA virus into the healthy host cell molecule's DNA and leads to the reproduction and production of virions. Thus, the treatment of AIDS can be done by hampering the activity of the discussed enzyme. Quercetin possesses strong antioxidant properties. As a result of this, it can be beneficial for human health as it is capable of improving heart rate and cancer risk becomes less. Antiallergic and anti-inflammatory properties are also some salient characteristics of Quercetin [100]. This wonderful flavonoid can cause inhibition of low-density lipoprotein (LDL) cholesterol. Lipoxygenase enzyme inhibition is another important function that may inhibit inflammatory mediators. Reports are there where Quercetin is shown to reduce the risk of various cancers, including colon, prostate, breast, uterine and tissue cancer. It is presumed to inhibit XOD.

In Fenton's reaction, harmful oxygen-free radicals get produced and interfere with healthy cells in the body and result in many unexpected disorders. Another important flavonoid, namely, Rutin has the potential to form chelates with metal ions and act as a strong antioxidant. This Rutin can

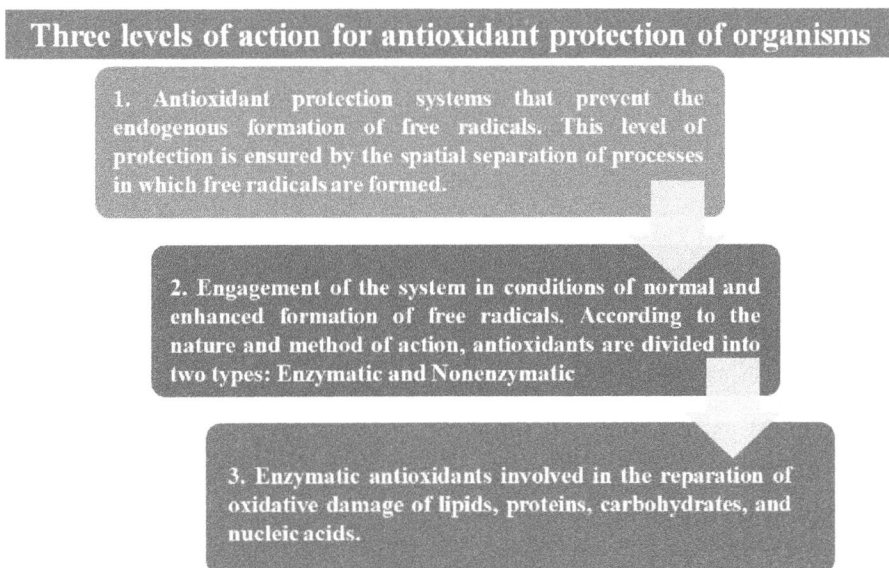

Three levels of action for antioxidant protection of organisms

1. Antioxidant protection systems that prevent the endogenous formation of free radicals. This level of protection is ensured by the spatial separation of processes in which free radicals are formed.

2. Engagement of the system in conditions of normal and enhanced formation of free radicals. According to the nature and method of action, antioxidants are divided into two types: Enzymatic and Nonenzymatic

3. Enzymatic antioxidants involved in the reparation of oxidative damage of lipids, proteins, carbohydrates, and nucleic acids.

SCHEME 6.34 Stepwise antioxidant protection of organisms.

be used to reduce Fenton's reaction. It stabilizes vitamin C. When consumed with vitamin C, rutin can increase the activity of ascorbic acid [101]. People with weak capillaries get easily bruised and bleed, rutin takes up the responsibility of strengthening the capillaries in the human body. Rutin is also capable of preventing atherogenesis, precancerous and carcinogenic conditions and has an anti-inflammatory effect. Kaempferol is another crucial chemopreventive flavonoid [102]. It can inhibit the formation of blood platelets and the oxidation of low-density lipoproteins thus resulting in the prevention of serious condition of arteriosclerosis. Studies show synergistic effect of quercetin in reducing the multiplication of malignant cells, so it is suggested that a combination of quercetin and kaempferol can be more fruitful in cancer treatments than their individual use.

Tangeretin may also be regarded as an important anticancer agent [103]. It has the potential to act against malignant cell forms. It carries out the strengthening mechanism of the cell wall and results in the protection of the cell wall from attack. It has the ability to carry out apoptosis of cells suffering from leukemia without damaging the healthy cells. Tangerine basically freezes the cancer cell growth and halts their duplication [104]. It has also been seen to have strong protective action against Parkinson's disease.

Flavonoids indeed are very essential for maintaining human health. With these many additional features, different flavonoids can be added to different surfactant formulations. Depending upon the requirement, be it an anti-dandruff shampoo, anti-acne face wash or any other problem-solving medicinal surfactant, benefits of flavonoids can be well utilized. As evident from this chapter, the amount of work done concerning flavonoids is immense. If the interest in incorporating renewable material-based surfactants keeps on increasing then it is expected that flavonoid-based surfactants will continue to grow as a widely available source for detergents in the coming years.

REFERENCES

1. Patel, M. (2004) Surfactants based on renewable raw materials. *J. Ind. Ecol.*, **7** (3), 47–62.
2. Deleu, M., and Paquot, M. (2004) From renewable vegetables resources to microorganisms: New trends in surfactants. *C.R. Chim.*, **7** (67), 641–646.
3. Makkar, R.S., and Rockne, K.J. (2003) Comparison of synthetic surfactants and biosurfactants in enhancing biodegradation of polycyclic aromatic hydrocarbons. *Environ. Toxicol. Chem.*, **22** (10), 2280–2292.
4. Halith, S.M., Abirami, A., Jayaprakash, S., Karthikeyini, C., Kulathuran Pillai, K., and Firthouse, P.U. Mohamed. (2009) Effect of *Ocimum sanctum* and *Azadirachta indica* on the formulation of antidandruff herbal shampoo powder. *Der Pharmacia Lettre*, **1** (2), 68–76.
5. Poojar, B., Ommurugan, B., Adiga, S., Thomas, H., Sori, R.K., Poojar, B., Hodlur, N., Tilak, A., Korde, R., Gandigawad, P., In, M., Sleep, R., Albino, D., Rats, W., Article, O., Schedule, P., Injury, C.C., Sori, R.K., Poojar, B., Hodlur, N., Tilak, A., Korde, R., and Gandigawad, P. (2017) Formulation and evaluation of licorice shampoo in comparison with commercial shampoo. *Asian J. Pharm. Clin. Res.*, **7** (10), 1–5.
6. Benincasa, M. (2007) Rhamnolipid produced from agroindustrial wastes enhances hydrocarbon biodegradation in contaminated soil. *Curr. Microbiol.*, **54** (6), 445–449.
7. Tawfik, S.M. (2015) Synthesis, surface, biological activity and mixed micellar phase properties of some biodegradable gemini cationic surfactants containing oxycarbonyl groups in the lipophilic part. *J. Ind. Eng. Chem.*, **28**, 171–183.
8. Sayed, G.H., Ghuiba, F.M., Abdou, M.I., Badr, E.A.A., Tawfik, S.M., and Negm, N.A.M. (2012) Synthesis, surface and thermodynamic parameters of some biodegradable nonionic surfactants derived from tannic acid. *Colloids Surf., A: Physicochem. Eng. Aspects*, **393** (1), 96–104.
9. Sayed, G.H., Ghuiba, F.M., Abdou, M.I., Badr, E.A.A., Tawfik, S.M., and Negm, N.A.M. (2012) Synthesis, surface, thermodynamic properties of some biodegradable vanillin-modified polyoxyethylene surfactants. *J. Surfactants Deterg.*, **15** (6), 735–743.
10. Tawfik, S.M., Abd-Elaal, A.A., and Aiad, I. (2016) Three gemini cationic surfactants as biodegradable corrosion inhibitors for carbon steel in HCl solution. *Res. Chem. Intermed.*, **42** (2), 1101–1123.
11. Tripathy, DB., and Mishra, A. (2016) Sustainable biosurfactants. *Sustainable Inorganic Chemistry.*, **1**, 175–192.

12. Middleton, E. (1998) Effect of plant flavonoids on immune and inflammatory cell function. *Adv. Exp. Med. Biol.*, **439**, 175–182.

13. Tripoli, E., Guardia, M.L., Giammanco, S., Majo, D.D., and Giammanco, M. (2007) Citrus flavonoids: Molecular structure, biological activity and nutritional properties: A review. *Food Chem.*, **104** (2), 466–479.

14. De Groot, H., and Rauen, U. (1998) Tissue injury by reactive oxygen species and the protective effects of flavonoids. *Fundam. Clin. Pharmacol.*, **12** (3), 249–255.

15. Zahoranová, A., Štefečka, M., Černák, M., and Šurda, V. (1999) Prebreakdown positive corona streamers and the streamer-cathode contact in hydrogen. *Czech. J. Phys.*, **49** (6), 941–956.

16. Wang, H.K., Xia, Y., Yang, Z.Y., Natschke, S.L., and Lee, K.H. (1998) Recent advances in the discovery and development of flavonoids and their analogues as antitumor and anti-HIV agents. *Adv. Exp. Med. Biol.*, **439**, 191–225.

17. Kaul, T.N., Middleton, E., and Ogra, P.L. (1985) Antiviral effect of flavonoids on human viruses. *J. Med. Virol.*, **15** (1), 71–79.

18. Wu, D., Kong, Y., Han, C., Chen, J., Hu, L., Jiang, H., and Shen, X. (2008) D-Alanine:D-alanine ligase as a new target for the flavonoids quercetin and apigenin. *Int. J. Antimicrob. Agents*, **32** (5), 421–426.

19. Li, M., and Xu, Z. (2008) Quercetin in a lotus leaves extract may be responsible for antibacterial activity. *Arch. Pharm. Res.*, **31** (5), 640–644.

20. Comalada, M., Camuesco, D., Sierra, S., Ballester, I., Xaus, J., Gálvez, J., and Zarzuelo, A. (2005) In vivo quercitrin anti-inflammatory effect involves release of quercetin, which inhibits inflammation through down-regulation of the NF-κB pathway. *Eur. J. Immunol.*, **35** (2), 584–592.

21. Hossain, H., Jahan, I.A., Nimmi, I., and Hasan, K. (2012) Evaluation of antinociceptive and antioxidant potential from the leaves of *Spilanthes paniculata* growing in Bangladesh. *Int. J. Pharm. Phytopharm. Res.*, **1** (4), 178–186.

22. De Stefani, E., Boffetta, P., Deneo-Pellegrini, H., Mendilaharsu, M., Carzoglio, J.C., Ronco, A., and Olivera, L. (1999) Dietary antioxidants and lung cancer risk: A case-control study in Uruguay. *Nutr. Cancer*, **34** (1), 100–110.

23. Knekt, P., Järvinen, R., Seppänen, R., Heliövaara, M., Teppo, L., Pukkala, E., and Aromaa, A. (1997) Dietary flavonoids and the risk of lung cancer and other malignant neoplasms. *Am. J. Epidemiol.*, **146** (3), 223–230.

24. Caltagirone, S., Rossi, C., Poggi, A., Ranelletti, F.O., Natali, P.G., Brunetti, M., Aiello, F.B., and Piantelli, M. (2000) Flavonoids apigenin and quercetin inhibit melanoma growth and metastatic potential. *Int. J. Cancer*, **87** (4), 595–600.

25. Paper, D.H. (1998) Natural products as angiogenesis inhibitors. *Planta Med.*, **64** (8), 686–695.

26. Osman, H.E., Maalej, N., Shanmuganayagam, D., and Folts, J.D. (1998) Grape juice but not orange or grapefruit juice inhibits platelet activity in dogs and monkeys (*Macaca fasciularis*). *J. Nutr.*, **128** (12), 2307–2312.

27. Alcaraz, M.J., and Ferrándiz, M.L. (1987) Modification of arachidonic metabolism by flavonoids. *J. Ethnopharmacol.*, **21** (3), 209–229.

28. Manach, C., Scalbert, A., Morand, C., Rémésy, C., and Jiménez, L. (2004) Polyphenols: Food sources and bioavailability. *Am. J. Clin. Nutr.*, **79** (5), 727–747.

29. Iwashina, T. (2013) Flavonoid properties of five families newly incorporated into the order Caryophyllales (review). *Bull. Natl. Mus. Nat. Sci. Ser. B, Bot.*, **39** (1), 25–51.

30. Matthies, A., Clavel, T., Gütschow, M., Engst, W., Haller, D., Blaut, M., and Braune, A. (2008) Conversion of daidzein and genistein by an anaerobic bacterium newly isolated from the mouse intestine. *Appl. Environ. Microbiol.*, **74** (15), 4847–4852.

31. Subapriya, R., and Nagini, S. (2005) Medicinal properties of neem leaves: A review. *Curr. Med. Chem. – Anticancer Agents*, **5** (2), 149–156.

32. Maithani, A., Parcha, V., Pant, G., Dhulia, I., and Kumar, D. (2011) *Azadirachta indica* (neem) leaf: A review. *J. Pharm. Res.*, **4** (6), 1824–1827.

33. Sithisarn, P., Supabphol, R., and Gritsanapan, W. (2006) Comparison of free radical scavenging activity of Siamese neem tree (*Azadirachta indica* A. Juss var. siamensis Valeton) leaf extracts prepared by different methods of extraction. *Med. Princ. Pract.*, **15** (3), 219–222.

34. Wang, Y.L., Xi, G.S., Zheng, Y.C., and Miao, F.S. (2010) Microwave-assisted extraction of flavonoids from Chinese herb Radix puerariae (Ge Gen). *J. Med. Plants Res.*, **4** (4), 304–308.

35. Paul, R., Prasad, M., and Sah, N.K. (2011) Anticancer biology of *Azadirachta indica* L. (neem): A mini review. *Cancer Biol. Ther.*, **12** (6), 467–476.

36. Reynolds, T., and Dweck, A.C. (1999) *Aloe vera* leaf gel: A review update. *J. Ethnopharmacol.*, **68** (1–3), 3–37.

37. Gantait, S., Sinniah, U.R., and Das, P.K. (2014) *Aloe vera*: A review update on advancement of in vitro culture. *Acta Agric. Scand. Sect. B Soil Plant Sci.*, **64** (1), 1–12.

38. Klein, A.D., and Penneys, N.S. (1988) *Aloe vera*. *J. Am. Acad. Dermatol.*, **18** (4), 714–720.

39. Hamman, J.H. (2008) Composition and applications of *Aloe vera* leaf gel. *Molecules*, **13** (8), 1599–1616.

40. Eshun, K., and He, Q. (2004) Aloe vera: A valuable ingredient for the food, pharmaceutical and cosmetic industries – a review. *Crit. Rev. Food Sci. Nutr.*, **44** (2), 91–96.

41. Ramachandra, C.T., and Srinivasa Rao, P. (2008) Processing of *Aloe vera* leaf gel: A review. *Am. J. Agric. Biol. Sci.*, **3** (2), 502–510.

42. Vogler, B.K., and Ernst, E. (1999) *Aloe vera*: A systematic review of its clinical effectiveness. *Br. J. Gen. Pract.*, **49** (447), 823–828.

43. Choi, S., and Chung, M.H. (2003) A review on the relationship between *Aloe vera* components and their biologic effects. *Semin. Integr. Med.*, **1** (1), 53–62.

44. Muthukumaran, P., Divya, R., Indhumathi, E., and Keerthika, C. (2018) Total phenolic and flavonoid content of membrane processed Aloe vera extract: A comparative study. *Int. Food Res. J.*, **25** (4), 1450–1456.

45. Nafis, A., Kasrati, A., Azmani, A., Ouhdouch, Y., and Hassani, L. (2018) Endophytic actinobacteria of medicinal plant Aloe vera: Isolation, antimicrobial, antioxidant, cytotoxicity assays and taxonomic study. *Asian Pac. J. Trop. Biomed.*, **8** (10), 513–518.

46. Kapoor, M.P., Suzuki, K., Derek, T., Ozeki, M., and Okubo, T. (2020) Clinical evaluation of *Emblica officinalis* Gatertn (Amla) in healthy human subjects: Health benefits and safety results from a randomized, double-blind, crossover placebo-controlled study. *Contemp. Clin. Trials Commun.*, **17**, 100499.

47. Bhandari, P., and Kamdod, M. (2012) *Emblica officinalis* (Amla): A review of potential therapeutic applications. *Int. J. Green Pharm.*, **6** (4), 257–269.

48. Liu, X., Zhao, M., Wang, J., Yang, B., and Jiang, Y. (2008) Antioxidant activity of methanolic extract of emblica fruit (*Phyllanthus emblica* L.) from six regions in China. *J. Food Compos. Anal.*, **21** (3), 219–228.

49. Anila, L., and Vijayalakshmi, N.R. (2002) Flavonoids from *Emblica officinalis* and *Mangifera indica* – Effectiveness for dyslipidemia. *J. Ethnopharmacol.*, **79** (1), 81–87.

50. Liu, X., Cui, C., Zhao, M., Wang, J., Luo, W., Yang, B., and Jiang, Y. (2008) Identification of phenolics in the fruit of emblica (*Phyllanthus emblica* L.) and their antioxidant activities. *Food Chem.*, **109** (4), 909–915.

51. Ur-Rehman, H., Yasin, K.A., Choudhary, M.A., Khaliq, N., Ur-Rahman, A., Choudhary, M.I., and Malik, S. (2007) Studies on the chemical constituents of *Phyllanthus emblica*. *Nat. Prod. Res.*, **21** (9), 775–781.

52. Zhao, T., Sun, Q., Marques, M., and Witcher, M. (2015) Anticancer properties of *phyllanthus emblica* (Indian gooseberry). *Oxid. Med. Cell. Longevity.*, **2015**, 1–7.

53. Yang, F., Yaseen, A., Chen, B., Li, F., Wang, L., Hu, W., and Wang, M. (2020) Chemical constituents from the fruits of *Phyllanthus emblica* L. *Biochem. Syst. Ecol.*, **92** (1), 104122–104128.

54. Dasaroju, S., and Gottumukkala, K.M. (2014) Review article current trends in the research of *Emblica officinalis* (Amla): A pharmacological perspective. *Int. J. Pharm. Sci. Rev. Res.*, **24** (2), 150–159.

55. Agarwal, M., Kumar, A., Gupta, R., and Upadhyaya, S. (2012) Extraction of polyphenol, flavonoid from *Emblica officinalis*, citrus limon, cucumis sativus and evaluation of their antioxidant activity. *Orient. J. Chem.*, **28** (2), 993–998.

56. Jirge, S.S., Tatke, P.A., and Gabhe, S.Y. (2014) Simultaneous estimation of kaempferol, rutin, and quercetin in various plant products and different dosage forms of bhuiamla and amla. *J. Planar Chromatogr. – Mod. TLC*, **27** (4), 267–273.

57. Prakash, D.V.S., and Vangalapati, M. (2015) Optimization of physico-chemical parameters for the extraction of quercetin from medicinal herbs. *J. Life Sci. Technol.*, **2** (2), 90–93.

58. Chen, C.Y., Kuo, P.L., Chen, Y.H., Huang, J.C., Ho, M.L., Lin, R.J., Chang, J.S., and Wang, H.M. (2010) Tyrosinase inhibition, free radical scavenging, antimicroorganism and anticancer proliferation activities of *Sapindus mukorossi* extracts. *J. Taiwan Inst. Chem. Eng.*, **41** (2), 129–135.

59. Upadhyay, A., and Singh, D.K. (2012) Efeitos farmacológicos do *Sapindus mukorossi*. *Rev. Inst. Med. Trop. Sao Paulo*, **54** (5), 273–280.

60. Wei, M.-P., Qiu, J.-D., Li, L., Xie, Y.-F., Yu, H., Guo, Y.-H., and Yao, W.-R. (2020) Saponin fraction from *Sapindus mukorossi* Gaertn as a novel cosmetic additive: Extraction, biological evaluation, analysis of anti-acne mechanism and toxicity prediction. *J. Ethnopharmacol.*, Mar 25, **268**, 113552. doi: 10.1016/j.jep.2020.113552.

61. Wang, X., Liu, J., Rui, X., Xu, Y., Zhao, G., Wang, L., Weng, X., Chen, Z., and Jia, L. (2021) Biogeographic divergence in leaf traits of *Sapindus mukorossi* and *Sapindus delavayi* and its relation to climate. *J. For. Res.*, **32**, 1445–1456.

62. Anjali, R.S., and Juyal, D. (2018) *Sapindus mukorossi*: A review article. *Pharma Innov. J.*, **7** (5), 470–472.

63. Umadevi, I., and Daniel, M. (1991) Chemosystematics of the Sapindaceae. *Feddes Repert.*, **102** (7–8), 607–612.

64. Arora, B., Bhadauria, P., Tripathi, D., and Sharma, A. (2012) Sapindus emarginatus: Phytochemistry and various biological activites. *Indo Global J. Pharm. Sci.*, **2** (3), 250–257.

65. Khanpara, K., and Harisha, C.R. (2015) A detailed investigation on shikakai (*Acacia Concinna* Linn.) – Fruit a detailed investigation on shikakai (Acacia). *J. Curr. Pharm. Res.*, **9** (1), 6–10.

66. Thakker, A.M. (2020) Sustainable processing of cotton fabrics with plant-based biomaterials *Sapindus mukorossi* and *Acacia concinna* for health-care applications. *J. Text. Inst.*, **112** (5), 718–726.

67. Kukhetpitakwong, R., Hahnvajanawong, C., Homchampa, P., Leelavatcharamas, V., Satra, J., and Khunkitti, W. (2006) Immunological adjuvant activities of saponin extracts from the pods of *Acacia concinna*. *Int. Immunopharmacol.*, **6** (11), 1729–1735.

68. Boonmee, S., and Kato-Noguchi, H. (2017) Allelopathic activity of *Acacia concinna* pod extracts. *Emirates J. Food Agric.*, **29** (4), 250–255.

69. Todkar, S.S., Chavan V.V., and Kulkarni, A.S. (2010) Screening of secondary metabolites and antibacterial activity of *Acacia concinna*. 974–979.

70. Khanpara, K., and Harisha, C.R. (2015) A detailed investigation on shikakai (*Acacia concinna* Linn.) – Fruit. *J. Curr. Pharm. Res.*, **9** (1), 6–10.

71. Kumar Dutta, A. (2019) Introductory chapter: Surfactants in household and personal care formulations – An overview. *Surfactants Deterg.*, **1**, 1–10.

72. De, S., Malik, S., Ghosh, A., Saha, R., and Saha, B. (2015) A review on natural surfactants. *RSC Adv.*, **5** (81), 65757–65767.

73. Gospel Ajuru, M. (2017) Qualitative and quantitative phytochemical screening of some plants used in ethnomedicine in the Niger Delta Region of Nigeria. *J. Food Nutr. Sci.*, **5** (5), 198.

74. Barros, R.G.C., Andrade, J.K.S., Pereira, U.C., de Oliveira, C.S., Rafaella Ribeiro Santos Rezende, Y., Oliveira Matos Silva, T., Pedreira Nogueira, J., Carvalho Gualberto, N., Caroline Santos Araujo, H., and Narain, N. (2020) Phytochemicals screening, antioxidant capacity and chemometric characterization of four edible flowers from Brazil. *Food Res. Int.*, **130**, 108899.

75. Gubitosa J., Rizza V., Fini P., and Cosma, P. (2019) Hair care cosmetics: From traditional shampoo to solid clay and herbal shampoo, a review. *Cosmetics*, **6** (1), 1–16.

76. Sharma, R.M., Shah, K., and Patel, J. (2011) Evaluation of prepared herbal shmapoo formulations and to compare formulated shampoo with marketed shampoos. *Int. J. Pharm. Pharm. Sci.*, **3** (4), 3–6.

77. Al Badi, K., and Khan, S.A. (2014) Formulation, evaluation and comparison of the herbal shampoo with the commercial shampoos. *Beni-Suef Univ. J. Basic Appl. Sci.*, **3** (4), 301–305.

78. Vijayalakshmi, A., Sangeetha, S., and Ranjith, N. (2018) Formulation and evaluation of herbal shampoo. *Asian J. Pharm. Clin. Res.*, **11** (Special Issue 4), 121–124.

79. Patel, I., and Talathi, A. (2016) Use of traditional Indian herbs for the formulation of shampoo and their comparative analysis. *Int. J. Pharm. Pharm. Sci.*, **8** (3), 28–32.

80. Heukelbach, J., Oliveira, F.A.S., and Speare, R. (2006) A new shampoo based on neem (*Azadirachta indica*) is highly effective against head lice in vitro. *Parasitol. Res.*, **99** (4), 353–356.

81. Ibrahim, M., Khaja, M.N., Aara, A., Khan, A.A., Habeeb, M.A., Devi, Y.P., Narasu, M.L., and Habibullah, C.M. (2008) Hepatoprotective activity of *Sapindus mukorossi* and *Rheum emodi* extracts: In vitro and in vivo studies. *World J. Gastroenterol.*, **14** (16), 2566–2571.

82. Khopde, S.M., Priyadarsini, K.I., Mohan, H., Gawandi, V.B., Satav, J.G., Yakhmi, J. V., Banavaliker, M.M., Biyani, M.K., and Mittal, J.P. (2001) Characterizing the antioxidant activity of amla (*Phyllanthus emblica*) extract. *Curr. Sci.*, **81** (2), 185–190.

83. Namita, and Nimisha (2013) Formulation and evaluation of herbal shampoo having antimicrobial potential. *Int. J. Pharm. Pharm. Sci.*, **5** (Suppl 3), 708–712.

84. Saeed, N., Khan, M.R., and Shabbir, M. (2012) Antioxidant activity, total phenolic and total flavonoid contents of whole plant extracts Torilis leptophylla L. *BMC Complement. Altern. Med.*, **12**, 221–233.

85. Solihah, M.A., Rosli, W.W.I., and Nurhanan, A.R. (2012) Phytochemicals screening and total phenolic content of Malaysian *Zea mays* hair extracts. *Int. Food Res. J.*, **19** (4), 1533–1538.

86. Mamillapalli, V., Katamaneni, M., Tiyyagura, V.M., Kanajam, P., Namagiri, A.P., Thondepu, H., Appikatla, B., Devangam, B., and Khantamneni, P. (2020) Formulation, phytochemical, physical, biological evaluation of polyherbal vanishing cream, and facewash. *Res. J. Pharm. Dos. Forms Technol.*, **12** (3), 139.

87. Brown, S.K., and Shalita, A.R. (1998) Acne vulgaris. *Lancet*, **351** (9119), 1871–1876.

88. Kanlayavattanakul, M., and Lourith, N. (2011) Therapeutic agents and herbs in topical application for acne treatment. *Int. J. Cosmet. Sci.*, **33** (4), 289–297.

89. Cong, T.X., Hao, D., Wen, X., Li, X.H., He, G., and Jiang, X. (2019) From pathogenesis of acne vulgaris to anti-acne agents. *Arch. Dermatol. Res.*, **311** (5), 337–349.

90. Sawarkar, H.A., Khadabadi, S.S., Mankar, D.M., Farooqui, I.A., and Jagtap, N.S. (2010) Development and biological evaluation of herbal anti-acne gel. *Int. J. PharmTech Res.*, **2** (3), 2028–2031.

91. Templeton, R.H. (2018) Reetha and shikakai as natural surfactants for cleaning of historic textiles. *Int. J. Res. Anal. Rev.*, **5** (2), 571–573.

92. Muntaha, S.T., and Khan, M.N. (2015) Natural surfactant extracted from *Sapindus mukurossi* as an eco-friendly alternate to synthetic surfactant – A dye surfactant interaction study. *J. Clean. Prod.*, **93**, 145–150.

93. Bolognesi, C. (2003) Genotoxicity of pesticides: A review of human biomonitoring studies. *Mutat. Res. – Rev. Mutat. Res.*, **543** (3), 251–272.

94. Storz, P. (2005) Reactive oxygen species in tumor progression. *Front. Biosci.*, **10** (2), 1881–1896.

95. Liou, G.Y., and Storz, P. (2010) Reactive oxygen species in cancer. *Free Radic. Res.*, **44** (5), 479–496.

96. Fang, C., Gu, L., Smerin, D., Mao, S., and Xiong, X. (2017) The Interrelation between reactive oxygen species and autophagy in neurological disorders. *Oxid. Med. Cell. Longevity*, **2017**, 1–16.

97. Mondal, S., and Tazib Rahaman, S. (2020) Flavonoids: A vital resource in healthcare and medicine. *Pharm. Pharmacol. Int. J.*, **8** (2), 91–104.

98. Millar, C.L., Duclos, Q., and Blesso, C.N. (2017) Effects of dietary flavonoids on reverse cholesterol transport, HDL metabolism, and HDL function. *Adv. Nutr.*, **8** (2), 226–239.

99. Mariee, A.D., Abd-Allah, G.M., and El-Beshbishy, H.A. (2012) Protective effect of dietary flavonoid quercetin against lipemic-oxidative hepatic injury in hypercholesterolemic rats. *Pharm. Biol.*, **50** (8), 1019–1025.

100. Anand David, A.V., Arulmoli, R., and Parasuraman, S. (2016) Overviews of biological importance of quercetin: A bioactive flavonoid. *Pharmacogn. Rev.*, **10** (20), 84–89.

101. Gullón, B., Lú-Chau, T.A., Moreira, M.T., Lema, J.M., and Eibes, G. (2017) Rutin: A review on extraction, identification and purification methods, biological activities and approaches to enhance its bioavailability. *Trends Food Sci. Technol.*, **67**, 220–235.

102. Imran, M., Salehi, B., Sharifi-Rad, J., Gondal, T.A., Saeed, F., Imran, A., Shahbaz, M., Fokou, P.V.T., Arshad, M.U., Khan, H., Guerreiro, S.G., Martins, N., and Estevinho, L.M. (2019) Kaempferol: A key emphasis to its anticancer potential. *Molecules*, **24** (12), 1–16.

103. Seleem, D., Pardi, V., and Murata, R.M. (2017) Review of flavonoids: A diverse group of natural compounds with anti-*Candida albicans* activity in vitro. *Arch. Oral Biol.*, **76**, 76–83.

104. Fang, Z., and Bhandari, B. (2010) Encapsulation of polyphenols – A review. *Trends Food Sci. Technol.*, **21** (10), 510–523.

7 Biosurfactants
Introduction, Types, and Industrial Applications

*Shashank Sharma[1], Nidhi Puri[2], Priyanka Dhingra[3]
and M. A. Quraishi[4]*

[1]Department of Chemistry, SBAS, Galgotias University,
Greater Noida, U.P., India

[2]Department of Applied Science & Humanities, I.T.S Engineering
College, Greater Noida 201307, India

[3]Department of Chemistry, JECRC University, Jaipur, Rajasthan, India

[4]Center of Research Excellence in Corrosion, Research Institute, King
Fahd University of Petroleum and Minerals, Dhahran 31261, Saudi Arabia

CONTENTS

DOI: 10.1201/9781003144878-7

7.1 INTRODUCTION

Public consciousness and seriousness about the environmental effects of different household goods has gradually grown over the last 50 years or so and has led to consumer choices when selecting, for example, soapy substances, detergents, cleaners, etc. In recent years, attention has turned to the global environmental effects of goods, and the "total carbon load" has become a concern. The economic significance of increased organic or bio-primary draft compounds for the chemical industry in conjunction with sharp price rises and competition for petroleum products. In the energy and fuel market, where the potential for the synthesis of reusable goods has risen significantly, this trend has been most noticeable. In recent decades, the detergent industry has drawn its notice to natural raw materials, either as hydrophilic or hydrophobic building blocks, to replace petrochemical products. From several sources, such as sugars, amino acids, cellulose, and other carbohydrates, hydrophilic building blocks have been picked. While the detergent industry's typical feedstocks are natural fats and their derivatives, there is an emerging need to discover novel hydrophobic components and normal hydrophobic materials may accumulate qualities which are not readily attained by traditional petrochemical synthesis. The utilization of unsaturated securities in unsaturated fats for fundamental compound change to get massiveness in the surfactant's hydrophobic moiety is a significant progression [1].

Surfactants are part of the most flexible category of chemicals used in different sectors, including cleansers, paints, paper items, drugs, beauty care products, oil, food, and water treatment. The activation potential of the surface forms surfactants suitable emulsifiers, dispersants, and foaming agents [2]. They minimize the interfacial tension structures of liquid-solid (e.g. wetting occurrences) or liquid-liquid (e.g. water-oil or oil-water) as well as the surface tension of air-water like aqueous media. In polar compounds, they help with the solubility of organic solvents. Surfactants are the functional materials used to extract oily matter from a specific medium in soaps and detergents and are widely used. Due to these qualities, surfactants are used in several production processes. They are of chemical or natural origin [3]. The usefulness of a surfactant in lowering surface tension is defined by its potency. The interfacial and surface tension of water against n-hexadecane water can be reduced by stable surfactants from 72 to 30 mN m^{-1} and from 40 to 1 mN m^{-1}, respectively [4].

Due to their less natural effect and their lower utilization of fossil assets, surfactants' popularity has caused a developing interest in inexhaustible surfactants over the most recent couple of years. Inexhaustible surfactants include mature and orchestrated surfactants from renewable crude materials (also referred to as biosurfactants or biological surfactants, i.e. bio-based or typical surfactants). The demand for green surfactants in 2017 amounted to 17.7 billion (13.5 billion for bio-based surfactants and 4.2 billion for biosurfactants) and it is expected to increase substantially 5% before 2022 [5].

Biosurfactants are surfactants synthesized by living cells [6]. With the same methods as synthetic surfactants, they seek to reduce interfacial and surface tension [7]. Emulsification, diffusion, solubilization, foaming, wetting, detergent capacity, and antimicrobial properties are the main functions of biosurfactants in some cases [8]. Exolipids are mainly biosurfactants, but they are cell-bound in some situations [6]. Biosurfactants are naturally derived from different substrates such as oils, sugars, alkanes, waste, etc. by bacteria, fungus, or yeast. For example, from substrates such as olive oil, succinate, C11 and C12 alkanes, pyruvate, glycerol, fructose, citrate, glucose, and mannitol, *Pseudomonas aeruginosa* may form rhamnolipids [9, 10]. Biosurfactants also have a range of industrial uses in the food processing industry, in the ecological area, in medicines and health care, personal care, farming, and agriculture [4, 11–21]. Table 7.1 gives various structural properties and descriptions of new types of surfactants.

TABLE 7.1

The Structural Properties and Descriptions of New Types of Surfactants

Class	Structure-Related Properties	Examples
Cationic	Surfactant mixture of equimolar cationic and anionic(absence of inorganic counterion)	n-dodecyltrimethylammonium n-dodecyl sulphate (DTADS) + $C_{12}H_{25}$ $(CH_3)_3$ N O_4S $C_{12}H_{25}$
Gemini or Dimeric	Two related surfactants connected by a spacer equal to or at the level of the head category	Propane-1,3-bis(dodecyldimethyl ammonium bromide) C_3H_6-1,3-bis[$(CH_3)_2$ N $C_{12}H_{25}$ Br]
Polymerizable	A surfactant which must be copolymerized or homopolymerized with several other components of the unit	11-(acryloyloxy)undecyltrimethyl ammonium bromide
Bolaform	Two charged head groups joined by a large polymethylene linear chain	Hexadecanediyl-1,16-bis(trimethyl ammonium bromide) Br⁻ $(CH_3)_3$ N⁺–$(CH_2)_{16}$–N⁺$(CH_3)_3$ Br⁻
Polymeric	Polymer with surface properties that are active	Copolymer of isobutylene and succinic anhydride

7.2 MAIN CHARACTERISTICS OF BIOSURFACTANTS

As compared to chemically synthesized counterparts, biosurfactants' properties and the large abundance of substrates have made them ideal for industrial applications. The surface movement, resistance to temperature, pH, ionic content, low toxic quality, biodegradability, antimicrobial activity, emulsifying and demulsifying capacity, and characterizing microbial surfactants are some of its properties [22]. The key highlights of all biosurfactant properties are schematically illustrated in Figure 7.1.

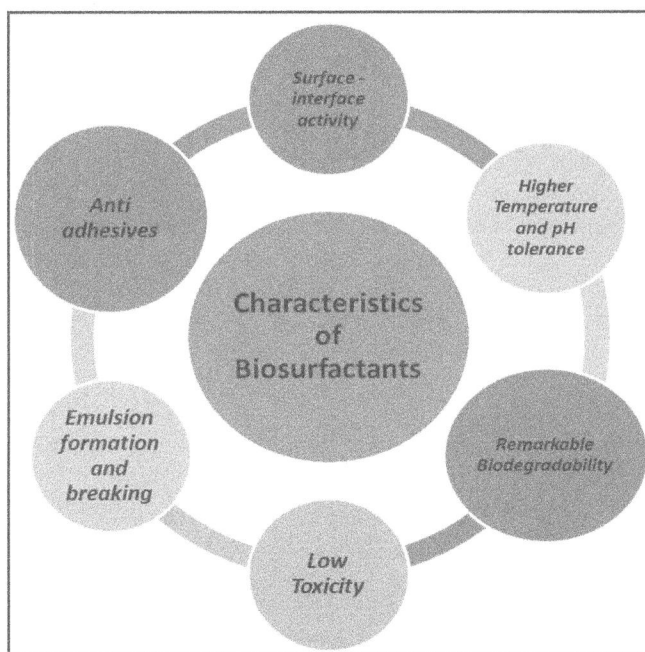

FIGURE 7.1 Schematic presentation of characteristics of biosurfactants.

7.2.1 SURFACE AND INTERFACE ACTIVITY

Joint forces are responsible for the phenomenon of surface tension within liquid molecules. It forms a "film" surface that, when it is wholly dipped, always makes it difficult to shift an object from the body rather than shifting it. The same condition also applies and is known as interfacial tension at the two fluids' interface, which does not get miscible into others like oil in water. A biosurfactant is able to reduce the surface tension of the output mixture which can be determined by its capacity, i.e. the water surface tension can be lowered by an effective biosurfactant from 72.0 to 35.0 mN m^{-1} [4]. According to the above criteria, such nominal degradation of surface tension to 8.0 mN m^{-1} is found to be a helpful microorganism-producing biosurfactant [23]. The surfactant developed by *Bacillus subtilis* m way minimize water/hexadecane interfacial strain to less than 1.0 mN m^{-1} and water surface tension to 25 mN m^{-1} [24]. *P. aeruginosa* produces *rhamnolipids* that minimize the tension of the water surface to 26 mN m^{-1} and the water/hexadecane interfacial pressure to be measured below 1.0 mN m^{-1} [25].

7.2.2 TEMPERATURE AND pH TOLERANCE

With a range of pH values from 2 to 12 and high temperatures, it is possible to use many biosurfactants. Up to 10% of salt concentration is also tolerated by biosurfactants, while 2% of NaCl is necessary to inactivate synthetic surfactants [26]. Due to their alleged profit-making interest, the processing of extremophile biosurfactants has received consideration in recent decades. Most biosurfactants are resistant to environmental factors and their surface movements, e.g. pH and temperature. McInerney et al. stated that temperature tolerance was up to 50°C, NaCl and Ca concentrations up to 50 and 25 g L^{-1}, respectively, pH between 4.5 and 9.0 in *B. licheniformis* that produces lichenysin [27]. One more biosurfactant generated by *Arthrobacter protophormiae* was established to be both stable pH (2–12) and thermostable (30–100°C) [28].

7.2.3 BIODEGRADABILITY

Biosurfactants are readily decomposed by water and soil microorganisms, turning these substances sufficient for bioremediation and waste management [26]. In contrast to synthetic surfactants, compounds produced from microbes can be severely weakened [29]. Mulligan et al. suggested that biosurfactants are compliant with different ecological purposes, including biosorption or bioremediation [30]. The emerging environmental challenge has contributed to the development of parallel alternatives, including biosurfactants, but there are ecological issues regarding synthetic or chemical surfactants [31]. Eco-friendly marine microorganism biosurfactants have also been linked to the biosorption of phenanthrene in water surfaces polluted with poorly soluble polycyclic aromatic hydrocarbons [32]. The eco-friendly biosurfactant sophorolipid was used in cochlodinium which provided 90% extraction efficiency in 30 minutes of therapy for controlling the blooms of marine algae [33].

7.2.4 LOW TOXICITY

Sambanthamoorthy et al. studied toxicity concerning human lung epithelial cell line biosurfactants obtained and tested by *Lactobacillus jensenii* and *L. rhamnosus* (A549) [34]. Biosurfactants, which have been produced from *L. rhamnosus* and *L. jensenii*, comprised a low level of contamination at nearly 200 mg mL^{-1}. At varying concentrations of 25–100 mg mL^{-1}, biosurfactants seem to have no contamination and are also used in foods, cosmetics, and pharmaceuticals. Less harmful content is of primary significance for environmental uses. It is possible to produce biosurfactants from mainly available raw materials and industrial waste [26].

7.2.5 EMULSION FORMATION AND BREAKING

A mixture of two or more than two typically undissolvable solutions is called an emulsion. It is also part of a more prevalent class called colloids of two-phase material structures. Milk and mayonnaise contain examples of emulsions in food preparations. In the Latin language, the term "emulsion" comes from the word for "milk". Emulsification occurs by reducing the interfacial tension between two points, according to the surface tension principle. In other words, emulsification is a process that produces a mixture of very tiny droplets of oil or fat stuck in a liquid, commonly known as emulsion water. Biosurfactants with high molecular weight are effective emulsifying agents which are also used to promote environmental bioaccumulation and eradication of oil content as an ingredient [35–45].

Many biosurfactants may be treated as emulsifiers or de-emulsifiers and are defined as a heterogeneous matrix of one insoluble liquid distributed as pellets, the distance and width of which are greater than 0.1 mm. Oil-in-water emulsions or water-in-oil emulsions have the least firmness and can be preserved by external substances, including biosurfactants, for a prolonged span of time as solid emulsions [46].

7.2.6 ANTI-ADHESIVES

A biofilm may be classified as an accumulated/colonized bacteria/other organic matter culture on every outside layer [47]. Several influences that include the form of microorganisms, electrical surface charges and hydrophobicity, ecological conditions, and the capability of microorganisms to create extracellular polymers that enable cells to bind to surfaces have affected bacterial surface adhesion [48]. It is possible to use biosurfactants to change surface hydrophobicity, which impacts the adhesion strength of surfaces of microbes. A surfactant reduces the rate of accumulation of other thermophilic strains of *Streptococcus* over the steel accountable for deterioration from *Streptococcus thermophilus*. The binding of *Listeria monocytogenes* to the steel surface was equally stopped by a *P. fluorescens*'s biosurfactant [25]. Biosurfactants have been shown to avoid pathogenic species from adhering to concrete surfaces or contamination sites. Initial biosurfactant adhesion to concrete surfaces can be a new and efficient way of fighting pathogenic microorganism colonization.

7.3 TYPES OF BIOSURFACTANTS

Biosurfactants use a large variety of structurally complex molecules which are primarily categorized by their microbial origin and chemical arrangement. These molecules are amphiphilic and involve (a) hydrophilic movement of monosaccharides, disaccharides, or polysaccharides, acids, cationic or anionic forms of peptide, and (b) hydrophobic, in other words, hydrocarbon chains comprising unsaturated or saturated forms of fatty acid. Such compounds can be generally categorized into two major groups: (a) most low-molecular-weight compounds, including glycolipids and certain short-chain lipopeptides, are typically biosurfactants, and (b) bio-emulsifiers, such as lipopolysaccharides or lipoproteins, are identified as high-molecular-weight compounds [39, 49–52]. Structural biosurfactants are broadly categorized into five classes: (a) polymers; (b) natural or phospholipids and fatty acids, (c) glycolipids; (d) lipoproteins or lipopeptides; and (e) biosurfactants in particulate form [2, 37, 53]. Figure 7.2 represents the broad view of types of biosurfactants.

7.3.1 GLYCOLIPIDS

They are carbohydrates linked by an ester group of aliphatic acids carrying long-chain or aliphatic hydroxy acids. Rhamnolipids, sophorolipids, and trehalolipids are among the most established glycolipids which are primarily biosurfactants. The origins and distribution of various glycolipids can be explained as follows:

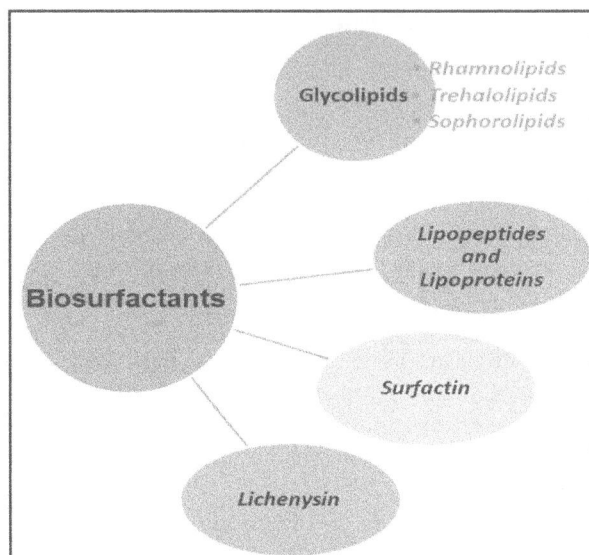

FIGURE 7.2   Schematic presentation of various types of biosurfactants.

7.3.1.1   Rhamnolipids

Edwards and Hayashi, 1965, reported that rhamnolipids are glycolipids which are linked to the same amount of hydroxydecanoic acid in one or two of the rhamnolipid molecules.

7.3.1.2   Trehalolipids

They are associated with many of the *Nocardia, Mycobacterium*, and *Corynebacterium* bacteria species. Asselineau and Asselineau reported that *Trehalose* lipids from *Arthrobacter* spp. and *Rhodococcus erythropolis* lowered the surface and interface stresses in culture broth of 1–5 and 25–40 mN m, respectively [54].

7.3.1.3   Sophorolipids

Yeast glycolipids containing dimeric carbohydrate sophorose are connected to a long-chain hydroxyl fatty acid by glycosidic connection. At least six to nine separate hydrophobic sophorolipids, normally sophorolipids, should be combined for many applications, and the sophorolipid lactone type [46, 55].

7.3.2   Lipopeptides and Lipoproteins

These comprise a lipid associated with a variety of polypeptides. Antimicrobial activity against diverse microbes, fungi, algae, and viruses has been demonstrated by many biosurfactants. The antifungal properties were noted by Ref. [56], and the antimicrobial properties of the *B. subtilis* – produced lipopeptide, iturin – were described by Singh and Cameotra [28]. With a shelf life of 6 months at −18°C and a pH of 5–11, iturin by *B. subtilis* was reported to be active also after autoclaving.

7.3.3   Surfactin

Cyclic lipopeptide (CLP) surfactin, consisting of a ring structure of seven amino acids, is one of the finest biosurfactants. Earlier studies have stated that surfactin from *B. subtilis* has different physicochemical properties. They discovered that surfactin is capable of reducing water surface stress (surface strain) and interfacial tension of the water. It has also been noticed that herpes and retroviruses are inactivated by surfactin.

7.3.4 LICHENYSIN

B. licheniformis contains a number of biosurfactants that show good durability in surfactin-related conditions of ambient temperatures, pH, and salt. McInerney et al. [27] suggested that water interfacial tension and surface tension could be decreased to 0.36 and 27 mN m^{-1}, respectively, by *B. licheniformis*.

7.4 APPLICATIONS OF BIOSURFACTANTS

At present, biosurfactants are very much useful in different applications. A broad view of their applications is given in Figure 7.3.

7.4.1 ANTIVIRAL APPLICATIONS

Evolutionary studies have indicated a possible theory, but the origin of microbial biosurfactants' production is still unsolved. This view acknowledges the comparative advantages presented by the development of biosurfactants, seeking to boost survival relative to many other microbes in areas, including resource acquisition and protection, which can be decreased as a consequence [57, 58]. Biosurfactant generation has usually started where species have less energy and can benefit from its antimicrobial properties. By extending the use of bioactive compounds to inactivate blanketed viruses, previous studies have investigated the protective nature of surfactants. *Tolypocladium inflatum* fungus–based biopeptide called cyclosporin A (CsA) is supposed to interact with the viral

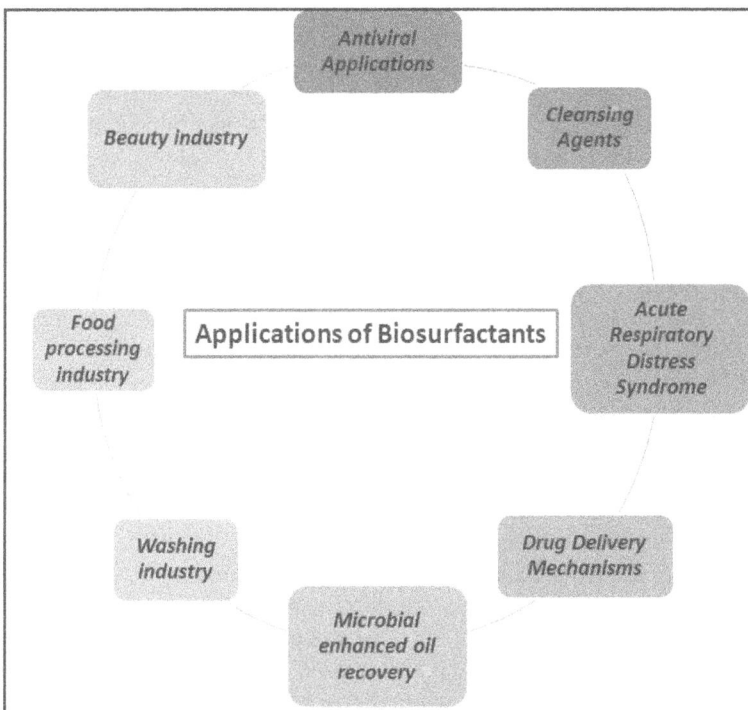

FIGURE 7.3 Applications of biosurfactants.

cycle to prevent replicating the influenza virus [59, 60]. CsA may not cause RNA replication or adsorption and therefore prevents actions, including integration or fading after protein synthesis [59]. This is necessary because sprouting allows infections to leave cells and attach to individual layers to enhance viral proteins that facilitate distribution and contamination [61]. By approaching subsequent life cycle activities of the virus, the problem of susceptibility to existing medications and limiting dissemination can be overcome. Lipopeptides used as adjuvants or associated with low-mass antigenic atoms were often used to introduce a safe framework to produce antibodies. Synthesized lipopeptide vaccine, say, causes cytotoxic *T-lymphocytes* relevant to the virus against the flu nucleoprotein antigenic determinant called epitope [62].

Related findings were reported in the T-cell and B-cell *in vivo* response to HIV-1 foot-and-mouth disease [63, 64]. In the development and production of new vaccines, this will have a very beneficial use. Sophorolipids (SL) are a class of yeast-based microbial glycolipids which have demonstrated anti-inflammatory, immunomodulatory, and improved survival features of sepsis in animal research systems. SL has been effective against herpes virus infection and HIV infection through sonophoresis head-collecting acetylation. In order to improve SL hydrophilicity, this improvement is supposed to promote its cytokine-stimulating and antiviral functions [65, 66]. SARS-CoV-2 problems are possible and can happen. This involves the screening of anticipated experts with new methods of operation to remove antagonistic hazardous impacts.

7.4.2 APPLICATIONS AS CLEANSING AGENTS

In worldwide disease outbreaks, effective antiviral implementation is critical, and immediate treatment of sick people is essential to managing suspected cases. The transmission rate of an infection could be reduced by extraordinary measures, including social distancing and shutting public spaces. However, when introduced particularly to SARS-CoV-2, only vaccines or antiviral drugs are useful in treatments and prevention of infection. Vaccine development is expected to be a sluggish process where an evolving and newly unindexed virus is encountered, whether the viral susceptibility correctly suits a commonly recognized virus for which treatments are available long before treatment. It is also vital to enhance successful and secure treatment approaches that completely remove residual aspects of the infection that may persist on grounds in public spaces, in clothing or dwellings, often after long periods of time prior to treatment. Anionic surfactant forms are frequently seen in household cleaners and detergents. When applied externally, the surfactant's fatty acid chains connect to the hydrophobic parts of microorganisms, although the surfactants of the hydrophilic field attach to water with a high affinity, causing solubilization of the microorganism simultaneously.

Such emulsification process thus encourages surface cleaning while depositing an effective surfactant surface. It extracts toxic contaminants from the surface and solubilizes them as the electrical charge of the anionic detergent particles is inserted into tiny droplets, resulting in detergent and dirt emulsification. The process of repulsion aims to clearly prevent the reintroduction to the surface of the same particle because surfactant molecules are regularly connected to the dangerous or dirt material. Yangxin et al. [67] reported that surfactants are the only most key components in household and laundry cleaners and generally responsible for 15–40% of the entire detergent formulation. As per increased experimental research, numerous combinations of surfactant forms and advancements in surfactant volume within current and emerging items have demonstrated successful surfactants over traditional products, including bleach. Analysis is reasonably clear because, with its active constituent, sodium hypochlorite, bleach has been an essential antibacterial agent that is efficient in killing bacteria, viruses, and fungi. Nevertheless, bleach may cause irritation to skin, mucus membranes, and airways, signaling that continued exposure is detrimental. Bleach, in addition, interfused under the action of temperature and light rapidly undergoes a reaction with other chemicals, decreasing its efficacy. Inadequate bleach applications can also affect quality; including variations from recommended dilutions and their excessive usage may cause environmental problems. Such negative environmental consequences may be increased when we wash public areas on a large scale to avoid viral spread.

Most of the above problems show how the use of bleach on its own for decontamination could even result in damage to healthcare professionals and enhance immune suppression in public spaces or high-risk areas, including hospitals/surgeries, that can result in increased vulnerability to SARS-CoV-2, particularly in areas where sensitivity to virus infections is much more probable [68]. Consequently, in combination with or even as a substitute for strong chemical cleaning products, the use of products having biosurfactants can be more effective for good decontamination. Biosurfactants have several advantages as an environmentally safe and not as much of hazardous alternative to conventional surfactants and, to date, glycolipids (sophoro-, rhamno- and mannosylerythritol lipids) are commonly marketed by various industries worldwide in cleaning applications. Saraya, Ecover, and Henkel companies generally use SL for purpose of dishwashing, clothing, and cleaning products for different organizations, while Evonik, BASF, Unilever, TeeGene market products that rely on lipopeptide and rhamnolipids biosurfactants [69–72].

7.4.3 Application of Biosurfactants to Acute Respiratory Distress Syndrome (ARDS)

Acute respiratory distress syndrome (ARDS) is an emergent medical disorder triggered by fluid accumulation in the patient's respiratory tract (alveoli), resulting in poor oxygen movement through alveolar membranes present in the blood [73]. In this case, Covid-19, ARDS is often the result of an already severe medical condition, causing oxygen loss to organs that contribute to high mortality rates observed in those who continue to experience effects [74]. Due to SARS-CoV-2 disease, surfactant brokenness is one driving explanation for this alveolar liquid growth, which affects the emulsification and, consequently, the fluid leeway from this specific territory [75].

A similar procedure's efficacy has played a significant role in socioeconomic reasons, with the availability of suitable ventilators and supplies that prove to be sufficient for each affected patient. The cost, location, and training needed to use them are challenging and, in some cases, impossible to take into account. With this in mind, biosurfactants are a promising area of research to identify novel ARDS therapies that could address socioeconomic obstacles that have hindered ventilation effectiveness to date. Thus, additional researches are able to give positive results by solubilization of the alveolar substrate for their applications as a cure for ARDS and hence are crucial in the future fight against Covid-19.

7.4.4 Biosurfactants in Drug Delivery Mechanisms

When determining likely helpful open doors for the Covid-19 pandemic that does not negotiate the drug's subatomic existence while also having the option of transmitting it viably to the region of interest, it is imperative to decide on a method of medication conveyance. Vaporized or tablet will be a possible conveyance strategy, with SARS-CoV-2 mainly affecting the upper gastrointestinal tract and respiratory framework. The micellar plan of biosurfactants enables them to be the perfect contender for any drug conveyance device, allowing them to develop a powerful liposome that encases the medication, shielding it from harm that could be broken somehow or for some cause [76].

When used in an aerosol, the physicochemical characteristics of biosurfactants enable them to retain their quality, because the main area of infection is inside the lungs, and this will be the probable mode of delivery of drugs. During this process, the solubility of biosurfactants effectively enhances the bioavailability of the drug after delivery. This self-solubilizing characteristic of biosurfactants thus enhances the proportionality of the dose of the medication, contributing to more consistent patient outcomes [77]. There is a clear limit for biosurfactants to intervene in the transmission of drugs, but the benefits of their use are often more prominent. In order to free surfactant dysfunction in the *alveoli*, the biosurfactants can also show appropriate antiviral characteristics at the infection point, another effect of SARS-CoV-2 virus, despite giving the drug a protected way to target.

Similarly, numerous viruses that occur around the acquired area can be inhibited. Although also trying to relieve symptoms directly, an important aspect is that it reduces their virulency and transmission among hosts. Likewise, this trademark has been reached out as specialists consider biosurfactants' capability to act in this manner as a treatment for themselves. The consideration of clinically affirmed biosurfactants in chewy candies or tablets is one such model; when biosurfactants are ingested, they can directly enter zones of the mouth and throat that can be undermined and subsequently give suggestive alleviation [78]. Regardless, a fume in the mouth that can be breathed in during the ingesting cycle could potentially frame the biosurfactant, allowing it to infiltrate regions of the respiratory tract to relieve there.

7.4.5   Applications in Microbial Enhanced Oil Recovery (MEOR)

Microbial enhanced oil recovery (MEOR) is a leading oil resurgence technique that persists in low-permeability reservoirs or high-viscosity crude oil. The residual oil in such areas is hard to reach, and capillary pressure holds the oil contained in the pores. Most of the biosurfactants diminish the interfacial stress amid oil/rock and oil/water. It lowers the capillary powers which prevent the movement of oil. Biosurfactants may also be tightly connected to the oil-water interface, resulting in emulsion, causing the water to settle and extract desorbed oil along with injection water [49].

7.4.6   Washing Industry

The majority of surfactants being used in cleaning are synthesized chemically as industrial detergents and have harmful effects on living organisms living in freshwater. Yet, many eco-friendly biosurfactants possess a good tendency to form emulsion when they are treated with vegetable oils. Also, many biosurfactants such as CLP are persistent within the pH range 7.0–12.0, and when they are subjected to higher temperatures, their surface-active properties are not affected. Because of such persistent nature and compatibility, various biosurfactants are being used in the preparation of laundry detergent.

7.4.7   Food Processing Industry

A number of food additives are comprised of biosurfactants, commonly known as emulsifiers, such as lectin and its derivatives, fatty acid esters containing glycerol, ethoxylated monoglyceride derivatives along with freshly synthesized oligopeptides, and sorbitan or ethylene glycol. Such emulsifiers boost the taste, benefit, and efficiency of food items with limited health hazards. Moreover, many other important applications of biosurfactants in food industry are popular, that is, including defoaming in sugar processing, fat stabilization in food oils, starch formation in instant potatoes, and coatings for fruits and vegetables which are protective, and improved solubility in instant drinks and soups, etc.

7.4.8   Beauty Industry

Biosurfactants have many roles in cosmetics industries and in the field of health care due to their skin-friendly characteristics. For example, sophorolipids are used as skin moisturizers and can be commonly found in personal care industries of Tokyo, Japan [79].

7.5   CONCLUSION

Despite several laboratory successes in the development of biosurfactants, industrial-scale production keeps a complicated problem. The commercialization of any product is based on its usability value, the easy availability of raw materials, and manufacturing costs. The most significant barrier to biosurfactants' production is the absence of proper knowledge of a bioreactor system; moreover, other factors, including low efficiency and expensive downstream processing, also inhibit the same. We expect super-active microbial strains to be developed using genetic engineering at the industrial level for future growth. The innovation of new methods and the development of more trustable resources are also expected in the future. It will then increase yields and minimize production costing and discover new biosurfactants and better understand these molecules' chemistry. So, the time is not long before the biosurfactants in the surfactant industry can start competition favorably with their synthetic counterparts. The rapid increase in the number of research articles and patents on the "biosurfactant" shows an eagerly growing interest in this topic.

REFERENCES

1. AAFI, www.herc.com
2. Desai, J. D., & Banat, I. M. (1997). Microbial production of surfactants and their commercial potential. *Microbiology and Molecular Biology Reviews*, *61*(1), 47–64.
3. Layman, P. L. (1985). Industrial surfactants set for strong growth. *Chemical & Engineering News*, *63*, 23–48.
4. Mulligan, C. N. (2005). Environmental applications for biosurfactants. *Environmental Pollution*, *133*(2), 183–198.
5. Markets and Markets. (2018). Natural surfactants market (bio-based surfactants) by product type (anionic, nonionic, cationic, and amphoteric), application (detergents, personal care, industrial & institutional cleaning, and oilfield chemicals) & region—Global forecast to 2022.
6. Syldatk, C., and Wagner, F. (1987). Production of biosurfactants, in *Biosurfactants and biotechnology*, ed. N. Kosaric, W. L. Cairns and N. C. C. Gray. Marcel Dekker, New York, NY, vol. 25, pp. 89–120.
7. Singh, A., Van Hamme, J. D., & Ward, O. P. (2007). Surfactants in microbiology and biotechnology: Part 2. Application aspects. *Biotechnology Advances*, *25*(1), 99–121.
8. Mukherjee, S., Das, P., & Sen, R. (2006). Towards commercial production of microbial surfactants. *TRENDS in Biotechnology*, *24*(11), 509–515.
9. Robert, M., Mercade, M. E., Bosch, M. P., Parra, J. L., Espuny, M. J., Manresa, M. A., & Guinea, J. (1989). Effect of the carbon source on biosurfactant production by *Pseudomonas aeruginosa* 44T1. *Biotechnology Letters*, *11*(12), 871–874.
10. Mulligan, C. N., & Gibbs, B. F. (2004). Types, production and applications of biosurfactants. *Proceedings-Indian National Science Academy Part B*, *70*(1), 31–56.
11. Hill, K., & Rhode, O. (1999). Sugar-based surfactants for consumer products and technical applications. *Lipid/Fett*, *101*(1), 25–33.
12. Banat, I. M., Makkar, R. S., & Cameotra, S. S. (2000). Potential commercial applications of microbial surfactants. *Applied Microbiology and Biotechnology*, *53*(5), 495–508.
13. von Rybinski, W., & Hill, K. (1998). Alkyl polyglycosides—properties and applications of a new class of surfactants. *Angewandte Chemie International Edition*, *37*(10), 1328–1345.
14. Rodrigues, L., Banat, I. M., Teixeira, J., & Oliveira, R. (2006). Biosurfactants: potential applications in medicine. *Journal of Antimicrobial Chemotherapy*, *57*(4), 609–618.
15. Nitschke, M., & Costa, S. G. V. A. O. (2007). Biosurfactants in food industry. *Trends in Food Science & Technology*, *18*(5), 252–259.
16. Lourith, N., & Kanlayavattanakul, M. (2009). Natural surfactants used in cosmetics: glycolipids. *International Journal of Cosmetic Science*, *31*(4), 255–261.
17. Gharaei-Fathabad, E. (2011). Biosurfactants in pharmaceutical industry: a mini-review. *American Journal of Drug Discovery and Development*, *1*(1), 58–69.
18. Szűts, A., & Szabó-Révész, P. (2012). Sucrose esters as natural surfactants in drug delivery systems—a mini-review. *International Journal of Pharmaceutics*, *433*(1–2), 1–9.
19. Marchant, R., & Banat, I. M. (2012). Microbial biosurfactants: challenges and opportunities for future exploitation. *Trends in Biotechnology*, *30*(11), 558–565.
20. Sachdev, D. P., & Cameotra, S. S. (2013). Biosurfactants in agriculture. *Applied Microbiology and Biotechnology*, *97*(3), 1005–1016.
21. Neta, N. S., Teixeira, J. A., & Rodrigues, L. R. (2015). Sugar ester surfactants: enzymatic synthesis and applications in food industry. *Critical Reviews in Food Science and Nutrition*, *55*(5), 595–610.
22. Chandran, P., & Das, N. (2010). Biosurfactant production and diesel oil degradation by yeast species *Trichosporon asahii* isolated from petroleum hydrocarbon contaminated soil. *International Journal of Engineering Science and Technology*, *2*(12), 6942–6953.

23. Van der Vegt, W., Van der Mei, H. C., Noordmans, J., & Busscher, H. J. (1991). Assessment of bacterial biosurfactant production through axisymmetric drop shape analysis by profile. *Applied Microbiology and Biotechnology*, *35*(6), 766–770.

24. Cavalero, D. A., & Cooper, D. G. (2003). The effect of medium composition on the structure and physical state of sophorolipids produced by *Candida bombicola* ATCC 22214. *Journal of Biotechnology*, *103*(1), 31–41.

25. Chakrabarti, S. (2012). *Bacterial biosurfactant: characterization, antimicrobial and metal remediation properties* (Doctoral dissertation).

26. Santos, D. K. F., Rufino, R. D., Luna, J. M., Santos, V. A., & Sarubbo, L. A. (2016). Biosurfactants: Multifunctional Biomolecules of the 21st Century. *International Journal of Molecular Sciences*, *17*(3), 401.

27. McInerney, M. J., Javaheri, M., & Nagle Jr, D. P. (1990). Properties of the biosurfactant produced by *Bacillus licheniformis* strain JF-2. *Journal of Industrial Microbiology and Biotechnology*, *5*(2–3), 95–101.

28. Singh, P., & Cameotra, S. S. (2004). Potential applications of microbial surfactants in biomedical sciences. *TRENDS in Biotechnology*, *22*(3), 142–146.

29. Mohan, P. K., Nakhla, G., & Yanful, E. K. (2006). Biokinetics of biodegradation of surfactants under aerobic, anoxic and anaerobic conditions. *Water Research*, *40*(3), 533–540.

30. Mulligan, C. N., Yong, R. N., & Gibbs, B. F. (2001). Remediation technologies for metal-contaminated soils and groundwater: an evaluation. *Engineering Geology*, *60*(1–4), 193–207.

31. Cameotra, S. S., & Makkar, R. S. (2004). Recent applications of biosurfactants as biological and immunological molecules. *Current Opinion in Microbiology*, *7*(3), 262–266.

32. Olivera, N. L., Commendatore, M. G., Delgado, O., & Esteves, J. L. (2003). Microbial characterization and hydrocarbon biodegradation potential of natural bilge waste microflora. *Journal of Industrial Microbiology and Biotechnology*, *30*(9), 542–548.

33. Lee, Y. J., Choi, J. K., Kim, E. K., Youn, S. H., & Yang, E. J. (2008). Field experiments on mitigation of harmful algal blooms using a Sophorolipid—yellow clay mixture and effects on marine plankton. *Harmful Algae*, *7*(2), 154–162.

34. Sambanthamoorthy, K., Feng, X., Patel, R., Patel, S., & Paranavitana, C. (2014). Antimicrobial and antibiofilm potential of biosurfactants isolated from lactobacilli against multi-drug-resistant pathogens. *BMC Microbiology*, *14*(1), 1–9.

35. Sajna, K. V., Sukumaran, R. K., Gottumukkala, L. D., & Pandey, A. (2015). Crude oil biodegradation aided by biosurfactants from Pseudozyma sp. NII 08165 or its culture broth. *Bioresource Technology*, *191*, 133–139.

36. Ron, E. Z., & Rosenberg, E. (2002). Biosurfactants and oil bioremediation. *Environmental Biotechnology*, *13*(3), 249–252.

37. Rahman, K. S., Rahman, T. J., Kourkoutas, Y., Petsas, I., Marchant, R., & Banat, I. M. (2003). Enhanced bioremediation of n-alkane in petroleum sludge using bacterial consortium amended with rhamnolipid and micronutrients. *Bioresource Technology*, *90*(2), 159–168.

38. Pacwa-Płociniczak, M., Płaza, G. A., Piotrowska-Seget, Z., & Cameotra, S. S. (2011). Environmental applications of biosurfactants: recent advances. *International Journal of Molecular Sciences*, *12*(1), 633–654.

39. Markande, A. R., Acharya, S. R., & Nerurkar, A. S. (2013). Physicochemical characterization of a thermostable glycoprotein bioemulsifier from Solibacillus silvestris AM1. *Process Biochemistry*, *48*(11), 1800–1808.

40. Leahy, J. G., & Colwell, R. R. (1990). Microbial degradation of hydrocarbons in the environment. *Microbiology and Molecular Biology Reviews*, *54*(3), 305–315.

41. Das, K., & Mukherjee, A. K. (2007). Crude petroleum-oil biodegradation efficiency of *Bacillus subtilis* and *Pseudomonas aeruginosa* strains isolated from a petroleum-oil contaminated soil from North-East India. *Bioresource Technology*, *98*(7), 1339–1345.

42. Cerqueira, V. S., Hollenbach, E. B., Maboni, F., Vainstein, M. H., Camargo, F. A., Maria do Carmo, R. P., & Bento, F. M. (2011). Biodegradation potential of oily sludge by pure and mixed bacterial cultures. *Bioresource Technology*, *102*(23), 11003–11010.

43. Atlas, R. M. (1981). Microbial degradation of petroleum hydrocarbons: an environmental perspective. *Microbiological Reviews*, *45*(1), 180.

44. Kavitha, V., Mandal, A. B., & Gnanamani, A. (2014). Microbial biosurfactant mediated removal and/or solubilization of crude oil contamination from soil and aqueous phase: an approach with *Bacillus licheniformis* MTCC 5514. *International Biodeterioration & Biodegradation*, *94*, 24–30.

45. Varjani, S. J. (2017). Microbial degradation of petroleum hydrocarbons. *Bioresource Technology*, *223*, 277–286.

46. Hu, Y., & Ju, L. K. (2001). Purification of lactonic sophorolipids by crystallization. *Journal of Biotechnology, 87*(3), 263–272.

47. Hood, S. K., & Zottola, E. A. (1995). Biofilms in food processing. *Food Control, 6*(1), 9–18.

48. Zottola, E. A. (1994). Microbial attachment and biofilms formation: a new problem for the food industry? *Food Technology, 48*, 107–114.

49. Banat, I. M. (1995). Biosurfactants production and possible uses in microbial enhanced oil recovery and oil pollution remediation: a review. *Bioresource Technology, 51*(1), 1–12.

50. Raza, Z. A., Khalid, Z. M., & Banat, I. M. (2009). Characterization of rhamnolipids produced by a *Pseudomonas aeruginosa* mutant strain grown on waste oils. *Journal of Environmental Science and Health, Part A, 44*(13), 1367–1373.

51. Yan, P., Lu, M., Yang, Q., Zhang, H. L., Zhang, Z. Z., & Chen, R. (2012). Oil recovery from refinery oily sludge using a rhamnolipid biosurfactant-producing Pseudomonas. *Bioresource Technology, 116*, 24–28.

52. Hošková, M., Schreiberová, O., Ježdík, R., Chudoba, J., Masák, J., Sigler, K., & Řezanka, T. (2013). Characterization of rhamnolipids produced by non-pathogenic Acinetobacter and Enterobacter bacteria. *Bioresource Technology, 130*, 510–516.

53. Jadhav, V. V., Yadav, A., Shouche, Y. S., Aphale, S., Moghe, A., Pillai, S., ... & Bhadekar, R. K. (2013). Studies on biosurfactant from *Oceanobacillus* sp. BRI 10 isolated from Antarctic sea water. *Desalination, 318*, 64–71.

54. Asselineau, C., & Asselineau, J. (1978). Trehalose-containing glycolipids. *Progress in the Chemistry of Fats and other Lipids, 16*, 59–99.

55. Gautam, K. K., & Tyagi, V. K. (2006). Microbial surfactants: a review. *Journal of Oleo Science, 55*(4), 155–166.

56. Besson, F., Peypoux, F., Michel, G., & Delcambe, L. (1976). Characterization of iturin A in antibiotics from various strains of *Bacillus subtilis*. *The Journal of Antibiotics, 29*(10), 1043–1049.

57. Kiran, G. S., Ninawe, A. S., Lipton, A. N., Pandian, V., & Selvin, J. (2016). Rhamnolipid biosurfactants: evolutionary implications, applications and future prospects from untapped marine resource. *Critical Reviews in Biotechnology, 36*(3), 399–415.

58. Cameotra, S. S., Makkar, R. S., Kaur, J., & Mehta, S. K. (2010). Synthesis of biosurfactants and their advantages to microorganisms and mankind. *Biosurfactants, 672*, 261–280.

59. Garoff, H., Hewson, R., & Opstelten, D. J. E. (1998). Virus maturation by budding. *Microbiology and Molecular Biology Reviews, 62*(4), 1171–1190.

60. Khan, T. N. (2017). Cyclosporin A production from *Tolipocladium inflatum*. *General Medicine Open Access, 5*(4), 1–3.

61. Hamamoto, I., Harazaki, K., Inase, N., Takaku, H., Tashiro, M., & Yamamoto, N. (2013). Cyclosporin A inhibits the propagation of influenza virus by interfering with a late event in the virus life cycle. *Japanese Journal of Infectious Diseases, 66*(4), 276–283.

62. Deres, K., Schild, H., Wiesmuller, K.-H., Jung, G., & Rammensee, H.-G. (1989). In vivo priming of virus-specific cytotoxic T lymphocytes with synthetic lipopeptide vaccine. *Nature, 342*, 561–564.

63. Loleit, M., Ihlenfeldt, H. G., Brünjes, J., Jung, G., Müller, B., Hoffmann, P., ... & Haas, G. (1996). Synthetic peptides coupled to the lipotripeptide P₃CSS induce in vivo B and thelper cell responses to HIV-1 reverse transcriptase. *Immunobiology, 195*(1), 61–76.

64. Wiesmüller, K. H., Jung, G., & Hess, G. (1989). Novel low-molecular-weight synthetic vaccine against foot-and-mouth disease containing a potent B-cell and macrophage activator. *Vaccine, 7*(1), 29–33.

65. Shah, V., Doncel, G. F., Seyoum, T., Eaton, K. M., Zalenskaya, I., Hagver, R., ... & Gross, R. (2005). Sophorolipids, microbial glycolipids with anti-human immunodeficiency virus and sperm-immobilizing activities. *Antimicrobial Agents and Chemotherapy, 49*(10), 4093–4100.

66. Gross, R. A., & Shah, V. (2007). Anti-herpes virus properties of various forms of sophorolipids. *Patent WO2007130738 A, 1*, 15.

67. Yangxin, Y. U., Jin, Z., & Bayly, A. E. (2008). Development of surfactants and builders in detergent formulations. *Chinese Journal of Chemical Engineering, 16*(4), 517–527.

68. World Health Organization. (2014). *Infection prevention and control of epidemic-and pandemic-prone acute respiratory infections in health care*. World Health Organization.

69. Singh, P., Patil, Y., & Rale, V. (2019). Biosurfactant production: emerging trends and promising strategies. *Journal of Applied Microbiology, 126*(1), 2–13.

70. Sekhon Randhawa, K. K., & Rahman, P. K. (2014). Rhamnolipid biosurfactants—past, present, and future scenario of global market. *Frontiers in Microbiology, 5*, 454.

71. Klosowska-Chomiczewska, I., Medrzycka, K., & Karpenko, E. (2011). Biosurfactants–biodegradability, toxicity, efficiency in comparison with synthetic surfactants. *Advanced Mechanical Engineering, 2*, 1–9.

72. Fracchia, L., Banat, J. J., Cavallo, M., & Banat, I. M. (2015). Potential therapeutic applications of microbial surface-active compounds. *AIMS Bioengineering*, *2*(3), 144–162.

73. Matthay, M. A., Zemans, R. L., Zimmerman, G. A., Arabi, Y. M., Beitler, J. R., Mercat, A., ... & Calfee, C. S. (2019). Acute respiratory distress syndrome. *Nature Reviews Disease Primers*, *5*(1), 1–22.

74. Matthay, M. A., Ware, L. B., & Zimmerman, G. A. (2012). The acute respiratory distress syndrome. *The Journal of Clinical Investigation*, *122*(8), 2731–2740.

75. Luks, A. M., Freer, L., Grissom, C. K., McIntosh, S. E., Schoene, R. B., Swenson, E. R., & Hackett, P. H. (2020). COVID-19 lung injury is not high altitude pulmonary edema. *High Altitude Medicine & Biology*, *21*(2), 192–193.

76. Sosnowski, T. R., & Gradon, L. E. O. N. (2009). Influence of a biosurfactant on reentrainment and deaggregation of powders. *Eng Chem App*, *48*, 176–177.

77. Tanawade, O., Shangrapawar, T., & Bhosale, A. (2020). Self-emulsifying drug delivery systems: an overview. *Journal of Current Pharma Research*, *10*(2), 3680–3693.

78. Vellingiri, B., Jayaramayya, K., Iyer, M., Narayanasamy, A., Govindasamy, V., Giridharan, B., ... & Subramaniam, M. D. (2020). COVID-19: a promising cure for the global panic. *Science of the Total Environment*, *725*, 138277.

79. www.kao.com/jp/

8 Prospects of Biosurfactants in Medicine

Gunjan Chauhan[1], Raman Singh[1], Anil K. Sharma[2] and Kuldeep Singh[1]

[1]Department of Chemistry, Maharishi Markandeshwar (Deemed to be University), Mullana, Haryana, India

[2]Department of Biotechnology, Maharishi Markandeshwar (Deemed to be University), Mullana, Haryana, India

CONTENTS

8.1 INTRODUCTION

Modern medicinal chemistry has been developed over the decades using interdisciplinary knowledge [1, 2]. Several chemical classes of bioactive molecules have played a pivotal role in providing suitable drugs for the pharmaceutical and medicinal use [3–9]. Distinct approaches such as conjugation, disjunction, and derivatizations have been used in the past [10–14]. Still, there is a wide scope for developing newer molecules and materials for pharmaceutical use. In recent literature, researchers from academia and industry have shown interest in the medicinal

DOI: 10.1201/9781003144878-8

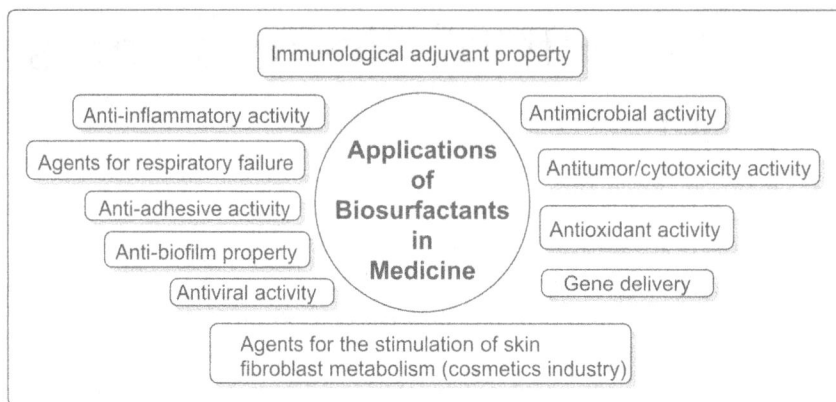

FIGURE 8.1 Applications of biosurfactants in medicine.

use of biosurfactants. In this article, the medicinal uses of biosurfactants have been discussed (Figure 8.1).

Natural surfactants or biosurfactants are secondary metabolites that various sources produce such as plants (e.g., saponins), microorganisms (e.g., glycolipids), and animals/humans (e.g., surface-active lipoprotein complex). These metabolites have a characteristic amphiphilic structure of a surfactant, by virtue of which they have diverse properties and applications to the environment as well as for therapeutic purposes. Biosurfactants have also been found to have utility in the inhibition of biofilm formation in bacteria by interrupting pathogenic organisms' adhesion to infection sites. Furthermore, biosurfactants have applications in diminishing antimicrobial resistance (AMR), which has appeared as a significant healthcare problem in today's context.

Kurt von Neergaard, a German Physiologist, in early 1929 proposed the existence of a surface-active substance in the lung [15]. He predicted the presence of an agent that can maintain alveolar surface tension, which confirms that lower surface tension would be useful for the respiratory mechanism to guarantee adequate alveolar ventilation throughout the respiratory cycle. Precisely 18 years later, Peter Gruenwald, a pathologist, repeated these experiments with stillborn infants' lungs. Gruenwald stated that surface tension causes resistance to aeration, and counteracts the entrance of air. He showed that surface-active substances reduce the pressure needed for aeration. Pattle and Clements found a substance in lung edema fluid and lung extracts that dramatically lowered surface tension. The material was composed of a phospholipid and a protein fraction and was termed surfactant (surface active agent). In 1959, Avery and Mead studied surfactant deficiency as the cause of infant respiratory distress syndrome [16].

8.2 BIOSURFACTANTS

Biosurfactants are bioproducts produced by microorganisms and play a critical role in reducing surface and interfacial tensions when present in aqueous solutions and as hydrocarbon mixtures. They are complex mixtures of heteropolysaccharides, lipopolysaccharides, lipoproteins, and proteins. They are amphipathic compounds with hydrophilic and hydrophobic moieties that preferentially partition between liquid interfaces with different degrees of polarity and hydrogen bridges, such as oil/water or air/water interfaces. They bind tightly to hydrocarbon surfaces and form stable emulsions by increasing kinetic stability in very low concentrations [17].

Biosurfactants have low critical micelle concentration (CMC) and high surface activity. They offer a wide range of diverse applications in practically all medical, industrial, environmental, and marine fields due to their immense structural diversity, complexity, and biochemical properties. Therefore, despite their high production cost, industrial and academic researchers showed their interest in isolation of novel biosurfactants to replace chemically synthesized counterparts, and efforts have been made to diminish the production cost of biosurfactants of amicrobial origin using alternative culture media [18–25].

8.3 CLASSIFICATION OF BIOSURFACTANTS

Biosurfactants are generally categorized by their molecular weight (high or low), type of secretion from the cell (intracellular, extracellular, or adherent to microbial cells), the ionic charge of the molecules (anionic, cationic, nonionic, or neutral), and chemical structure [26–29].

According to molecular weights, biosurfactants are divided into high-molecular-mass compounds such as amphipathic polysaccharides, proteins, lipoproteins, lipopolysaccharides/complex mixtures [30, 31], and low-molecular-mass biosurfactants containing phospholipids, glycolipids, and lipopeptides. The biosurfactants with low molecular mass are efficient in lowering surface and interfacial tensions [32], whereas high-molecular-mass biosurfactants are more effective at stabilizing oil-in-water emulsions.

The chemical structure is considered the main base for the classification in lipopolysaccharide, lipoprotein, biopolymer, phospholipid, lipopeptide, or glycolipid [33]. The glycolipids and lipopeptides are the most widely explored groups in the literature due to the high production yields and numerous applications potential in the most diverse fields [34].

The surfactants are generally categorized based on ionic charges. The hydrophilic group of a surfactant is usually referred to as the "head group" and is either strongly polar or charged, whereas the hydrophobic group of surfactant is usually called "tail" and is mostly a simple hydrocarbon group. Anionic surfactants include the traditional soaps and early synthetic detergents, sulfonate, and sulfates. Cationic surfactants are usually quaternary ammonium, imidazolinium, or alkyl pyridinium compounds. Zwitterionic/Amphoteric surfactants are used in the form of betaines or sulfobetaines. These compounds are milder on the skin than the anionic and have an especially low eye-sting effect, so these are used in toilets and baby shampoos. Ethoxylates usually dominate nonionic surfactants. This surfactant class also includes several semipolar compounds such as amine oxides, sulfoxide, and phospholine oxides. Some types of surfactants usually contain both nonionic and anionic groups, such as alkyl ethoxysulfates. These surfactants are mild on the skin, and skin contact cannot be avoided, such as dishwashing liquids and shampoos.

8.4 SYNTHESIS OF BIOSURFACTANTS

By using a variety of microorganisms, biosurfactants can be synthesized having distinct molecular structures. Most biosurfactants are produced from solid/liquid hydrocarbons, oils, and fats that are water-insoluble substrates, although many are obtained through soluble substrates or a combination of both types of substrates [35]. Most commonly reported microorganisms for the production of biosurfactants are *Arthrobacter*, *Pseudomonas aeruginosa*, *Acinetobacter calcoaceticus*, *Bacillus subtilis*, *Candida lipolytica*, and *Torulopsis bombicola* (Figure 8.2).

Biosurfactants can be synthesized through distinct metabolic pathways in the stationary growth phase of the microorganism termed as "*de novo*" synthesis, i.e., microorganisms cannot synthesize them through the main metabolic pathways. It is also possible to synthesize biosurfactants under optimal growth conditions, provided there is a previous kinetic study to determine the variables that need to be adjusted for this purpose [36, 37]. To favor production, insoluble substrates are added to water such as oils, fats, and/or solid and liquid hydrocarbons, which also assist in the survival of the microorganisms in highly contaminated or nutrient-poor environments [38, 39]. Moreover, productivity can be varied by evaluating the conversion rate of the substrate into the product in accordance with the metabolic pathway employed by the microorganism [40, 41]. The substrate used in the fabrication of biosurfactants directly affects the growth of the microorganism and the yield of the desired product.

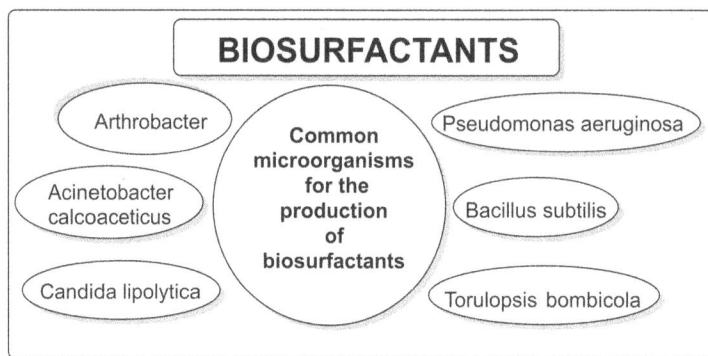

FIGURE 8.2 Commonly reported microorganisms for the production of biosurfactants.

Furthermore, the yield is affected by physiological state, cell density, and nutrient availability. The same carbon source can be used to synthesize different groups such as mannoproteins, glycolipids, and complexes containing a mixture of carbohydrates, lipids, and proteins, depending on the microorganism used. Therefore, the study of the carbon source is extremely important as different substrates lead to the production of different chemical structures as well as different properties and applications. Thus, depending on the desired application, variations in the culture conditions can significantly affect a biosurfactant structure [21]. The structure of biosurfactants is also influenced by the microorganism as well.

8.5 MECHANISMS OF INTERACTIONS

Biosurfactants are complex mixtures of heteropolysaccharides, lipopolysaccharides, lipoproteins, and proteins. They are amphipathic compounds with both hydrophilic and hydrophobic moieties that preferentially partition between liquid interfaces with different degrees of polarity and hydrogen bridges, such as oil/water or air/water interfaces. They bind tightly to hydrocarbon surfaces and form stable emulsions by increasing kinetic stability in very low concentrations [17].

8.6 APPLICATIONS OF BIOSURFACTANT

8.6.1 In Medicine

Biosurfactants have been considered multifunctional biomolecules due to their functional abilities and eco-friendly properties. Biosurfactants have several advantages over the chemical surfactants, such as lower toxicity, higher biodegradability, better environmental compatibility, higher foaming, high selectivity, and specific activity at extreme conditions such as temperatures, pH, and salinity. Recently, attention towards biosurfactants has doubled mainly due to their wide range of functional properties and the diverse synthetic capabilities of the microbes. Biosurfactants have tremendous applications in various industries such as petroleum, food, medicine, pharmaceutical [42], chemical, paper and pulp [43], textile, and cosmetics [44–46] and are also considered "green molecules" because of their wide applications in bioremediation of soil.

The biosurfactants could have a wide range of applications in pharmaceutical fields such as gene delivery, agents for respiratory failure, immunological adjuvants, anti-adhesive agents in surgical, inhibition of the adhesion of pathogenic organisms to solid surfaces, recovery of intracellular products, antimicrobial activity, antiviral activity, anticancer activity, and agents for the stimulation of skin fibroblast metabolism [47, 48].

8.6.2 ANTIMICROBIAL ACTIVITY

For the ability of their molecules to self-associate and form a pore-bearing channel, or micellar aggregation, inside a lipid membrane, upon which their antibiotic activity depends, lipopeptides are considered potentially useful antimicrobial agents (Figure 8.3) [49, 50]. The sequence of the hydrocarbon chains is affected, and its activity varies with the membrane thickness in penetrating into the membrane through hydrophobic interactions [51]. In 2003, the FDA approved daptomycin (CubicinR) to treat skin infections, which is one of the antimicrobial lipopeptides under commercial development. This practical lipopeptide is synthesized from *Streptomyces roseosporus* and has excellent performance against multiresistant bacteria, such as Methicillin-resistant *Staphylococcus aureus* (MRSA) [52]. Hence, biosurfactants can possibly be used as drugs in antimicrobial chemotherapy because of their nonhemolytic nature [53]. These activities are nonspecific modes of action and are useful for acting on different cell membranes, either Gram-positive or Gram-negative bacteria [54]. This means such performance of surfactin as a lipopeptide is on membrane integrity, rather than other important cellular processes and could make up the next antibiotic generations [55]. Abalos et al. identified six rhamnolipids in cultures of *P. aeruginosa* [56]. AT10 has exhibited excellent antifungal properties against different fungi. *C. bombicola*-derived sophorolipids inhibited the growth of both Gram-negative and Gram-positive bacteria [57].

The biosurfactant rufisan produced by *C. lipolytica* demonstrated antimicrobial activities against *Streptococcus agalactiae*, *S. mutans*, *S. mutans* NS, *S. mutans* HG, *S. sanguis* 12, and *S. oralis* J22 [58]. The biosurfactant from *Anser indicus* M6 has antimicrobial and biofilm activity against MRSA and can be used in dental treatment such as in root canal treatment, in mouth washes to remove dental flora, and for wound healing [59]. The biosurfactant produced by marine *B. circulans* has a potent antimicrobial activity against Gram-positive or Gram-negative pathogenic, and semipathogenic microbial strains including multidrug-resistant (MDR) strains [60]. Fernandes et al. investigated the antimicrobial activity of biosurfactants from *B. subtilis* R14 against 29 bacterial strains [61]. Their results demonstrated that lipopeptides have a broad spectrum of action, including antimicrobial activity against microorganisms with MDR profiles [61]. Iturin is a lipopeptide produced by *B. subtilis* that has demonstrated antifungal activity by affecting the morphology and structure of the cell membrane of yeasts. Rhamnolipid produced by *P. aeruginosa*, lipopeptides produced by *B. subtilis* [62], or *B. licheniformis* [63], and mannosylerythritol from *C. antarctica* [64] have shown antimicrobial activities.

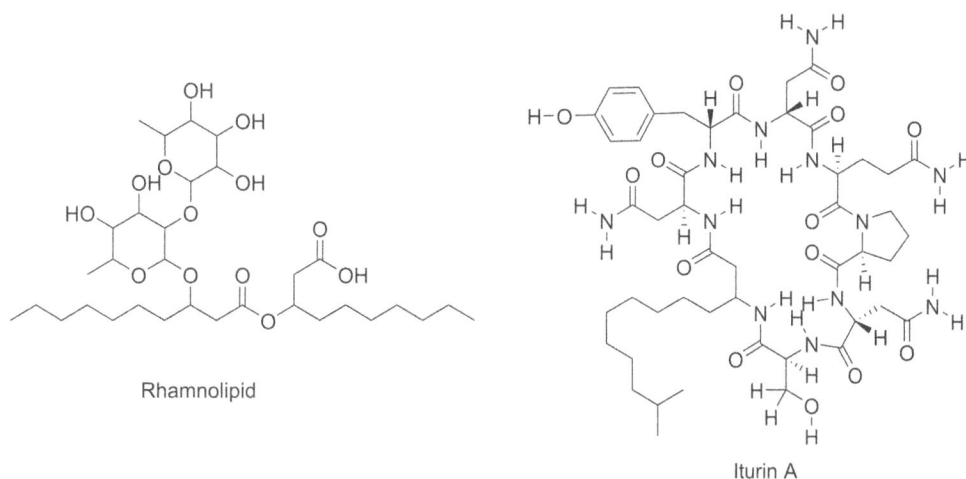

Rhamnolipid

Iturin A

FIGURE 8.3 Surfactants with antimicrobial activity.

Iturin A has been investigated as a useful antifungal molecule for deep mycosis. Other variants of the iturin group, including bacillomycin D and bacillomycin Lc, also present antimicrobial activity against *Aspergillus flavus*. Various pathogenic yeasts, along with human mycoses, including *Candida* spp., *Colletotrichum gloeosporioides*, *Corynespora cassiicola*, *Cryptococcus neoformans*, *Fusarium* spp., *Fusarium oxysporum*, *Rhizoctonia* spp., *Trichophyton rubrum*, and *Trichosporon asahii*, have been greatly inhibited *in vitro*, depicting the antifungal performance by biosurfactants [65, 66]. The major mode of action of biosurfactants on fungal pathogens is the direct lysis of zoospores generated by the intercalation of biosurfactants via plasma membranes of the zoospore that are independent of a cell wall [67].

8.6.3 ANTIVIRAL ACTIVITY

Surfactin, which is considered the best in medical field applications, can perform the effective antiviral activity [68]. From experiments in vitro, surfactin and fengycin from *B. subtilis* could significantly inhibit the activities of cell-free virus stocks of many viruses, such as the porcine parvovirus, the pseudorabies virus, the Newcastle disease virus, and the infectious bursal disease virus (Figure 8.4). They have also been able to inhibit infections and the replication of these viruses [69].

In vitro experiments have demonstrated that surfactin can effectively inactivate the virus that causes herpes as well as the retrovirus and other compact RNA and DNA viruses. The antiviral activity of surfactin has been determined for a broad spectrum of viruses. Surfactin has been suggested as showing antiviral action because of physicochemical interactions between the virus lipid membrane and the membrane-active surfactant. This causes permeability changes and finally results in the disintegration of the membrane system through a micelle effect [70]. Moreover, the feline calicivirus, herpes simplex virus, murine encephalomyocarditis virus, semliki forest virus, simian immunodeficiency virus, suid herpes virus, and vesicular stomatitis virus were found to be affected by surfactin. *Pumilacidin* has significant antiviral performance against herpes simplex virus 1 (HSV-1).

Sophorolipids are also reported to perform against the human immunodeficiency virus [71]. Herpes simplex virus types 1 and 2 were affected by rhamnolipid alginate complex, especially the herpes virus cytopathic effect in the Madin-Darby bovine kidney cell line, by a concentration lower than the CMC [72]. Antibiotic effects and growth inhibition of human immunodeficiency virus in leucocytes by biosurfactants have also been reported [26]. Sophorolipids surfactants from *C. bombicola* and its structural analogs such as sophorolipid diacetate ethyl ester is the most potent

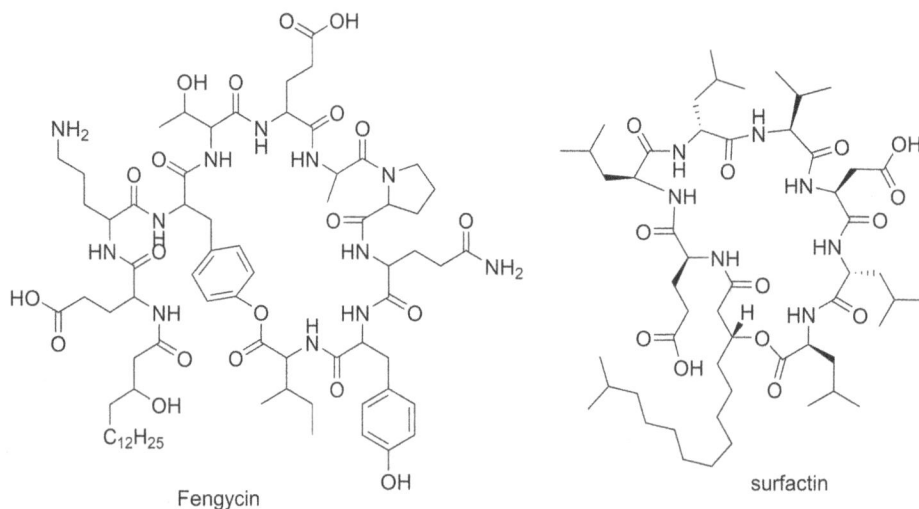

FIGURE 8.4 Structure of Fengycin and surfactin.

spermicidal and virucidal agent. It was also reported that this substance has a virucidal activity similar to nonoxynol-9 against the human semen.

8.6.4 ANTI-ADHESIVE ACTIVITY

Anti-adherent activity, which is the ability to inhibit the adherence of pathogenic microorganisms to solid surfaces or infectious sites, has been reported for biosurfactants, leading to a reduction in hospital infections with no need for drugs or synthetic chemical agents [18]. Meylheuc et al. studied a biosurfactant obtained from *P. fluorescens* with inhibitory properties regarding the adherence of Listeria monocytogenes to stainless steel and polytetrafluoroethylene surfaces. *C. sphaerica*-derived lunasan inhibited the adherence of *P. aeruginosa*, *S. agalactiae*, and *S. sanguis* [58]. Pretreatment of silicone rubber with surfactant produced by *S. thermophilus* inhibited the adhesion of *C. albicans* (by 85%) whereas surfactants obtained from *Lactobacillus fermentum* and *L. acidophilus* adsorbed on glass, reduced the number of adhering uropathogenic cells of *Enterococcus faecalis* by 77% [73]. The biosurfactant obtained [74] from *L. fermentum* inhibited *S. aureus* infection and adhered to surgical implants [75]. Surfactin decreased the amount of biofilm formation by *Salmonella typhimurium*, *Salmonella enterica*, *Escherichia coli*, and *Proteus mirabilis* in PVC plates and vinyl urethral catheters [76]. *C. sphaerica* UCP 0995 produces a biosurfactant known as lunasan, when using residual refined soy oil as a carbon source, and it has shown growth inhibition as well as anti-adhesive properties on pathogens such as *S. aureus*, *C. albicans*, and *S. agalactiae* [58].

8.6.5 ANTITUMOR/CYTOTOXICITY ACTIVITY

Some microbial compounds or biosurfactants, considered by cell differentiation, cell immune response, and signal transduction, could control various mammalian cell functions [77]. Sophorolipids from *C. bombicola* have been studied due to their spermicidal and cytotoxic activities as well as anti-HIV action that can reduce the proliferation of acquired immunodeficiency syndrome (AIDS). The biosurfactant produced by *A. indicus* M6 inhibited cell proliferation of lung cancer cells (A549) at G1 phase, and it is nontoxic in nature [78]. It was reported that apoptosis was induced in human breast cancer MCF-7 cells through a reactive oxygen species (ROS)/c-Jun N-terminal kinase (JNK)-mediated mitochondrial/caspase pathway by surfactin [79]. The ROS and Ca^{2+} impact on mitochondria permeability transition pore (MPTP) activity and MCF-7 cell apoptosis induced by surfactin have recently been investigated [80]. The ROS formation, which made the MPTP open and the mitochondrial membrane potentially collapse, was initially induced by surfactin. This caused an increase in the cytoplasmic Ca^{2+} concentration. Cytochrome *c*, which activated caspase-9 and eventually induced apoptosis, was also released from mitochondria to the cytoplasm through the MPTP [81]. Novel cytotoxic activity against cancer cell lines was revealed by lipopeptides, isoforms of surfactins, and fengycins produced by *B. circulans* DMS2, a marine microorganism [82].

8.6.6 GENE DELIVERY

Gene transfection such as lipofection using cationic liposomes is a promising way to deliver foreign genes to the target cells without any side effects [83]. Kitamoto et al. (2002) demonstrated that in comparison with commercially available cationic liposomes, liposomes based on biosurfactants show increasing efficiency of a gene [84]. Gene isolation for protein molecules in this surfactant and cloning in bacteria allow fermentative production for medical applications [85].

8.6.6.1 Agents for Respiratory Failure

The deficiency of lung surfactant, a protein-phospholipid complex, is responsible for respiratory failure in premature infants [86]. Isolation of the genes for protein molecules of this surfactant and cloning in bacteria has made possible its fermentative production for medical applications [85].

8.6.7 IMMUNOLOGICAL ADJUVANT PROPERTY

Bacterial lipopeptides constitute potent nontoxic and non-pyrogenic immunological adjuvants when mixed with conventional antigens. A marked enhancement of humoral immune response was obtained with the low-molecular-mass antigens Iturin AL, herbicolin A, and microcystin (MLR) coupled to poly-L-lysine (MLR-PLL) in rabbits and chickens [87].

8.6.8 ANTI-BIOFILM PROPERTY

Biofilms are complex microbial communities adhered on solid-liquid, liquid-air, liquid-liquid, or solid-air interfaces, embedded in their own microbial-originated matrix of protective and adhesive extracellular polymeric substances (EPSs), which are mainly polysaccharides, lipids, and proteins [88, 89]. In the biomedical field, biosurfactants are getting more attention for their anti-biofilm and antimicrobial activity because of lower toxicity for plants and animals, high biodegradability, low irritancy, and compatibility with human skin [90]. Loss of lipopolysaccharides (LPSs) was observed in *P. aeruginosa* strains treated with rhamnolipids (RLs) and this resulted in increased changes in cell surface hydrophobicity and decrease in the outer membrane proteins of bacteria [91]. Because of their tensoactive properties, many surfactants can cause membrane disruption and cellular lysis by increasing membrane permeability, causing metabolite leakage by altering the physical membrane structure or disrupting proteins and interfering with important membrane functions such as energy generation and transport [92]. The use of biosurfactants as alternatives to anti-biofilm antimicrobial agents has extensively been explored [93]. Surfactin induces nanoripples in lipid bilayers, which may explain the biosurfactant action on biofilm permeability or integrity, probably through the formation of some kind of channels within the biofilm increasing penetrability [94].

The mixture of iturin and fengycin produced by *B. amyloliquefaciens* strain AR2 inhibited *Candida* biofilm formation as well as dislodging preformed biofilm [95]. Complexes of glycolipids from *Brevibacterium casei* MSA19 have been reported to disrupt and significantly inhibit single and mixed biofilms of pathogenic and nonpathogenic bacteria of humans and fish [96]. A glycolipid biosurfactant from *Lysinibacillus fusiformis* S9 restricted the biofilm formation of *E. coli* and *S. mutans* completely. The biosurfactant inhibited bacterial attachment and biofilm formation equally on hydrophilic and hydrophobic surfaces like glass and catheter tubing [97]. *Serratia marcescens* is capable of producing a glycolipid composed of glucose and palmitic acid that prevented the adhesion of *C. albicans*, *P. aeruginosa* and *B. pumilus*, as well as interrupting the formation of biofilms of these cultures in microtitration plates [98]. Biosurfactants produced by another probiotic strain, *L. acidophilus*, was able to reduce the biofilm formation of *S. mutans*, a primary dental cariogen, by inhibiting its attachment [99]. A glycolipid-type biosurfactant produced by the yeast *T. montevideense* CLOA72 diminished biofilm formation by *C. albicans* obtained from the apical tooth canal by up to 85% in polystyrene microplates [93].

8.6.9 ANTIOXIDANT ACTIVITY

Biosurfactants show some potential as antioxidant agents; mannosylerythritol lipids (MELs) has the highest antioxidant and protective effects in cells and suggest potential use as antiaging skin care ingredients. Similar observations were reported for a biosurfactant obtained from *B. subtilis* RW-I showing antioxidant capacity to scavenge free radicals and suggest potential as alternative natural antioxidants [100]. Polysaccharide emulsifier from *Klebsiella* was also shown to have potent inhibition of the autoxidation of soybean oil [101]. The biosurfactant synthesized by *A. junii* B6 through ROS has been shown free radical scavenging property. Performing three different antioxidant tests on a biosurfactant from *C. utilis*, researchers found that the biomolecule had greater complex-reducing capacity than radical-sequestration capacity [102]. Moreover, a linear relation was found between the concentration of biosurfactant and antioxidant activity. Regarding radical sequestering activity, biosurfactants from species of *Lactobacillus* achieve the best results when submitted to the 2,2-diphenyl-1-picrylhydrazyl (DPPH) test [103].

8.6.10   Anti-Inflammatory Activity

LPS-induced nitric oxide production in RAW264.7 cells or primary macrophages by inhibiting NF-κB activation could be downregulated by such bioactive surfactants [104]. The significant anti-inflammatory performances were the work of surfactin isomers produced by the mangrove bacterium *Bacillus* sp. (No. 061341) especially [105]. The overproduction of nitric oxide and the release of IL-6 in the LPS-induced murine macrophage cell RAW264.7 could be effectively inhibited by the group of cyclic lipopeptides. The existence of the free carboxyl group in the structure of surfactin isomer was important to the anti-inflammatory performances. Anti-inflammatory performance in the context of periodontitis caused by *Porphyromonas gingivalis*, the major pathogen of periodontal disease, was recently explored, which was further induced by the mechanisms which are responsible for surfactin [106]. Sophorolipids have also been studied as anti-inflammatory agents for patients with immune diseases [107].

8.6.11   Agents for the Stimulation of Skin Fibroblast Metabolism (Cosmetics Industry)

A broad potential application area is the cosmetic industry, where surface-active substances are found in daily use products [108]. Many biosurfactant properties such as emulsification, demulsification, foaming, water binding capacity, spreading, and wetting properties affect viscosity on product consistency, which the cosmetic industry can efficiently utilize. Cosmetic products using surfactants including bath products, acne pads, antidandruff products, contact lens solution, hair colors and care products, deodorants, nail care, body massage accessories, lipsticks, lip makers, eyeshades, soap, toothpaste and polishes, denture cleansers, adhesives, antiperspirants, baby products, foot care, mousses, antiseptics, shampoos, conditioners, shampoos, conditioners, shave and depilatory products, moisturizers, health and beauty products creams, lotions, liquids, pastes, powders, sticks, gels, films, and sprays could be used and may be replaced by biosurfactants [109, 110]. Monoglyceride, widely used surfactants in the cosmetic industry, have been reported to be produced from glycerol tallow using *P. fluorescens* lipase treatment [111].

The use of sophorolipids in lactone form comprises a major part of diacetyl lactones as agents for stimulating skin dermal fibroblast cell metabolism and mainly as agents for the stimulation of collagen neosynthesis. This can be applied in cosmetology and also in dermatology. The purified lactone sophorolipid product is important in the formulation of dermis antiaging products because of its effect on the stimulation of cells of the dermis. By encouraging the production of new collagen fibers, purified lactone sophorolipids may be used both as a preventive measure against ageing of the skin and used in creams for the body and the body milks, lotions, and gels that are used for the skin.

8.6.12   Their Medicinal Applications

There are also many other biomedical-related properties of biosurfactants that can be applied for effective use. The aggregation of amyloid β-peptide [(AB (1-40)] into fibrils, the main pathological process associated with Alzheimer's disease, was affected by a high surfactin micelle concentration [112]. The surfactin and its synthetic analogs can change the nanoscale organization of supported bilayers and induce nanoripple structures with intriguing perspectives for biotechnological and biomedical applications [113, 114]. Eeman et al. studied the ability of fengycin and found that it interacted with the lipid constituents of the stratum corneum extracellular matrix and with cholesterol [115]. Deleu et al. reported that fengycin could also cause membrane perturbations [50]. It can be concluded that the hemolysis of human erythrocytes was the consequence of trehalose lipid through a colloid-osmotic mechanism, especially the formation of enhanced permeability domains, or "pores", strengthened by biosurfactant in the erythrocyte membrane. The *P. aeruginosa* dirham-nolipid could induce the permeabilization of biological and artificial membranes [116]. The usual disc shape of erythrocytes could be changed with biosurfactant addition into that of spheroechino-cytes, which was shown through scanning electron microscopy. Moreover, surfactin has been found to affect insulin absorption in the lungs of laboratory rats [26].

8.6.13   As Growth Inhibitors

The biosurfactant synthesized by *Acinetobacter* sp. ACMS25 has been reported to inhibit the growth of *Xanthomonas oryzae* pv. *Oryzae* XAV24. This biosurfactant has potential application as a biocontrol agent against the rice blight disease. It also improved the germination and vigor index in rice plants. Surfactants are being used as adjuvant with fungicides, insecticides, and herbicides [117]. Biosurfactants are additionally utilized as a part of agribusiness. They play a critical part in helping biocontrol components of microorganisms such as parasitism, antibiosis, rivalry, instigated absolute protection, and hypo harmfulness [118].

8.6.14   In Food Supplements, Drug Delivery, and Formulation

Biosurfactants have been used in nanoparticle synthesis which is an emerging field as a part of green chemistry [119, 120]. Reddy et al. found that silver nanoparticle synthesis could be stabilized for two months using surfactin, which is a biodegradable, and renewable stabilizing agent with lower toxicity [27, 121]. A biosurfactant produced by *P. Aeruginosa* grown in a low-cost medium has been employed to stabilize silver nanoparticles in the liquid phase [122]. New applications for biosurfac-tants stabilize the nanoparticles by biosurfactants before addition during remediation processes or use in the development of drug delivery systems [21]. Biosurfactants have been used to develop new formulations to replace synthetic additives with plant-based compounds, such as lecithin and gum Arabic [123]. Thus, novel additives with thickening, stabilizing, and emulsifying properties, as well as antioxidant, anti-biofilm, and antimicrobial properties, have been developed [124, 125].

8.6.15   Biosurfactants as Emulsifying and Stabilizing Agents

For successful emulsion, emulsifiers play a fundamental role in facilitating the initial formation of lipid droplets during homogenization and increasing the stability of these droplets. Emulsifiers increase the useful life of products and minimize the separation of the phases due to their capac-ity to control the clustering of globules, favoring the stability of heterogeneous systems [126]. Considering the demand for novel natural products to diminish the use of emulsifiers derived from genetically modified plants, such as soy, which can alter the natural flavour and texture of foods, biosurfactants can replace these compounds and enhance these factors [127, 128]. Lecithin and its derivatives, fatty acid esters containing glycerol, sorbitol, or ethylene glycol, and ethoxylated derivatives of monoglycerides including synthesized oligopeptides are now used as emulsifiers in the food industries [129–131].

8.7   CONCLUSIONS

Due to the unique features of biosurfactants, they are used in several fields, including medicine. Biosurfactants have antibacterial, antifungal, and antiviral properties. Moreover, they have been potentially used as immunomodulatory molecules, adhesive agents, as adjuvants, in vaccines, as inhibitors for fibrin clot formation, gene therapy, and help to reduce hospital-based infections as well. High production cost is an inhibitory factor for their wide use in pharmaceutical and medicine fields. Further research is required to improve production and fully exploit their use in different medicinal areas.

ACKNOWLEDGMENTS

The authors acknowledge the support from MM(DU).

REFERENCES

1. Wermuth, C. G. (2008). *The Practice of Medicinal Chemistry*. Elsevier. https://doi.org/10.1016/B978-0-12-374194-3.X0001-7
2. Silverman, R. B., & Holladay, M. W. (2014). *The Organic Chemistry of Drug Design and Drug Action: Third Edition*. Elsevier. https://doi.org/10.1016/C2009-0-64537-2
3. Silakari, O. (2018). *Key Heterocycle Cores for Designing Multitargeting Molecules*. Elsevier Science. https://doi.org/10.1016/c2016-0-01252-4
4. Costantino, L., & Barlocco, D. (2006). Privileged structures as leads in medicinal chemistry. *Current Medicinal Chemistry*, *13*(1), 65–85.
5. Yet, L. (2018). *Privileged Structures in Drug Discovery*. Wiley. https://doi.org/10.1002/9781118686263
6. Halve, A. K., Bhadauria, D., Bhaskar, B., Dubey, R., & Bhadauria, R. (2007). Design, synthesis and in vitro antibacterial studies of some biologically significant N-3-chloro-4-[2′-hydroxy-5′-(phenylazo)phenyl]azetidin-2-ones. *Journal of the Indian Chemical Society*, *84*(2), 193–196. https://doi.org/10.1002/chin.200742095
7. Halve, A. K., Bhashkar, B., Sharma, V., Bhadauria, R., Kankoriya, A., Soni, A., & Tiwari, K. (2008). Synthesis and in vitro antimicrobial studies of some new 3-[phenyldiazenyl] benzaldehyde N-phenyl thiosemicarbazones. *Journal of Enzyme Inhibition and Medicinal Chemistry*, *23*(1), 77–81. https://doi.org/10.1080/14756360701408614
8. Halve, A. K., Dubey, R., Bhadauria, D., Bhaskar, B., & Bhadauria, R. (2006). Synthesis, antimicrobial screening and structure-activity relationship of some novel 2-hydroxy-5-(nitro-substituted phenylazo) benzylidine anilines. *Indian Journal of Pharmaceutical Sciences*, *68*(4), 510–514. https://doi.org/10.4103/0250-474X.27831
9. Halve, A. K., Bhaskar, B., Sharma, V., Bhadauria, D., & Bhadauria, R. (2007). Facile synthesis and antimicrobial screening of some biorelevant thiosemicarbazone and its analogues. *Journal of the Indian Chemical Society*, *84*(10), 1032–1034.
10. Kaur, R., Singh, R., Kumar, A., Kaur, S., Priyadarshi, N., Singhal, N. K., & Singh, K. (2020). 1,2,3-Triazole β-lactam conjugates as antimicrobial agents. *Heliyon*, *6*(6), e04241. https://doi.org/10.1016/j.heliyon.2020.e04241
11. Tanwer, N., Kaur, R., Rana, D., Singh, R., & Singh, K. (2015). Synthesis and characterization of pyrazoline derivatives. *Journal of Integrated Science and Technology*, *3*(2), 39–41.
12. Singh, K., Kaur, R., & Singh, R. (2020). *Synthesis of 1,2,3-Triazole Carboxamides and Derivatives Thereof*. India: IPO.
13. Kaur, R., Singh, K., & Singh, R. (2016). 1,5-Benzothiazepine: Bioactivity and targets. *Chemical Biology Letters*, *3*(1), 18–31.
14. Kaur, R., Singh, R., Ahlawat, P., Kaushik, P., & Singh, K. (2020). Contemporary advances in therapeutic portfolio of 2-Azetidinones. *Chemical Biology Letters*, *7*(1), 13–26.
15. Parmigiani, S., & Solari, E. (2003). The era of pulmonary surfactant from Laplace to nowadays. *Acta Biomedica de l'Ateneo Parmense*, *74*(2), 69–75+108.
16. Avery, M. E. (2000). Surfactant deficiency in hyaline membrane disease: The story of discovery. *American Journal of Respiratory and Critical Care Medicine*, *161*(4 Pt 1), 1074–1075. https://doi.org/10.1164/ajrccm.161.4.16142
17. Uzoigwe, C., Burgess, J. G., Ennis, C. J., & Rahman, P. K. S. M. (2015). Bioemulsifiers are not biosurfactants and require different screening approaches. *Frontiers in Microbiology*. https://doi.org/10.3389/fmicb.2015.00245
18. Rufino, R. D., Luna, J. M., Sarubbo, L. A., Rodrigues, L. R. M., Teixeira, J. A. C., & Campos-Takaki, G. M. (2011). Antimicrobial and anti-adhesive potential of a biosurfactant Rufisan produced by *Candida lipolytica* UCP 0988. *Colloids and Surfaces B: Biointerfaces*, *84*(1), 1–5. https://doi.org/10.1016/j.colsurfb.2010.10.045
19. Freitas, B. G., Brito, J. G. M., Brasileiro, P. P. F., Rufino, R. D., Luna, J. M., Santos, V. A., & Sarubbo, L. A. (2016). Formulation of a commercial biosurfactant for application as a dispersant of petroleum and by-products spilled in oceans. *Frontiers in Microbiology*, *7*(OCT). https://doi.org/10.3389/fmicb.2016.01646
20. Luna, J. M., Filho, A. S. S., Rufino, R. D., & Sarubbo, L. A. (2016). Production of biosurfactant from *Candida bombicola* URM 3718 for environmental applications. *Chemical Engineering Transactions*, *49*, 583–588. https://doi.org/10.3303/CET1649098

21. Santos, D. K. F., Rufino, R. D., Luna, J. M., Santos, V. A., & Sarubbo, L. A. (2016). Biosurfactants: Multifunctional biomolecules of the 21st century. *International Journal of Molecular Sciences*. https://doi.org/10.3390/ijms17030401

22. Resende, A. H. M., Da Rocha E Silva, N. M. P., Rufin, R. D., De Luna, J. M., & Sarubbo, L. A. (2017). Biosurfactant production by bacteria isolated from seawater for remediation of environments contaminated with oil products. *Chemical Engineering Transactions*, *57*, 1555–1560. https://doi.org/10.3303/CET1757260

23. Bezerra, K. G. O., Rufino, R. D., Luna, J. M., & Sarubbo, L. A. (2018, November 1). Saponins and microbial biosurfactants: Potential raw materials for the formulation of cosmetics. In *Biotechnology Progress*. John Wiley and Sons Inc. https://doi.org/10.1002/btpr.2682

24. Silva, E. J., Correa, P. F., Almeida, D. G., Luna, J. M., Rufino, R. D., & Sarubbo, L. A. (2018). Recovery of contaminated marine environments by biosurfactant-enhanced bioremediation. *Colloids and Surfaces B: Biointerfaces*, *172*, 127–135. https://doi.org/10.1016/j.colsurfb.2018.08.034

25. Da Silva, I. A., Resende, A. H. M., da Rocha e Silva, N. M. P., Brasileiro, P. P. F., de Amorim, J. D. P., de Luna, J. M., … & Sarubbo, L. A. (2018). Application of biosurfactants produced by bacillus cereus and *Candida sphaerica* in the bioremediation of petroleum derivative in soil and water. *Chemical Engineering Transactions*, *64*, 553–558. https://doi.org/10.3303/CET1864093

26. Muthusamy, K., Gopalakrishnan, S., Ravi, T. K., & Sivachidambaram, P. (2008). Biosurfactants: Properties, commercial production and application. *Current Science*, *94*(6), 736–747.

27. Banat, I. M., Franzetti, A., Gandolfi, I., Bestetti, G., Martinotti, M. G., Fracchia, L., … & Marchant, R. (2010, June). Microbial biosurfactants production, applications and future potential. *Applied Microbiology and Biotechnology*. https://doi.org/10.1007/s00253-010-2589-0

28. Rahman, P. K. S. M., & Gakpe, E. (2008). Production, characterisation and applications of biosurfactants – Review. *Biotechnology*, *7*(2), 360–370. https://doi.org/10.3923/biotech.2008.360.370

29. Mnif, I., Ellouz-Chaabouni, S., & Ghribi, D. (2018). Glycolipid biosurfactants, main classes, functional properties and related potential applications in environmental biotechnology. *Journal of Polymers and the Environment*, *26*(5), 2192–2206. https://doi.org/10.1007/s10924-017-1076-4

30. Smyth, T. J. P., Perfumo, A., McClean, S., Marchant, R., & Banat, I. M. (2010). Isolation and analysis of lipopeptides and high molecular weight biosurfactants. In *Handbook of Hydrocarbon and Lipid Microbiology* (pp. 3687–3704). Berlin Heidelberg: Springer. https://doi.org/10.1007/978-3-540-77587-4_290

31. Smyth, T. J. P., Perfumo, A., Marchant, R., & Banat, I. M. (2010). Isolation and analysis of low molecular weight microbial glycolipids. In *Handbook of Hydrocarbon and Lipid Microbiology* (pp. 3705–3723). Berlin Heidelberg: Springer. https://doi.org/10.1007/978-3-540-77587-4_291

32. Calvo, C., Manzanera, M., Silva-Castro, G. A., Uad, I., & González-López, J. (2009, June 1). Application of bioemulsifiers in soil oil bioremediation processes. Future prospects. *Science of the Total Environment*. https://doi.org/10.1016/j.scitotenv.2008.07.008

33. Najmi, Z., Ebrahimipour, G., Franzetti, A., & Banat, I. M. (2018, July 1). In situ downstream strategies for cost-effective bio/surfactant recovery. *Biotechnology and Applied Biochemistry*. Wiley-Blackwell Publishing Ltd. https://doi.org/10.1002/bab.1641

34. Souza, K. S. T., Gudiña, E. J., Azevedo, Z., de Freitas, V., Schwan, R. F., Rodrigues, L. R., … & Teixeira, J. A. (2017). New glycolipid biosurfactants produced by the yeast strain *Wickerhamomyces anomalus* CCMA 0358. *Colloids and Surfaces B: Biointerfaces*, *154*, 373–382. https://doi.org/10.1016/j.colsurfb.2017.03.041

35. Van Hamme, J. D., Singh, A., & Ward, O. P. (2006, November). Physiological aspects. Part 1 in a series of papers devoted to surfactants in microbiology and biotechnology. *Biotechnology Advances*. https://doi.org/10.1016/j.biotechadv.2006.08.001

36. Jahan, R., Bodratti, A. M., Tsianou, M., & Alexandridis, P. (2020, January 1). Biosurfactants, natural alternatives to synthetic surfactants: Physicochemical properties and applications. *Advances in Colloid and Interface Science*. Elsevier B.V. https://doi.org/10.1016/j.cis.2019.102061

37. Campos, J. M., Stamford, T. L. M., & Sarubbo, L. A. (2014). Production of a bioemulsifier with potential application in the food industry. *Applied Biochemistry and Biotechnology*, *172*(6), 3234–3252. https://doi.org/10.1007/s12010-014-0761-1

38. Clements, T., Ndlovu, T., Khan, S., & Khan, W. (2019, January 18). Biosurfactants produced by Serratia species: Classification, biosynthesis, production and application. *Applied Microbiology and Biotechnology*. Springer Verlag. https://doi.org/10.1007/s00253-018-9520-5

39. Bhardwaj, G., Cameotra, S. S., & Chopra, H. K. (2013). Utilization of oleo-chemical industry by-products for biosurfactant production. *AMB Express*. Springer Verlag. https://doi.org/10.1186/2191-0855-3-68

40. Elshafie, A. E., Joshi, S. J., Al-Wahaibi, Y. M., Al-Bemani, A. S., Al-Bahry, S. N., Al-Maqbali, D., & Banat, I. M. (2015). Sophorolipids production by *Candida bombicola* ATCC 22214 and its potential application in microbial enhanced oil recovery. *Frontiers in Microbiology*, 6(NOV). https://doi.org/10.3389/fmicb.2015.01324

41. Campos, J. M., Stamford, T. L. M., Rufino, R. D., Luna, J. M., Stamford, T. C. M., & Sarubbo, L. A. (2015). Formulation of mayonnaise with the addition of a bioemulsifier isolated from *Candida utilis*. *Toxicology Reports*, 2, 1164–1170. https://doi.org/10.1016/j.toxrep.2015.08.009

42. Maier, R. M., & Soberón-Chávez, G. (2000). *Pseudomonas aeruginosa* rhamnolipids: Biosynthesis and potential applications. *Applied Microbiology and Biotechnology*, 54(5), 625–633. https://doi.org/10.1007/s002530000443

43. Deleu, M., & Paquot, M. (2004). From renewable vegetables resources to microorganisms: New trends in surfactants. *Comptes Rendus Chimie*, 7(6–7), 641–646. https://doi.org/10.1016/j.crci.2004.04.002

44. Banat, I. M. (1995). Characterization of biosurfactants and their use in pollution removal – State of the art. (Review). *Acta Biotechnologica*, 15(3), 251–267. https://doi.org/10.1002/abio.370150302

45. Haferburg, D., Hommel, R., Claus, R., & Kleber, H.-P. (2005). Extracellular microbial lipids as biosurfactants. In *Bioproducts* (pp. 53–93). Springer-Verlag. https://doi.org/10.1007/bfb0002453

46. Banat, I. M., Makkar, R. S., & Cameotra, S. S. (2000). Potential commercial applications of microbial surfactants. *Applied Microbiology and Biotechnology*. Springer Verlag. https://doi.org/10.1007/s002530051648

47. Kakugawa, K., Tamai, M., Imamura, K., Miyamoto, K., Miyoshi, S., Morinaga, Y., ... & Miyakawa, T. (2002). Isolation of yeast *Kurtzmanomyces* sp. I-11, novel producer of mannosylerythritol lipid. *Bioscience, Biotechnology and Biochemistry*. https://doi.org/10.1271/bbb.66.188

48. Mukherjee, S., Das, P., & Sen, R. (2006). Towards commercial production of microbial surfactants. *Trends in Biotechnology*. https://doi.org/10.1016/j.tibtech.2006.09.005

49. Biniarz, P., Łukaszewicz, M., & Janek, T. (2017). Screening concepts, characterization and structural analysis of microbial-derived bioactive lipopeptides: A review. *Critical Reviews in Biotechnology*, 37(3), 393–410. https://doi.org/10.3109/07388551.2016.1163324

50. Deleu, M., Paquot, M., & Nylander, T. (2008). Effect of fengycin, a lipopeptide produced by *Bacillus subtilis*, on model biomembranes. *Biophysical Journal*, 94(7), 2667–2679. https://doi.org/10.1529/biophysj.107.114090

51. Bonmatin, J.-M., Laprevote, O., & Peypoux, F. (2012). Diversity among microbial cyclic lipopeptides: Iturins and surfactins. Activity-structure relationships to design new bioactive agents. *Combinatorial Chemistry & High Throughput Screening*, 6(6), 541–556. https://doi.org/10.2174/138620703106298716

52. Liu, Y., Ding, S., Dietrich, R., Märtlbauer, E., & Zhu, K. (2017). A biosurfactant-inspired heptapeptide with improved specificity to kill MRSA. *Angewandte Chemie International Edition*, 56(21), 1486–1490. https://doi.org/10.1002/anie.201703383

53. Fracchia, L., Cavallo, M., Giovanna, M., & M., I. (2012). Biosurfactants and bioemulsifiers biomedical and related applications – Present status and future potentials. In *Biomedical Science, Engineering and Technology*. InTech. https://doi.org/10.5772/23821

54. Lu, J. R., Zhao, X. B., & Yaseen, M. (2007, April). Biomimetic amphiphiles: Biosurfactants. *Current Opinion in Colloid and Interface Science*. https://doi.org/10.1016/j.cocis.2007.05.004

55. Rodrigues, L. R., & Teixeira, J. A. (2010). Biomedical and therapeutic applications of biosurfactants. *Advances in Experimental Medicine and Biology*, 672, 75–87. https://doi.org/10.1007/978-1-4419-5979-9_6

56. Abalos, A., Pinazo, A., Infante, M. R., Casals, M., García, F., & Manresa, A. (2001). Physicochemical and antimicrobial properties of new rhamnolipids produced by *Pseudomonas aeruginosa* AT10 from soybean oil refinery wastes. *Langmuir*, 17(5), 1367–1371. https://doi.org/10.1021/la0011735

57. Joshi-Navare, K., & Prabhune, A. (2013). A biosurfactant-sophorolipid acts in synergy with antibiotics to enhance their efficiency. *BioMed Research International*, 2013, 1–8. https://doi.org/10.1155/2013/512495

58. Luna, J. M., Rufino, R. D., Sarubbo, L. A., Rodrigues, L. R. M., Teixeira, J. A. C., & De Campos-Takaki, G. M. (2011). Evaluation antimicrobial and antiadhesive properties of the biosurfactant Lunasan produced by *Candida sphaerica* UCP 0995. *Current Microbiology*, 62(5), 1527–1534. https://doi.org/10.1007/s00284-011-9889-1

59. Karlapudi, A. P., T. C., V., Srirama, K., Kota, R. K., Mikkili, I., & Kodali, V. P. (2020). Evaluation of anti-cancer, anti-microbial and anti-biofilm potential of biosurfactant extracted from an Acinetobacter M6 strain. *Journal of King Saud University – Science*, 32(1), 223–227. https://doi.org/10.1016/j.jksus.2018.04.007

60. Das, P., Mukherjee, S., & Sen, R. (2009). Antiadhesive action of a marine microbial surfactant. *Colloids and Surfaces B: Biointerfaces*, 71(2), 183–186. https://doi.org/10.1016/j.colsurfb.2009.02.004

61. Fernandes, P. A. V., De Arruda, I. R., Dos Santos, A. F. A. B., De Araújo, A. A., Maior, A. M. S., & Ximenes, E. A. (2007). Antimicrobial activity of surfactants produced by *Bacillus subtilis* R14 against multidrug-resistant bacteria. *Brazilian Journal of Microbiology*, *38*(4), 704–709. https://doi.org/10.1590/S1517-83822007000400022

62. Sandrin, C., Peypoux, F., & Michel, G. (1990). Coproduction of surfactin and iturin A, lipopeptides with surfactant and antifungal properties, by *Bacillus subtilis*. *Biotechnology and Applied Biochemistry*, *12*(4), 370–375. https://doi.org/10.1111/j.1470-8744.1990.tb00109.x

63. Yakimov, M. M., Timmis, K. N., Wray, V., & Fredrickson, H. L. (1995). Characterization of a new lipopeptide surfactant produced by thermotolerant and halotolerant subsurface *Bacillus licheniformis* BAS50. *Applied and Environmental Microbiology*, *61*(5), 1706–1713. https://doi.org/10.1128/AEM.61.5.1706-1713.1995

64. Singh, M., & Desai, J. D. (1989). Hydrocarbon emulsification by *Candida tropicalis* and *Debaryomyces polymorphus*. *Indian Journal of Experimental Biology*, *27*(3), 224–226.

65. Mnif, I., & Ghribi, D. (2016). Glycolipid biosurfactants: Main properties and potential applications in agriculture and food industry. *Journal of the Science of Food and Agriculture*, *96*(13), 4310–4320. https://doi.org/10.1002/jsfa.7759

66. Fariq, A., & Saeed, A. (2016). Production and biomedical applications of probiotic biosurfactants. *Current Microbiology*, *72*(4), 489–495. https://doi.org/10.1007/s00284-015-0978-4

67. Vatsa, P., Sanchez, L., Clement, C., Baillieul, F., & Dorey, S. (2010). Rhamnolipid biosurfactants as new players in animal and plant defense against microbes. *International Journal of Molecular Sciences*, *11*(12), 5095–5108. https://doi.org/10.3390/ijms11125095

68. Naruse, N., Tenmyo, O., Kobaru, S., Kamei, H., Miyaki, T., Konishi, M., & Oki, T. (1990). Pumilacidin, a complex of new antiviral antibiotics. Production, isolation, chemical properties, structure and biological activity. *The Journal of Antibiotics*, *43*(3), 267–280. https://doi.org/10.7164/antibiotics.43.267

69. Huang, X., Lu, Z., Zhao, H., Bie, X., Lü, F. X., & Yang, S. (2006). Antiviral activity of antimicrobial lipopeptide from *Bacillus subtilis* fmbj against pseudorabies virus, porcine parvovirus, Newcastle disease virus and infectious bursal disease virus in vitro. *International Journal of Peptide Research and Therapeutics*, *12*(4), 373–377. https://doi.org/10.1007/s10989-006-9041-4

70. Vollenbroich, D., Özel, M., Vater, J., Kamp, R. M., & Pauli, G. (1997). Mechanism of inactivation of enveloped viruses by the biosurfactant surfactin from *Bacillus subtilis*. *Biologicals*, *25*(3), 289–297. https://doi.org/10.1006/biol.1997.0099

71. Shah, V., Doncel, G. F., Seyoum, T., Eaton, K. M., Zalenskaya, I., Hagver, R., … & Gross, R. (2005). Sophorolipids, microbial glycolipids with anti-human immunodeficiency virus and sperm-immobilizing activities. *Antimicrobial Agents and Chemotherapy*, *49*(10), 4093–4100. https://doi.org/10.1128/AAC.49.10.4093-4100.2005

72. Remichkova, M., Galabova, D., Roeva, I., Karpenko, E., Shulga, A., & Galabov, A. S. (2008). Anti-herpesvirus activities of *Pseudomonas* sp. S-17 rhamnolipid and its complex with alginate. *Zeitschrift für Naturforschung C*, *63*(1–2), 75–81. https://doi.org/10.1515/znc-2008-1-214

73. Busscher, H. J., Van Hoogmoed, C. G., Geertsema-Doornbusch, G. I., Van Der Kuijl-Booij, M., & Van Der Mei, H. C. (1997). *Streptococcus thermophilus* and its biosurfactants inhibit adhesion by *Candida* spp. on silicone rubber. *Applied and Environmental Microbiology*, *63*(10), 3810–3817. https://doi.org/10.1128/aem.63.10.3810-3817.1997

74. Meylheuc, T., van Oss, C. J., & Bellon-Fontaine, M.-N. (2001). Adsorption of biosurfactant on solid surfaces and consequences regarding the bioadhesion of Listeria monocytogenes LO28. *Journal of Applied Microbiology*, *91*(5), 822–832. https://doi.org/10.1046/j.1365-2672.2001.01455.x

75. Bing, S. G., Kim, J., Reid, G., Cadieux, P., & Howard, J. C. (2002). *Lactobacillus fermentum* RC-14 inhibits *Staphylococcus aureus* infection of surgical implants in rats. *Journal of Infectious Diseases*, *185*(9), 1369–1372. https://doi.org/10.1086/340126

76. Mireles, J. R., Toguchi, A., & Harshey, R. M. (2001). *Salmonella enterica* serovar typhimurium swarming mutants with altered biofilm-forming abilities: Surfactin inhibits biofilm formation. *Journal of Bacteriology*, *183*(20), 5848–5854. https://doi.org/10.1128/JB.183.20.5848-5854.2001

77. Osada, H. (1998). Bioprobes for investigating mammalian cell cycle control. *Journal of Antibiotics*, *51*(11), 973–982. https://doi.org/10.7164/antibiotics.51.973

78. Shalini, D., Benson, A., Gomathi, R., John Henry, A., Jerritta, S., & Melvin Joe, M. (2017). Isolation, characterization of glycolipid type biosurfactant from endophytic *Acinetobacter* sp. ACMS25 and evaluation of its biocontrol efficiency against *Xanthomonas oryzae*. *Biocatalysis and Agricultural Biotechnology*, *11*, 252–258. https://doi.org/10.1016/j.bcab.2017.07.013

79. Cao, X. H., Wang, A. H., Wang, C. L., Mao, D. Z., Lu, M. F., Cui, Y. Q., & Jiao, R. Z. (2010). Surfactin induces apoptosis in human breast cancer MCF-7 cells through a ROS/JNK-mediated mitochondrial/caspase pathway. *Chemico-Biological Interactions*, *183*(3), 357–362. https://doi.org/10.1016/j.cbi.2009.11.027

80. Cao, X. H., Zhao, S. S., Liu, D. Y., Wang, Z., Niu, L. L., Hou, L. H., & Wang, C. L. (2011). ROS-Ca(2+) is associated with mitochondria permeability transition pore involved in surfactin-induced MCF-7 cells apoptosis. *Chemico-Biological Interactions*, *190*(1), 16–27. https://doi.org/10.1016/j.cbi.2011.01.010

81. Singh, B. R., Singh, B. N., Khan, W., Singh, H. B., & Naqvi, A. H. (2012). ROS-mediated apoptotic cell death in prostate cancer LNCaP cells induced by biosurfactant stabilized CdS quantum dots. *Biomaterials*, *33*(23), 5753–5767. https://doi.org/10.1016/j.biomaterials.2012.04.045

82. Sivapathasekaran, C., Das, P., Mukherjee, S., Saravanakumar, J., Mandal, M., & Sen, R. (2010). Marine bacterium derived lipopeptides: Characterization and cytotoxic activity against cancer cell lines. *International Journal of Peptide Research and Therapeutics*, *16*(4), 215–222. https://doi.org/10.1007/s10989-010-9212-1

83. Zhang, Y., Li, H., Sun, J., Gao, J., Liu, W., Li, B., … & Chen, J. (2010). DC-Chol/DOPE cationic liposomes: A comparative study of the influence factors on plasmid pDNA and siRNA gene delivery. *International Journal of Pharmaceutics*, *390*(2), 198–207. https://doi.org/10.1016/j.ijpharm.2010.01.035

84. Kitamoto, D., Isoda, H., & Nakahara, T. (2002). Functions and potential applications of glycolipid biosurfactants—From energy-saving materials to gene delivery carriers. *Journal of Bioscience and Bioengineering*, *94*(3), 187–201. https://doi.org/10.1016/s1389-1723(02)80149-9

85. Gautam, K. K., & Tyagi, V. K. (2006). Microbial surfactants: A review. *Journal of Oleo Science*, *55*(4), 155–166. https://doi.org/10.5650/jos.55.155

86. Chakraborty, M., & Kotecha, S. (2013). Pulmonary surfactant in newborn infants and children. *Breathe*, *9*(6), 476–488. https://doi.org/10.1183/20734735.006513

87. Rodrigues, L., van der Mei, H., Banat, I. M., Teixeira, J., & Oliveira, R. (2006). Inhibition of microbial adhesion to silicone rubber treated with biosurfactant from *Streptococcus thermophilus* A. *FEMS Immunology & Medical Microbiology*, *46*(1), 107–112. https://doi.org/10.1111/j.1574-695X.2005.00006.x

88. Flemming, H. C., & Wingender, J. (2010). The biofilm matrix. *Nature Reviews Microbiology*, *8*(9), 623–633. https://doi.org/10.1038/nrmicro2415

89. Branda, S. S., Vik, Å., Friedman, L., & Kolter, R. (2005). Biofilms: The matrix revisited. *Trends in Microbiology*, *13*(1), 20–26. https://doi.org/10.1016/j.tim.2004.11.006

90. Cameotra, S. S., & Makkar, R. S. (2004). Recent applications of biosurfactants as biological and immunological molecules. *Current Opinion in Microbiology*, *7*(3), 262–266. https://doi.org/10.1016/j.mib.2004.04.006

91. Sotirova, A., Spasova, D., Vasileva-Tonkova, E., & Galabova, D. (2009). Effects of rhamnolipid-biosurfactant on cell surface of *Pseudomonas aeruginosa*. *Microbiological Research*, *164*(3), 297–303. https://doi.org/10.1016/j.micres.2007.01.005

92. Thimon, L. (1995). Effect of the lipopeptide antibiotic, iturin A, on morphology and membrane ultrastructure of yeast cells. *FEMS Microbiology Letters*, *128*(2), 101–106. https://doi.org/10.1016/0378-1097(95)00090-R

93. Monteiro, A. S., Miranda, T. T., Lula, I., Denadai, Â. M. L., Sinisterra, R. D., Santoro, M. M., & Santos, V. L. (2011). Inhibition of *Candida albicans* CC biofilms formation in polystyrene plate surfaces by biosurfactant produced by *Trichosporon montevideense* CLOA72. *Colloids and Surfaces B: Biointerfaces*, *84*(2), 467–476. https://doi.org/10.1016/j.colsurfb.2011.02.001

94. Brasseur, R., Braun, N., El Kirat, K., Deleu, M., Mingeot-Leclercq, M.-P., & Dufrêne, Y. F. (2007). The biologically important surfactin lipopeptide induces nanoripples in supported lipid bilayers. *Langmuir*, *23*(19), 9769–9772. https://doi.org/10.1021/la7014868

95. Rautela, R., Singh, A. K., Shukla, A., & Cameotra, S. S. (2014). Lipopeptides from Bacillus strain AR2 inhibits biofilm formation by *Candida albicans*. *Antonie van Leeuwenhoek, International Journal of General and Molecular Microbiology*, *105*(5), 809–821. https://doi.org/10.1007/s10482-014-0135-2

96. Kiran, G. S., Sabarathnam, B., & Selvin, J. (2010). Biofilm disruption potential of a glycolipid biosurfactant from marine *Brevibacterium casei*. *FEMS Immunology & Medical Microbiology*, *59*(3), 432–438. https://doi.org/10.1111/j.1574-695X.2010.00698.x

97. Pradhan, A. K., Pradhan, N., Sukla, L. B., Panda, P. K., & Mishra, B. K. (2014). Inhibition of pathogenic bacterial biofilm by biosurfactant produced by *Lysinibacillus fusiformis* S9. *Bioprocess and Biosystems Engineering*, *37*(2), 139–149. https://doi.org/10.1007/s00449-013-0976-5

98. Dusane, D. H., Pawar, V. S., Nancharaiah, Y. V., Venugopalan, V. P., Kumar, A. R., & Zinjarde, S. S. (2011). Anti-biofilm potential of a glycolipid surfactant produced by a tropical marine strain of *Serratia marcescens*. *Biofouling*, *27*(6), 645–654. https://doi.org/10.1080/08927014.2011.594883

99. Tahmourespour, A., Salehi, R., & Kermanshahi, R. K. (2011). *Lactobacillus acidophilus*-derived biosurfactant effect on GTFB and GTFC expression level in *Streptococcus mutans* biofilm cells. *Brazilian Journal of Microbiology, 42*(1), 330–339. https://doi.org/10.1590/S1517-83822011000100042

100. Yalçin, E., & Çavuşoğlu, K. (2010). Structural analysis and antioxidant activity of a biosurfactant obtained from *Bacillus subtilis* RW-I. *Turkish Journal of Biochemistry, 35*(3), 243–247.

101. Kawaguchi, K., Satomi, K., Yokoyama, M., & Ishida, Y. (1996). Antioxidative properties of an extracellular polysaccharide produced by a bacterium *Klebsiella* sp. isolated from river water. *Nippon Suisan Gakkaishi, 62*(1), 123–128. https://doi.org/10.2331/suisan.62.123

102. Ribeiro, B. G., de Veras, B. O., dos Santos Aguiar, J., Medeiros Campos Guerra, J., & Sarubbo, L. A. (2020). Biosurfactant produced by *Candida utilis* UFPEDA1009 with potential application in cookie formulation. *Electronic Journal of Biotechnology, 46*, 14–21. https://doi.org/10.1016/j.ejbt.2020.05.001

103. Merghni, A., Dallel, I., Noumi, E., Kadmi, Y., Hentati, H., Tobji, S., … & Mastouri, M. (2017). Antioxidant and antiproliferative potential of biosurfactants isolated from *Lactobacillus casei* and their anti-biofilm effect in oral *Staphylococcus aureus* strains. *Microbial Pathogenesis, 104*, 84–89. https://doi.org/10.1016/j.micpath.2017.01.017

104. Byeon, S. E., Lee, Y. G., Kim, B. H., Shen, T., Lee, S. Y., Park, H. J., … & Cho, J. Y. (2008). Surfactin blocks NO production in lipopolysaccharide-activated macrophages by inhibiting NF-κB activation. *Journal of Microbiology and Biotechnology, 18*(12), 1984–1989. https://doi.org/10.4014/jmb.0800.189

105. Tang, J.-S., Zhao, F., Gao, H., Dai, Y., Yao, Z.-H., Hong, K., … & Yao, X.-S. (2010). Characterization and online detection of surfactin isomers based on HPLC-MSn analyses and their inhibitory effects on the overproduction of nitric oxide and the release of TNF-α and IL-6 in LPS-induced macrophages. *Marine Drugs, 8*(10), 2605–2618. https://doi.org/10.3390/md8102605

106. Park, S. Y., Kim, Y. H., Kim, E. K., Ryu, E. Y., & Lee, S. J. (2010). Heme oxygenase-1 signals are involved in preferential inhibition of pro-inflammatory cytokine release by surfactin in cells activated with *Porphyromonas gingivalis* lipopolysaccharide. *Chemico-Biological Interactions, 188*(3), 437–445. https://doi.org/10.1016/j.cbi.2010.09.007

107. Kitamoto, D., Morita, T., Fukuoka, T., Konishi, M. A., & Imura, T. (2009, October). Self-assembling properties of glycolipid biosurfactants and their potential applications. *Current Opinion in Colloid and Interface Science*. https://doi.org/10.1016/j.cocis.2009.05.009

108. Fiechter, A. (1992). Biosurfactants: Moving towards industrial application. *Trends in Biotechnology, 10*(C), 208–217. https://doi.org/10.1016/0167-7799(92)90215-H

109. Ueno, Y., Hirashima, N., Inoh, Y., Furuno, T., & Nakanishi, M. (2007). Characterization of biosurfactant-containing liposomes and their efficiency for gene transfection. *Biological and Pharmaceutical Bulletin, 30*(1), 169–172. https://doi.org/10.1248/bpb.30.169

110. Villeneuve, P. (2007). Lipases in lipophilization reactions. *Biotechnology Advances, 25*(6), 515–536. https://doi.org/10.1016/j.biotechadv.2007.06.001

111. McNeill, G. P., & Yamane, T. (1991). Further improvements in the yield of monoglycerides during enzymatic glycerolysis of fats and oils. *Journal of the American Oil Chemists' Society, 68*(1), 6–10. https://doi.org/10.1007/BF02660299

112. Han, Y., Huang, X., Cao, M., & Wang, Y. (2008). Micellization of surfactin and its effect on the aggregate conformation of amyloid β(1-40). *The Journal of Physical Chemistry B, 112*(47), 15195–15201. https://doi.org/10.1021/jp805966x

113. Bouffioux, O., Berquand, A., Eeman, M., Paquot, M., Dufrêne, Y. F., Brasseur, R., & Deleu, M. (2007). Molecular organization of surfactin-phospholipid monolayers: Effect of phospholipid chain length and polar head. *Biochimica et Biophysica Acta – Biomembranes, 1768*(7), 1758–1768. https://doi.org/10.1016/j.bbamem.2007.04.015

114. Francius, G., Dufour, S., Deleu, M., Paquot, M., Mingeot-Leclercq, M. P., & Dufrêne, Y. F. (2008). Nanoscale membrane activity of surfactins: Influence of geometry, charge and hydrophobicity. *Biochimica et Biophysica Acta - Biomembranes, 1778*(10), 2058–2068. https://doi.org/10.1016/j.bbamem.2008.03.023

115. Eeman, M., Francius, G., Dufrêne, Y. F., Nott, K., Paquot, M., & Deleu, M. (2009). Effect of cholesterol and fatty acids on the molecular interactions of Fengycin with Stratum corneum mimicking lipid monolayers. *Langmuir, 25*(5), 3029–3039. https://doi.org/10.1021/la803439n

116. Sánchez, M., Aranda, F. J., Teruel, J. A., Espuny, M. J., Marqués, A., Manresa, Á., & Ortiz, A. (2010). Permeabilization of biological and artificial membranes by a bacterial dirhamnolipid produced by *Pseudomonas aeruginosa*. *Journal of Colloid and Interface Science, 341*(2), 240–247. https://doi.org/10.1016/j.jcis.2009.09.042

117. Rostás, M., & Blassmann, K. (2009). Insects had it first: Surfactants as a defence against predators. *Proceedings of the Royal Society B: Biological Sciences*, *276*(1657), 633–638. https://doi.org/10.1098/rspb.2008.1281

118. Roy, A. (2018). A review on the biosurfactants: Properties, types and its applications. *Journal of Fundamentals of Renewable Energy and Applications*, *08*(01). https://doi.org/10.4172/2090-4541.1000248

119. Kiran, G. S., Sabu, A., & Selvin, J. (2010). Synthesis of silver nanoparticles by glycolipid biosurfactant produced from marine *Brevibacterium casei* MSA19. *Journal of Biotechnology*, *148*(4), 221–225. https://doi.org/10.1016/j.jbiotec.2010.06.012

120. Sabatini, D. A., Knox, R. C., Harwell, J. H., & Wu, B. (2000). Integrated design of surfactant enhanced DNAPL remediation: Efficient supersolubilization and gradient systems. *Journal of Contaminant Hydrology*, *45*(1–2), 99–121. https://doi.org/10.1016/S0169-7722(00)00121-2

121. Reddy, A. S., Chen, C.-Y., Baker, S. C., Chen, C.-C., Jean, J.-S., Fan, C.-W., … & Wang, J.-C. (2009). Synthesis of silver nanoparticles using surfactin: A biosurfactant as stabilizing agent. *Materials Letters*, *63*(15), 1227–1230. https://doi.org/10.1016/j.matlet.2009.02.028

122. Farias, C. B. B., Silva, A. F., Rufino, R. D., Luna, J. M., Gomes Souza, J. E., & Sarubbo, L. A. (2014). Synthesis of silver nanoparticles using a biosurfactant produced in low-cost mediums as stabilizing agent. *Electronic Journal of Biotechnology*, *17*(3), 122–125. https://doi.org/10.1016/j.ejbt.2014.04.003

123. Hasenhuettl, G. L. (2019). Synthesis and commercial preparation of food emulsifiers. In G. L. Hasenhuettl & R. W. Hartel (Eds.), *Food Emulsifiers and Their Applications* (pp. 11–39). Cham: Springer International Publishing. https://doi.org/10.1007/978-3-030-29187-7_2

124. Nitschke, M., & Costa, S. G. V. A. O. (2007). Biosurfactants in food industry. *Trends in Food Science and Technology*, *18*(5), 252–259. https://doi.org/10.1016/j.tifs.2007.01.002

125. Faustino, M., Veiga, M., Sousa, P., Costa, E., Silva, S., & Pintado, M. (2019). Agro-food byproducts as a new source of natural food additives. *Molecules*, *24*(6), 1056. https://doi.org/10.3390/molecules24061056

126. McClements, D. J., & Gumus, C. E. (2016). Natural emulsifiers—Biosurfactants, phospholipids, biopolymers, and colloidal particles: Molecular and physicochemical basis of functional performance. *Advances in Colloid and Interface Science*, *234*, 3–26. https://doi.org/10.1016/j.cis.2016.03.002

127. Egolf, A., Hartmann, C., & Siegrist, M. (2019). When evolution works against the future: Disgust's contributions to the acceptance of new food technologies. *Risk Analysis*, *39*(7), 1546–1559. https://doi.org/10.1111/risa.13279

128. Nitschke, M., & Silva, S. S. E. (2018). Recent food applications of microbial surfactants. *Critical Reviews in Food Science and Nutrition*, *58*(4), 631–638. https://doi.org/10.1080/10408398.2016.1208635

129. Liu, J., Zou, A., & Mu, B. (2010). Surfactin effect on the physicochemical property of PC liposome. *Colloids and Surfaces A: Physicochemical and Engineering Aspects*, *361*(1–3), 90–95. https://doi.org/10.1016/j.colsurfa.2010.03.021

130. Pornsunthorntawee, O., Wongpanit, P., Chavadej, S., Abe, M., & Rujiravanit, R. (2008). Structural and physicochemical characterization of crude biosurfactant produced by *Pseudomonas aeruginosa* SP4 isolated from petroleum-contaminated soil. *Bioresource Technology*, *99*(6), 1589–1595. https://doi.org/10.1016/j.biortech.2007.04.020

131. Bajpai Tripathy, D., & Mishra, A. (2011). Sustainable biosurfactants. *Encyclopedia of Inorganic and Bioinorganic Chemistry*, 1–17. https://doi.org/10.1002/9781119951438.eibc2433

9 Synthesis and Properties of Nano-Surfactants

Shipra Mital Gupta[1], Priyanka Yadav[1] and S. K. Sharma[2]
[1]University School of Basic and Applied Sciences, Guru Gobind Singh Indraprastha University, New Delhi, India
[2]University School of Chemical Technology, Guru Gobind Singh Indraprastha University, New Delhi, India

CONTENTS

9.1 INTRODUCTION

The word nano-surfactant represents nanoparticle-based surface active agent. It is a combination of nanoparticle and surfactant. Nanoparticle is defined as a nano-sized solid particle of varying shape. The term surfactant can be used as an umbrella term for different types of surface modifiers. It includes compounds which change the miscibility between two incompatible phases and thereby increasing their miscibility like in water-oil emulsion. The range of surfactants encompasses polymers, peptides, DNA, ligands and bioconjugates, etc. that alter the properties of the surface. Nano-surfactant or nanoparticle-based surfactants are the nano-range molecules which show the synergistic effect of both nanoparticle and surfactant. They have a large activated specific surface area that corresponds to nanoparticles and shows amphiphilic behavior that corresponds to surfactants. Sometimes they are also called as amphiphilic nanoparticles because of the amphiphilic behavior of surfactants. Nano-surfactants overcome the various shortcomings associated with traditional surfactants and nanoparticles. Unlike bulk surfactants, they are stable at high temperatures and highly saline conditions [1]. Unlike nanoparticles, nano-surfactants get dispersed in various fluids and the rate of agglomeration of nano-surfactants is slower compared to pristine nanoparticles [2]. Therefore, nano-surfactants open a wide window for many applications like stabilization of two or more phase systems, for enhanced hydrocarbon recovery, lubricants, etc.

9.2 NANO-SURFACTANTS CLASSIFICATION

Nano-surfactants can be broadly classified as natural and synthetic based on the origin of surfactants. A brief description is given in subsequent subsections.

9.2.1 NATURAL NANO-SURFACTANT

Naturally occurring substances, such as graphene, that show amphiphilic behavior come under this category. Pristine graphene at a size below 1 μm shows amphiphilic behavior. Graphene flake attracts water at its edges but repels it on its surface thus acting as a surfactant [3]. Since graphene serves the purpose of both surfactant and nanoparticle, therefore, it eliminates the need of adding surfactant externally to the system. This is advantageous as the addition of surfactant externally may produce foam when being heated and contaminate the system in applications such as heat-transfer [4].

Zeng et al. [5] prepared printing ink by using graphene quantum dots as nano-surfactants. Graphene quantum dots serve the purpose of both ink material and surfactant. Therefore, preparing ink by this method is economical and eco-friendly. Utilizing such substances for industrial purposes like in heat transfer fluids, printing inks, etc. could be beneficial for instrument performance. These serve both as nanoparticles and surfactants. Therefore, there is no need of adding surfactant externally [6]. Graphite oxide is a unique two-dimensional amphiphile that can be adhered to gas-water, liquid-water and solid-water interfaces and lowers the interfacial energy. The amphiphilicity of graphite oxide is tunable by pH value of solution and sheet size etc. Graphite oxide can be readily dispersed in water because of the presence of ionizable –COOH groups on their edges. The basal plane of graphite oxide contains many polyaromatic rings due to which it is dispersible in oil [3]. Materials like molybdenum disulfide [7], aluminosilicate clays, fullerene [8] and certain quantum dots are capable of lowering the interfacial tension and show properties of a surfactant upon designing their surface appropriately [6].

9.2.2 SYNTHETIC NANO-SURFACTANTS

These are man-made nano-surfactants and consist of two parts. First part is the nanoparticle, also known as the base portion. It can be organic or inorganic and their combination. Various organic and inorganic nanoparticles are used for synthesizing nano-surfactants. A summary of nanoparticles used for nano-surfactant synthesis is presented in Table 9.1. Second part is surfactant which is either hydrophilic, hydophobic or zwitterionic in nature [9]. Examples of synthetic nano-surfactants are janus nanoparticles, covalent surface modified nanoparticles and non-covalent surface modified nanoparticles [5, 10].

TABLE 9.1

Nanoparticles Used for Nano-Surfactant Synthesis

Nanoparticles	References
Organic nanoparticle	
Graphene	[16]
CNT	[27,36]
Inorganic nanoparticles	
ZrO_2	[23]
SiO_2	[25,30,31]
TiO_2	[26,38]
TiO_2/Graphene	[32]
Al_2O_3	[38]
Fe_2O_3	[42]
Al_2O_3/TiO_2	[46]
Au	[47]
Ni	[53]
CuO	[55]

9.3 SYNTHESIS OF NANO-SURFACTANTS

Synthesis of nano-surfactants includes surface-functionalized nanoparticles using electrostatic forces or covalent bonds. It provides them the capability of reducing interfacial tension and stabilizing various colloidal systems [5]. The act of modifying the surface of a nanoparticle either by covalent or by non-covalent interactions is known as surface modification or functionalization. The main goal of functionalizing nanoparticles is to cover their surface with a molecule or functional group compatible according to usage. The modified nanoparticles show remarkable physical properties like anti-agglomeration, anticorrosion and noninvasive characteristics [11]. Subsequent sections discuss surface modifications of nanoparticles by using covalent and non-covalent methods of functionalization which are used to synthesize nano-surfactants. A summary of various approaches used for synthesis of nano-surfactants is shown in Figure 9.1. The appropriate designing of nanoparticles makes them capable of lowering interfacial tension and show surfactancy.

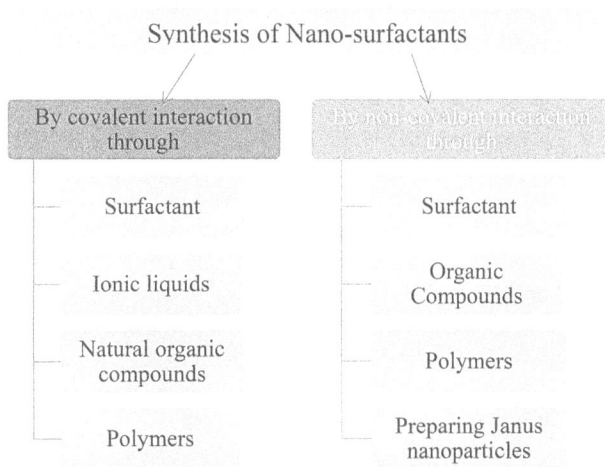

Synthesis of Nano-surfactants

By covalent interaction through
- Surfactant
- Ionic liquids
- Natural organic compounds
- Polymers

By non-covalent interaction through
- Surfactant
- Organic Compounds
- Polymers
- Preparing Janus nanoparticles

FIGURE 9.1 Various approaches used for synthesis of nano-surfactants.

9.3.1 Nano-Surfactants Synthesis Using Non-Covalent Interactions

Non-covalent surface modifications demonstrate weak interactions which include electrostatic attraction and repulsions, hydrogen bonding and π-π, anion-π, cation-π interactions, etc. Usually, physical methods like magnetic stirring, ultra-sonication, homogenization, etc. are used for non-covalent functionalization of nanoparticles [12]. Adsorption of surfactants at nanoparticle surface is strongly influenced by (a) type of nanoparticle (metallic, nonmetallic and their oxides), (b) nature of the structural groups on the surface of nanoparticle, (c) ionic or nonionic nature of surfactant that interact with nanoparticle, (d) hydrophobic groups that are present on surfactant (long or short), (e) tail part of surfactant (straight-chained or branched, aliphatic or aromatic, etc.) and (f) environment of the basefluid (polar or nonpolar).

These factors determine the amount of surfactant needed to induce a given amount of decrease in surface tension efficiency and the maximum surface tension reduction possible for a given surfactant, regardless of concentration (effectiveness) [13]. Mechanism of surfactant adsorption on the surface of a nanoparticle is also determined by these factors. Mechanisms of absorption in the formation of nano-surfactants are discussed next.

9.3.1.1 Absorption through Electrostatic Attraction

Ion pairs are formed at the surface of a nanoparticle which is in contact with a surfactant solution. Ion pairs consist of a positive ion and a negative ion. When the nanoparticle is positively charged, then a negatively charged surfactant molecule forms an ion pair with the nanoparticle and when the nanoparticle is negatively charged, then a positively charged surfactant molecule forms an ion pair with the nanoparticle. The system as a whole remains neutral during the procedure, whereas the average locations of anions and cations are changing respectively to the nanoparticle surface [14, 15].

9.3.1.2 Absorption through Hydrogen Bonding

Hydrogen bonding also contributes to surfactant adsorption on the surface of nanoparticles. It is the result of attractive force between covalently bonded hydrogen atom and electronegative atoms such as nitrogen, oxygen, fluorine, etc. [14]. The electronegative groups and hydrogen atoms that are present on the surface of nanoparticles and surfactant and vice-versa undergo absorption via hydrogen bonding [15].

9.3.1.3 Absorption through Polarization of π Bond

The nanoparticles having π electron density such as graphite, carbon nanotube (CNT), etc. interact with cationic or anionic groups of surfactants through cation-π or anion-π interactions. If both the nanoparticle and surfactant have π electron density then π-π interaction will occur [14, 15]. Graphene nanoparticles can be dispersed in aqueous-organic solvents by adding surfactant to this system. The surfactant forms non-covalent interactions (π-π interactions) with graphene which increases its dispersibility in aqueous-organic solvents [16]. All these mechanisms play a significant role in surfactant adsorption on nanoparticle surfaces. The individual contributions of the above mentioned non-covalent adsorption mechanisms can be ascertained by the use of existing chemical knowledge, measuring zeta potential, substituting functional groups in surfactants, changing surface type, and using different solvents and molecular simulations [14].

Various types of substances such as surfactants, ionic liquids, polymers, etc. can be utilized for non-covalent functionalization of nanoparticles as described in subsequent paragraphs.

9.3.1.3.1 By Using Surfactants

Surfactants are classified as cationic, anionic, nonionic and zwitterionic on the basis of the charge of their head groups. Cationic surfactant have positive charge on their head group e.g. cetyl trimethyl ammonium bromide (CTAB), cetylpyridinium chloride (CPC), benzalkonium chloride (BAC), benzethonium chloride (BZT), dioctadecyldimethylammonium bromide (DODAB), dimethyl-dioctadecylammonium chloride (DDAC), etc. Anionic surfactants have negatively charged head groups e.g. sodium dodecylbenzene sulfonate (SDBS), sodium dodecyl sulfate (SDS), sodium stearate, etc. Nonionic surfactant has no charge on hydrophilic head e.g. tween 80, tween X-100, rokacet O7, rokanol K7, stearyl alcohol, etc. Zwitterionic surfactant has both charges on hydrophilic heads thereby making overall net charge zero e.g. sodium lauroamphoacetate, cocamidopropyl betaine, hydroxysultaine, etc. [17, 18].

The adsorption of ionic surfactants on nanoparticle surfaces is governed by Coulombic attractions. In the case of interaction with cationic surfactants, Coulombic attractions exist between the positive head group of surfactant and the negative surface of nanoparticles. In the case of interaction with anionic surfactants, Coulombic attractions exist between the negative head group of surfactant and the positive surface of the nanoparticle. Strong hydrophobic attraction between the solid nanoparticle surface and the hydrophobic tail of surfactant governs the adsorption of nonionic surfactants [19, 20].

After the adsorption of surfactant molecules on the surface, the surfactant self-assembles into aggregative structures called at a concentration above critical micelle concentration (CMC) [20, 21]. Surfactant adsorption on nanoparticle surfaces alters the surface wettability of nanoparticles [22]. The addition of these surface-modified nanoparticles results in lowering of interfacial tension among different phases.

Non-covalent interactions between surfactant and nanoparticle surfaces include electrostatic attraction and electrostatic repulsions. Electrostatic attractions exist between negative nanoparticle surfaces and positive surfactants and vice versa. Electrostatic repulsions exist between similarly charged nanoparticle surfaces and surfactants.

Esmaeilzadeh et al. [23] modified ZrO_2 (negatively charged) nanoparticles' surface by the use of different types of surfactant SDS (anionic), CTAB (cationic) and Lauryl alcohol 7 (LA7) (nonionic). SDS, CTAB and LA7 interacted with ZrO_2 through electrostatic repulsions, electrostatic attraction and hydrophobic absorption, respectively. The inclusion of nanoparticles showed an increment in surfactant surface activity and reduced the interfacial tension at heptane-water interface.

Suleimanov et al. [24] demonstrated the effect of nanofluids for enhanced oil recovery (EOR). Anionic surfactant sulfanole and nonferrous metal nanoparticles were used for this study. Addition of nonferrous metal nanofluid reduced surface tension by 70–90% compared to aqueous solution of surfactant on an oil boundary. This experimental study showed that the presence of nanoparticles enhances the oil recovery process in comparison to surfactant solution.

Vatanparast et al. [25] studied the modification of silica nanoparticles by two anionic surfactants (SDS and dodecyl benzene sulfonic acid) for stabilization of heptane-water system. The presence of electrostatic repulsion between similarly charged nanoparticles and surfactants resulted in more lowering of interfacial tension compared to surfactant alone.

In some cases, nanoparticles are modified with multiple surfactants simultaneously. Bogunovic et al. [26] utilized oleylamine and pluronic as dispersants for modifying TiO_2 nanoparticle surface. It was found that both surfactants interact with TiO_2 nanoparticles. The surfactant-modified TiO_2 nanoparticle showed good tribiological properties and can be utilized as a lubricant. Madni et al. [27] used mixed surfactant sodium octanoate (SOCT) and dodecyltrimethylammoniumbromide (DTAB) to modify multi-walled carbon nanotube (MWCNT) surface. This resulted in lowering of interfacial tension at nanoparticle/water interface and these modified nanoparticles can be utilized for heat transfer applications.

Non-covalent interaction among surfactant molecules and nanoparticles provides a stable dispersion of nanoparticles without damaging the surface of the nanoparticle. This is an easy and economical way to produce nano-surfactants.

9.3.1.3.2   By Using Ionic Liquids

Ionic liquid consists of a large asymmetric organic cationic group and an inorganic anionic group. Large molecule size and nature of anionic groups are responsible for the presence of diffuse charges on the ions of these salts. In these salts, regular crystalline structure is not formed due to the reduced electrostatic interaction between anion and cation. Therefore, ionic liquids are liquid at room temperature [28–30].

del Río et al. [16] demonstrated the synergistic relationship in an ionic liquid and nanoparticle as an additive in lubricants. Graphene nanoplatelets and hexagonal boron nitride nanoparticles were used for this purpose. Tributyl(ethyl)phosphonium diethyl phosphate was used as ionic liquid and triisotridecyltrimellitate (TTM) was used as an ester-type basefluid. Lubricants were prepared with and without the inclusion of nanoparticles in ionic liquid. It was found that nanoparticles modified with ionic liquids showed better results compared to TTM-ionic liquid-based lubricant.

Seymour et al. [31] modified hairy silica nanoparticles by using phosphonium-organophosphate ionic liquid [P8888][DEHP] and utilized these modified nanoparticles in a polyalphaolefin (PAO)-based lubricant. The phosphate anions of ionic liquid can interact with silica nanoparticles to form linkages. The attachment of ionic liquid on the nanoparticle surface provides amphiphilic characteristics to the nanoparticle and improves its dispersion in PAO. Therefore, improvement in lubricating properties is observed due to the combined effect of hairy silica nanoparticles and phosphonium-phosphate ionic liquid.

Kheireddin et al. [32] used ionic liquid 1-butyl-3-methylimidazolium (trifluoromethysulfony) imide) for surface modification of SiO_2 nanoparticles through electrostatic interactions. These nanoparticles were dispersed in a basefluid estisol. These dispersions were then utilized as lubricants. Silica nanoparticle addition to ionic liquid-based lubricants showed improvement in their lubrication property [32].

More investigations on using nanoparticles with green solvents such as ionic liquids should be encouraged with a view for environmental protection and prolonged lifetime of machinery.

9.3.1.3.3 By Using Natural Organic Compounds

An increase in interest in utilizing naturally occurring organic compounds as surfactants occurs due to their remarkable properties like biodegradability and low toxicity. Due to their diverse nature, these molecules show application potential in various fields such as petroleum, agriculture, medicine, cosmetics, food, etc. Due to their biodegradable nature, higher stability under varied physicochemical environments and low toxicity, these are considered 'green' material.

Tavakoli et al. [33] added nanoparticles such as graphene, TiO_2 and TiO_2/graphene composite to henna-tragacanth extract to produce nano-surfactants useful in EOR. Henna extract and tragacanth extract were utilized as natural surfactant for modifying the surface of nanoparticles. Tragacanth is discharged from Astragalus Asiatic species as dried gum [34]. Gum tragacanth is a carbohydrate mainly consisting of water-soluble tragacanthin and insoluble bassorin which swells in water. It interacts with nanoparticle through electrostatic interactions. Henna is widely known as *Lawsonia inermis* L. gallic acid, lawsone, tannic acid and dextrose, etc. are the main components of henna. The compounds like gallic acid and dextrose are helpful in stabilizing interactions with nanoparticles [35]. Results proved that addition of graphene and TiO_2 nanoparticles as well as nanocomposite of TiO_2/graphene in henna-tragacanth extract reduced the interfacial tension of kerosene oil-water compared to surfactant alone. Emadi et al. [36] used Cedr extract and nano silica for preparation of nano-surfactant. Cedr extract can be economically usable for EOR because it is environmentally safe and low-cost compound. Cedr extract is a polynucleotide surfactant found in nature and biologically degradable. An increase in concentration of surfactant from 0 to 10 wt% at a fixed nanoparticle concentration reduced the interfacial tension of kerosene-aqueous phase from 35 to 11.9 mN/m. Addition of silica nanoparticles to Cedr extract enhances the oil recovery up to 74%.

Natural products like green tea impart amphiphilic character to carbon-based nanoparticles like CNT. Uddin et al. [37] reported green tea showed good dispersion in both aqueous media as well as organic solvents. Water-soluble polyphenols consisting of four primary catechins: epicatechin, epicatechin gallate, epigallocatechin and epigallocatechin gallate are found in green tea. Catechin is composed of phenol groups through which π-π stacking interactions occur with CNT graphitic lattice. Non-covalent interaction between CNT and catechin provides amphiphilic character to CNT. The modified CNT showed good dispersion in polar and nonpolar solvents. Utilizing naturally occurring organic products for preparation of nanoparticle-based surfactant is an economic and eco-friendly way.

9.3.1.3.4 By Using Polymer

Polymer can be introduced both covalently and non-covalently on nanoparticle surfaces known as polymer grafting and polymer wrapping, respectively. During polymer wrapping, polymer envelopes the surface of a nanoparticle through physical absorption. Non-covalent interactions like π-π, CH-π and cation-π are responsible for absorption. This is also known as polymer wrapping or coating [38]. Hendraningra et al. [39] prepared nano-surfactants by using this technique. Hydrophilic oxide nanoparticles Al_2O_3, TiO_2 and SiO_2 were used as nanoparticles and polyvinylpyrrolidone (PVP) was used as a polymer for wrapping nanoparticles. PVP has solubility in both hydrophilic and hydrophobic solvents [40] and is nontoxic in nature and environmentally safe. The wrapping of PVP on nanoparticles modified the surface properties of nanoparticles and these lead to a reduction in interfacial tension. These nanoparticles were then utilized for EOR. Metal oxide combined with PVP EOR to a higher extent than nanoparticle-based systems alone.

Polymer coating imparts properties of polymer to the nanoparticle surface. Lee et al. [41] used oligothiophene-terminated poly(ethylene glycol) (TN-PEG) for coating the surface of CNT. TN-PEG is an amphiphilic molecule. TN-PEG was adsorbed on the nanotube surface by strong π-π interaction. The prepared nano-surfactant can be utilized for various applications like heat transfer and EOR. Yousefvand et al. [42] added nano-silica to anionic hydrolyzed polyacrylamide (PAM) (HPAM) polymer and investigated their effect on EOR. An enhancement of 10% in oil recovery was observed. Khan [43] observed the result of adding nanoparticles to the polymer system for enhancing the oil recovery. TiO_2, SiO_2 and Fe_2O_3 were used as nanoparticles, CTAB as surfactant and PAM was used as polymer. It was observed that polymer chains get absorbed at the nanoparticle surface. The TiO_2-based system reduced the interfacial tension at the oil-water interface more compared to SiO_2 and Fe_2O_3.

Developing polymer-based nano-surfactants ensures the applicability of nanoparticles in various applications as well as enhances the properties like anticorrosion and anti-abrasion, etc. [2].

9.3.2 NANO-SURFACTANTS SYNTHESIS USING COVALENT INTERACTIONS

Covalent chemical modification can be done by adding functional groups to the nanoparticle surface covalent bonding. There are two strategies for the addition of chemical functionality on the nanoparticles surface: direct functionalization and by post-functionalization [44]. Strategies for the addition of chemical functionality on the nanoparticles surface are shown in Figure 9.2 and described next.

9.3.2.1 Direct Functionalization

This method involves functionalization of nanoparticles in a single step. A large number of functional groups can be established on the nanoparticle surface by direct functionalization. Functional groups containing reactive species can attach effectively on the surface of metal and metal oxide nanoparticles. Different functional groups can be utilized for nanoparticle functionalization depending on requirement [11]. Vidal-Vidal et al. [45] prepared magnetic nanoparticles by using ferrous and ferric salts with oleylamine in water-oil microemulsion. Oleylamine-functionalized maghemite nanoparticles produced stable colloidal dispersion in oil-water mixture.

Surface of metal oxide nanoparticles can be modified by coupling with alkoxysilane compounds [46]. Luo et al. [47] modified Al_2O_3/TiO_2 composite with silane derivative 3-glycidoxypropyltrimethoxysilane. Modified Al_2O_3/TiO_2 nanoparticles showed enhanced dispersion stability in lubricating oil.

(A) Direct functionalization

Nanoparticle   Functionalized Nanoparticle

(B) Post functionalization

Nanoparticle

Functionalized Nanoparticle

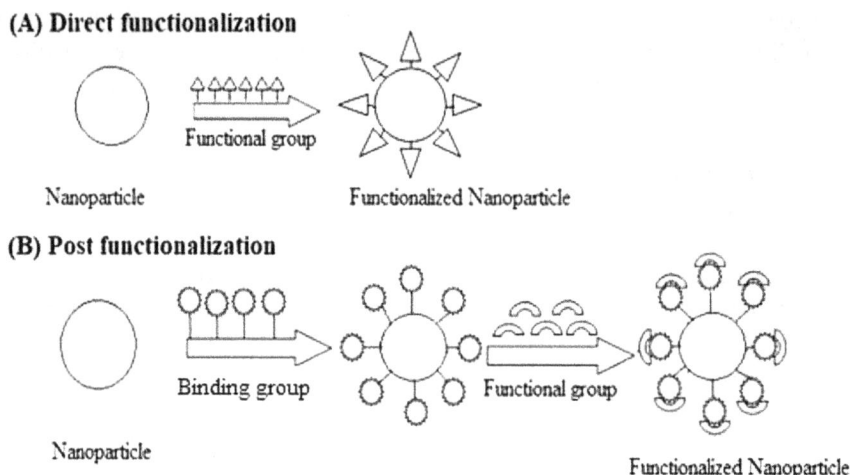

FIGURE 9.2 Strategies for addition of chemical functionality on the nanoparticles surface (a) direct functionalization (b) post-functionalization.

Surface modification of noble metals like Ag, Au, Pt, etc. can be done by organosulfur derivatives [44]. Andryszewski et al. [48] synthesized gold nanoparticles functionalized with aminothiolate which displays both hydrophilic and hydrophobic properties. These functionalized gold nanoparticles acted as nano-surfactants and lowered oil-water interfacial tension. Functionalization of carbon-containing nanoparticles like CNTs, graphite, etc. is generally done for improving their dispersibility in organic and polar solvents. It can be done by using free radical reactions, oxidation in presence of strong acids, fluorination, etc. [49, 50].

9.3.2.2 Post-functionalization

This functionalization occurs in two steps. In the first step, the reactive functional group of a bifunctional compound is introduced on the nanoparticle surface. In the next step, this functional group is used for post-functionalization of the ligand with required property on the nanoparticle surface [51].

Post-functionalization is a more versatile method compared to direct functionalization. The major advantage of this method is that many bifunctional agents are easily available and desirable functional groups can be introduced easily. Functional groups like amine, carboxylic acid, thiol, etc. are very often associated with post-functionalization of the nanoparticle surface. Due to their rich chemistry, they are suitable for further reactions [51]. Cao et al. [52] used 3-aminopropyltriethoxysilane (APTES) to functionalize nanosilica. It was observed that the amine group strengthened the interaction between nanoparticles and polymers (2-acrylamido-2-methyl-1-propane sulfonic acid). Polymer-amine-functionalized nanoparticle systems enhanced the oil recovery factor (16.30%) compared to polymer-nanoparticle systems (10.84%).

Different types of surfactants, biomolecules, metals and polymers are attached to the nanoparticle surface via covalent bonding by using either direct or post-functionalization methods. The following sections summarize covalent functionalization using different materials.

9.3.2.2.1 By Using Surfactant

Surfactant molecules attach onto the surface of nanoparticles via chemical bonds. Interaction between different nanoparticles and surfactants is due to varying reaction mechanisms. Surfactants can be introduced directly on the surface of nanoparticles by addition and mixing of nanoparticles and surfactants to solvent at high temperature. Normally, oppositely charged surfactant and nanoparticle pairs were selected for functionalization of nanoparticles. For example, the −COOH group of surfactants may interact with the −OH group of the oxide nanoparticles via esterification.

Cationic surfactants form ionic bonds with the nanoparticle surface. Treating the metal oxide nanoparticle surface by acid or oxidant improves the binding between nanoparticle surface and surfactant, which leads to an increase in the dispersion of nanoparticles [20, 41, 46, 53].

Ma et al. [53] modified the surface of silica nanoparticles by covalent functionalization. Silica nanoparticles and surfactant CTAB were put in distilled water by stirring and heating. During this process, the oppositely charged surfaces of surfactant and nanoparticle interact via electrostatic attraction. CTAB chains bind around silica nanoparticles. Finally, surfactants are grafted on the nanoparticle surface by covalent bonding with the functional group present on the nanoparticle surface.

Functionalization of nanoparticles with surfactant molecules increases the size of nanoparticles, which increases steric hindrance among them. As a result, particles become monodispersed. Grafting of surfactant molecules on nanoparticles increases their amphiphilic character making them suitable for use as nano-surfactants.

9.3.2.2.2 By Using Organic Compounds

Various organic compounds like oleylamine (OAM), oleic acid (OA), etc. are used for modification of nanoparticle surfaces. Chen et al. [54] prepared nickel-based nano-surfactant by covalent functionalization of nickel nanoparticles with OAM in the presence of OA to obtain lubricant. Nickel nano-surfactant was obtained by adding nickel source to OAM and the resultant solution was heated and stirred. The addition of nickel nano-surfactant to basefluid PAO oil and OA enhances the lubricating properties of PAO-based lubricants.

Li and Zhu [55] prepared surface-modified SiO_2 nano-surfactant. n-hexane and OA were mixed with an appropriate amount of SiO_2 nanoparticles. The mixture was heated under vigorous stirring. OA was covalently functionalized on the surface of SiO_2 nanoparticles to get surface-modified nanoparticles of SiO_2. These acted as additive for lubricants and heavy oils. Li and Chang [56] prepared CuO-modified nanoparticles by two methods. In the first method, CuO nanoparticles were mixed with OA in alkane as a basefluid and the resultant was stirred and ultrasonicated. In the second method, OA was mixed with NaOH followed by the addition of CuO nanoparticles and the mixture was heated and stirred. The resultant oxalic acid functionalized (covalently functionalized)-CuO nanoparticles were then added to alkane as basefluid for application in lubricants. The absorption of OA on the surface of CuO increased its solubility in various long-chain alkanes and lubricants. The functionalization of nanoparticles with different organic functional groups enhances the properties of nanoparticles, rendering them suitable for use in nano-surfactants.

9.3.2.2.3 By Using Polymer

Introduction of polymers to the surface of nanoparticles covalently is known as polymer grafting. Polymer-grafted nanoparticles are also known as hairy nanoparticles (HNP) [57]. The gigantic polymeric molecules provide steric stabilization to the colloidal system of nanoparticles due to its web-like structure [58]. Normally 'grafting to' and 'grafting from' techniques are used. In the 'grafting to' approach, polymers react with functional groups on the surface of the nanoparticle. In the 'grafting from' approach, the nanoparticle surface is first attached to an initiator where polymerization takes place producing polymer chains [57]. Figure 9.3 shows the approaches used for polymer grafting of nanoparticle surfaces.

Niu et al. [59] grafted acid-functionalized MWCNT by using 'grafting to' technique with PEG to obtain MWCNT-g-PEG (g is grafted). Grafting occurred by condensation between carboxylic groups on MWCNT and hydroxyl groups on PEG. The agent used for condensing was N,N'-dicyclohexylcarbodiimide (DCC). MWCNT-g-PEG contains many amphiphilic groups. The presence of amphiphilic groups enables MWCNT to remain colloidal stable in various solvents and to act as a dispersant between different phases. Li et al. [60] prepared silica nanoparticles functionalized by 'grafting from' technique. The prepared nanoparticles were amphiphilic by nature and found to form stable suspensions in $CHCl_3$ and methanol mixture. This type of nano-surfactants show properties of both nanoparticles and polymers and open a wide variety of applications.

(A) 'Grafting To'

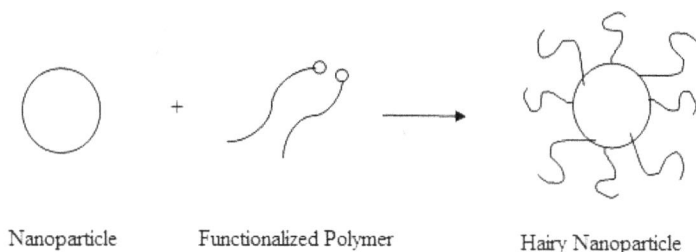

Nanoparticle   Functionalized Polymer   Hairy Nanoparticle

(B) 'Grafting From'

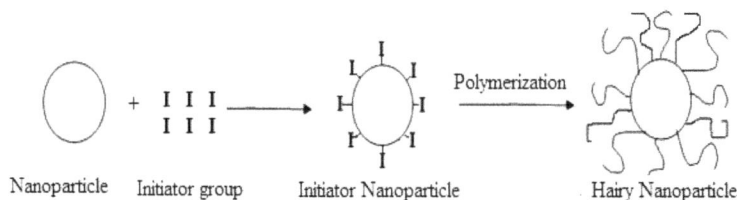

Nanoparticle   Initiator group   Initiator Nanoparticle   Hairy Nanoparticle

FIGURE 9.3 Approaches used for grafting the polymer on nanoparticle surface (a) 'grafting to' (b) 'grafting from'.

9.3.2.2.4 *Preparing Janus Nanoparticles*

Janus nanoparticles were first synthesized by C. Casagrande and M. Veyssie. Janus nanoparticles are named after Roman God Janus who had two faces. In the same way, Janus nanoparticles posse two sides, one hydrophilic and other hydrophobic or one anionic and the other cationic. Thus, they have certain features that are common with surfactants [10, 61–63].

For preparing Janus nanoparticles, the focus of surface modification is to alter the unprotected region and protect certain regions of homogeneous particles. Geometric constraints imposed by neighboring particles, templates, masking, etc. are used for protecting nanoparticle surfaces. A perfect Janus nanoparticle is composed of two perfect hemispheres that showed miscibility in immiscible phases [10]. Luo et al. [47] designed Janus kaolinite nanoplates to accumulate at the oil-water interface. The prepared nanoparticles can be utilized to lower interfacial tension between oil-water phases, which indicates that they can be utilized as nano-surfactants in future. Andryszewski et al. [48] also synthesized Janus nanoparticles of gold. Such nanoparticles displayed amphiphilic (Janus-like) structure when adsorbed at the oil-water interface. These nanoparticles may be used for nano-fluid preparation and interfacial studies. Janus nanoparticle combines the amphiphilicity inherent to surfactants with nanoparticle character. Therefore, these can be used as interfacial stabilizers.

9.4 PROPERTIES OF NANO-SURFACTANT AND FACTORS AFFECTING THEIR FUNCTIONING

Nano-surfactants possess properties of both nanoparticle and surfactant i.e. they have a small size and are amphiphilic in nature. They show synergetic behavior of both nanoparticle and surfactant and find applications in various fields such as in EOR, lubricants, printing inks, etc. For all these applications, colloidal stability of nano-surfactants in various solvents and emulsions is a

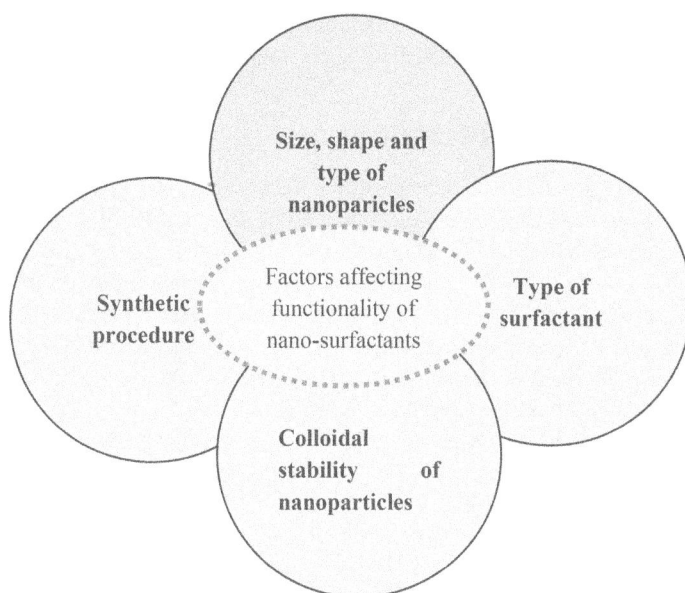

FIGURE 9.4 Factors affecting the functionality of nano-surfacants.

highly desirable characteristic. Factors like type, size and shape of nanoparticle, type of surfactant attached to the surface of nanoparticle, synthetic procedure adopted for nano-surfactant preparation and colloidal stability of nanoparticles (as shown in Figure 9.4), which affects the functioning of nano-surfactants, are discussed here.

9.4.1 Colloidal Stability

Colloidal stability of nano-surfactants is very important for their functionality in various applications like oil recovery, nanofluids, lubricants, etc. Colloidal stability means the nanoparticle would not aggregate and settle out at a considerable rate [64, 65]. If nano-surfactants get agglomerated and settle out, then the functioning of nano-surfactants gets adversely affected. It restricts the commercialization and industrial usage of nano-surfactants. Therefore, long-term stable dispersion of nano-surfactants is a must.

The DLVO theory, which is named after four scientists, Derjaguin, Landau, Verwey and Overbeek, describes the stability of colloids in suspension. This theory focuses on the relation between van der Waals attraction (electrostatic stabilization) and electrostatic repulsion (steric stabilization), which are opposing forces [66, 67]. Nano-surfactant consists of a surfactant part and a nanoparticle part. Surfactant part of a nanoparticle has the capability to prevent the nanoparticles from aggregation, keeping them as dispersed basefluid. Surfactant part on the nanoparticle surface stabilizes the colloidal dispersion through the steric stabilization mechanism. Surfactant molecules act as a barrier among the nanoparticles, which screens the van der Waals attraction among them. In simple terms, surfactant surrounds the nanoparticle surface with its tail, which stretches into the basefluid and remains suspended within the basefluid through the mechanism of steric stabilization [68, 69].

The presence of similar charges on the nanoparticle surface causes electrostatic stabilization of the colloidal system. Surface charge on nanoparticles could be introduced by (a) adsorption of ions, (b) physical adsorption of charged species, (c) substitution of ions and (d) collection or depletion of electrons at surfaces, etc. [17]. Colloidal stability of nano-surfactants also depends upon the type of nanoparticle, its shape and size, type of surfactant/amphiphilic group attached on nanoparticle and the synthetic procedure adopted to prepare nano-surfactant.

9.4.2 SIZE, SHAPE AND TYPE OF NANOPARTICLES

The stability of dispersion of surfactant-modified nanoparticles is associated with their size. Nanoparticles having size less than 10 nm displayed more stability [46]. The shape of the nanoparticles also affects their functioning. Neighboring plated or rod-shaped nanoparticles (like CNT, graphene, etc.) have a large contact area, therefore, stronger tendency to form aggregates compared to spherical nanoparticles. The size of particles increases after aggregation and sedimentation becomes dominant, which affects their dispersive behavior and promotes settling. Type of nanoparticle used for synthesizing nano-surfactants also play a crucial role in stability and functioning of nano-surfactants. Tavakoli et al. [33] prepared nano-surfactants by using TiO_2 and graphene as nanoparticles and heena-tragacanth as surfactant. It was found that TiO_2 nano-surfactant reduced the interfacial tension of the oil-water system for EOR application to a greater extent in comparison to graphene-based nano-surfactant.

9.4.3 TYPE OF SURFACTANT

Colloidal stability of nano-surfactant is also affected by the type of surfactant group or amphiphile attached to a nanoparticle. Esmaeilzadeh et al. [23] prepared ZrO_2 nano-surfactant by using SDS, CTAB and LA7 surfactants. It was found that among all LA7-based nano-surfactants, it has a substantial impact on reducing heptane-water interface.

9.4.4 SYNTHETIC PROCEDURE

Synthetic procedure adopted for nano-surfactant synthesis also affects the functioning of nano-surfactant. Covalent as well as non-covalent functionalization of nanoparticles is utilized to prepare nano-surfactants. Generally, covalent modifications are significantly better than non-covalent with regards to stability of functionalization [38]. Lee et al. [41] compared the interfacial behavior of silica nanoparticles modified by covalent and non-covalent procedure. It was found that the interfacial tension of the oil-water system was decreased to a greater extent when nano-surfactants were prepared by covalent method.

9.5   CHARACTERIZATION

In the process of formation of nano-surfactant from a nanoparticle and surfactant, nanoparticles may lose or gain some properties. The surface of nanoparticles may be distorted while introducing surfactant groups on the surface. Functionalization with acids, oxidizing agents, etc. are carried out so that the surfactant group can be introduced easily to the nanoparticle surface. The alterations in the surface morphology can be studied by using characterization techniques like scanning electron microscopy-energy dispersive X-ray (SEM-EDX), transmission electron microscopy (TEM), etc. The mode of anchoring of the surfactant part on the nanoparticle surface can be covalent or non-covalent. This can be analyzed by Fourier transform-infrared (FTIR), nuclear magnetic resonance (NMR), Raman, etc. The colloidal stability of prepared nano-surfactants can be studied by dynamic light scattering (DLS) techniques and zeta potential values. These characterization techniques are summarized in Figure 9.5.

9.5.1   SEM-EDX

SEM gives a high-resolution image of the material under study when a rastering electron beam is focused onto its surface and the secondary electrons or backscattered electrons are detected. Energy

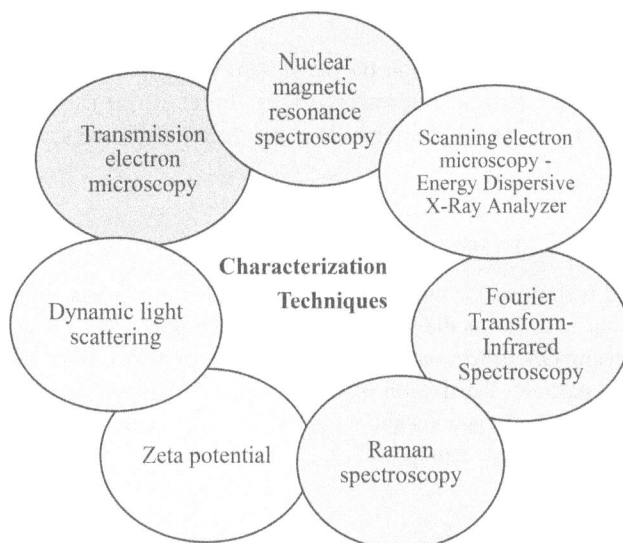

FIGURE 9.5 Techniques used for characterization of nano-surfactants.

Dispersive X-Ray Analyzer (EDX or EDA) helps to identify elemental composition and provides quantitative information [70]. Seymour et al. [31] used SEM-EDX for analyzing the composition and surface morphology of synthesized hairy silica nanoparticles in ionic liquid and used this as a nano-surfactant in PAO-based lubricants.

9.5.2 TRANSMISSION ELECTRON MICROSCOPY

TEM is one of the most widely used techniques for characterizing nanomaterials. TEM provides the images and chemical information of nanomaterials at a spatial resolution of atomic dimensions [28, 71]. Zhang et al. [45] prepared an iron oxide-based nano-surfactant by using OA. The absorbance of OA on the iron oxide nanoparticle surface and size was observed by using TEM analysis.

9.5.3 FOURIER TRANSFORM-INFRARED SPECTROSCOPY

FTIR spectroscopy is a method used to determine functional groups present on the surface of both organic and inorganic nanoparticles [72]. Absorption in the IR region occurs due to the excitation of molecules from lower to higher vibrational level. This technique provides the graphical data of absorption of IR radiation by the material versus wavenumber. The band in infrared absorption determines the structures and molecular components. Li et al. [60] used FTIR spectroscopy for determination of a functional group attached to a silica nano-surfactant.

9.5.4 NUCLEAR MAGNETIC RESONANCE SPECTROSCOPY

NMR spectroscopy determines the molecular structure of the material by measurement of the interaction of nuclear spins under the influence of a magnetic field [73]. Li et al. [60] prepared hairy silica nano-surfactants by introducing poly-(acrylic acid) (PAA) and polystyrene (PS) polymer brushes on silica nanoparticles. NMR showed the attachment and dispersion of mixed PAA-PS particles in chloroform and methanol.

9.5.5 RAMAN SPECTROSCOPY

Raman spectroscopy provides information on the structure, composition and homogeneity of the material and surface groups attached on material [72]. Raj et al. [7] use Raman spectroscopy to determine the polymorphic nature of amphiphilic molybdenum disulfide sheets for their utilization in EOR.

9.5.6 DYNAMIC LIGHT SCATTERING

DLS is the most used technique for measurement of particles in nanometer size range. Brownian motion of dispersed particles forms the basis of the DLS technique [74]. Zhang et al. [75] prepared nano-surfactants by using OA-functionalized iron oxide nanoparticles. DLS was performed for particle size analysis and particle distribution of OA-functionalized iron oxide nanoparticles. It was found that the solution was homogenous and an increment in particle size indicated the absorption of OA on the surface of iron oxide nanoparticles.

9.5.7 ZETA POTENTIAL

The determination of Zeta potential is an important technique used to estimate the charge on the surface of a particle. Zeta potential characterization provides an understanding of the physical stability of nano-surfactants. The magnitude of the Zeta potential can be used to estimate the potential stability of the system. Solutions having a Zeta potential value more positive than +30 mV or more negative than −30 mV are usually taken to be stable [76]. Luo et al. [47] used zeta potential study for finding the dispersive stability of modified Al_2O_3/TiO_2 nanocomposite. Prepared nano-surfactants showed good dispersion stability in lubricants. Cao et al. [52] used zeta potential for analyzing the dispersion stability of polymer-amino-functionalized silica nanoparticles so that they can be used effectively as nano-surfactants in EOR.

9.6 CONCLUSION

There are various shortcomings of using a surfactant such as foaming, reduced efficiency at high temperature and saline conditions, etc. for various industrial applications. The attachment of nanoparticles to surfactants either via non-covalent or covalent functionalization leads to formation of nano-surfactants. Nano-surfactants show qualities of both the surfactant and nanoparticle and overcome the shortcomings of both surfactant and nanoparticle. Unlike surfactants, nano-surfactants show functionality under high temperature and saline conditions. Unlike nanoparticles, nano-surfactants can form stable colloidal dispersions. Therefore, it opens a window for their usage in the field of nanofluids, lubricants, EOR, etc. Various types of nanoparticles (organic, inorganic) and surfactants (ionic, nonionic, etc.) can be utilized according to our need for synthesizing nano-surfactants. The techniques like SEM-EDX, TEM, FTIR, NMR, Zeta, DLS, etc. can be utilized for characterization of nano-surfactants. The factors like colloidal stability of nano-surfactant, nanoparticle used for the preparation of nano-surfactant, type of surfactant utilized for nano-surfactant preparation, etc. should be considered before preparing a nano-surfactant as these factors affect their functionality.

REFERENCES

1. Ahmadi, M. A., & Shadizadeh, S. R. (2017). Nano-surfactant flooding in carbonate reservoirs: A mechanistic study. *European Physical Journal Plus*, 132(6), 1–13.
2. Kango, S., Kalia, S., Celli, A., Njuguna, J., Habibi, Y., & Kumar, R. (2013). Surface modification of inorganic nanoparticles for development of organic-inorganic nanocomposites – A review. *Progress in Polymer Science*, 38(8), 1232–1261.

3. Cote, L. J., Kim, J., Tung, V. C., Luo, J., Kim, F., & Huang, J. (2011). Graphene oxide as surfactant sheets. *Pure and Applied Chemistry*, 83(1), 95–110.
4. Zhang, Z. J., Simionesie, D., & Schaschke, C. (2014). Graphite and hybrid nanomaterials as lubricant additives. *Lubricants*, 2(2), 44–65.
5. Zeng, M., Kuang, W., Khan, I., Huang, D., Du, Y., Saeidi-Javash, M., Zhang, L., Cheng, Z., Hoffman, A. J., & Zhang, Y. (2020). Colloidal nanosurfactants for 3D conformal printing of 2D van der Waals materials. *Advanced Materials*, 32(39), 1–8.
6. Zeng, M., & Zhang, Y. (2019). Colloidal nanoparticle inks for printing functional devices: Emerging trends and future prospects. *Journal of Materials Chemistry A*, 7(41), 23301–23336.
7. Raj, I., Qu, M., Xiao, L., Hou, J., Li, Y., Liang, T., Yang, T., & Zhao, M. (2019). Ultralow concentration of molybdenum disulfide nanosheets for enhanced oil recovery. *Fuel*, 251(December 2018), 514–522.
8. Hirsch, A. (2008). Amphiphilic architectures based on fullerene and calixarene platforms: From buckysomes to shape-persistent micelles. *Pure and Applied Chemistry*, 80(3), 571–587.
9. Suresh. (2017). Methods of using nano-surfactants for enhanced oil recovery. United States Patent. Patent no. US9,708,525B2.
10. Walther, A., & Müller, A. H. E. (2013). Janus particles: Synthesis, self-assembly, physical properties, and applications. *Chemical Reviews*, 113(7), 5194–5261.
11. Thiruppathi, R., Mishra, S., Ganapathy, M., Padmanabhan, P., & Gulyás, B. (2017). Nanoparticle functionalization and its potentials for molecular imaging. *Advanced Science*, 4(3), 1–14.
12. Rubio, J. I. (2012). *Nanofluids: Thermophysical Analysis and Heat Transfer Performance Master of Science* Thesis.
13. Eastoe, J., & Tabor, R. F. (2014). Surfactants and nanoscience. In *Colloidal Foundations of Nanoscience*. Elsevier B.V.
14. Heinz, H., Pramanik, C., Heinz, O., Ding, Y., Mishra, R. K., Marchon, D., Flatt, R. J., Estrela-Lopis, I., Llop, J., Moya, S., & Ziolo, R. F. (2017). Nanoparticle decoration with surfactants: Molecular interactions, assembly, and applications. *Surface Science Reports*, 72(1), 1–58.
15. Rosen, M. J., & College, B. (1975). Relationship of structure to properties in surfactant: III. Adsorption at the solid-liquid interface from aqueous solution. *Journal of the American Oil Chemists Society*, 52(11), 431–435.
16. del Rio, J.M., Lopez, E.R., & Fernandez, J. (2020). Synergy between boron nitride or graphene nanoplatelets and tri(butyl)ethylphosphonium diethylphosphate ionic liquid as lubricant additives of triisotridecyltrimellitate oil. *journal of molecular liquids*, 301, 1–11.
17. Sharma, B., Sharma, S. K., & Gupta, S. M. (2016). Preparation and evaluation of stable nanofluids for heat transfer application: A review. *Experimental Thermal and Fluid Science*, 79, 202–212.
18. Borode, A. O., Ahmed, N. A., & Olubambi, P. A. (2019). Surfactant-aided dispersion of carbon nanomaterials in aqueous solution. *Physics of Fluids*, 31(7), 071301-1-24.
19. Sidik, N.A.C., Mohammed, H.A., Alwai, O.A., Samion, S. (2014). A review on preparation methods ad challenges of nannofluids. *International Communications in Heat and Mass Transfer*, 54, 115–125.
20. Ma, P. C., Siddiqui, N. A., Marom, G., & Kim, J. K. (2010). Dispersion and functionalization of carbon nanotubes for polymer-based nanocomposites: A review. *Composites Part A: Applied Science and Manufacturing*, 41(10), 1345–1367.
21. Vaisman, L., Wagner, H. D., & Marom, G. (2006). The role of surfactants in dispersion of carbon nanotubes. *Advances in Colloid and Interface Science*, 128–130(2006), 37–46
22. Ravera, F., Santini, E., Loglio, G., Ferrari, M., & Liggieri, L. (2006). Effect of nanoparticles on the interfacial properties of liquid/liquid and liquid/air surface layers. *Journal of Physical Chemistry B*, 110(39), 19543–19551.
23. Esmaeilzadeh, P., Hosseinpour, N., Bahramian, A., Fakhroueian, Z., & Arya, S. (2014). Effect of ZrO_2 nanoparticles on the interfacial behavior of surfactant solutions at air-water and *n*-heptane-water interfaces. *Fluid Phase Equilibria*, 361, 289–295.
24. Suleimanov, B. A., Ismailov, F. S., & Veliyev, E. F. (2011). Nanofluid for enhanced oil recovery. *Journal of Petroleum Science and Engineering*, 78(2), 431–437.
25. Vatanparast, H., Shahabi, F., Bahramian, A., Javadi, A., & Miller, R. (2018). The role of electrostatic repulsion on increasing surface activity of anionic surfactants in the presence of hydrophilic silica nanoparticles. *Scientific Reports*, 8(1), 1–11.
26. Bogunovic, L., Zuenkeler, S., Toensing, K., & Anselmetti, D. (2015). An oil-based lubrication system based on nanoparticular TiO_2 with superior friction and wear properties. *Tribology Letters*, 59(2), 1–12.

27. Madni, I., Hwang, C. Y., Park, S. D., Choa, Y. H., & Kim, H. T. (2010). Mixed surfactant system for stable suspension of multiwalled carbon nanotubes. *Colloids and Surfaces A: Physicochemical and Engineering Aspects*, 358(1–3), 101–107.

28. Somers, A. E., Howlett, P. C., MacFarlane, D. R., & Forsyth, M. (2013). A review of ionic liquid lubricants. *Lubricants*, 1(1), 3–21.

29. Ohno, H. (2009). Ionic liquids. *ChemInform*, 34(9), 153–159.

30. Murshed, S. M. S., & Nieto de Castro, C. A. (2016). Conduction and convection heat transfer characteristics of ethylene glycol based nanofluids – A review. *Applied Energy*, 184, 681–695.

31. Seymour, B. T., Fu, W., Wright, R. A. E., Luo, H., Qu, J., Dai, S., & Zhao, B. (2018). Improved lubricating performance by combining oil-soluble hairy silica nanoparticles and an ionic liquid as an additive for a synthetic base oil. *ACS Applied Materials and Interfaces*, 10(17), 15129–15139.

32. Kheireddin, B. A., Lu, W., Chen, I. C., & Akbulut, M. (2013). Inorganic nanoparticle-based ionic liquid lubricants. *Wear*, 303(1–2), 185–190.

33. Tavakoli, P., Shadizadeh, S. R., Hayati, F., & Fattahi, M. (2020). Effects of synthesized nanoparticles and Henna-Tragacanth solutions on oil/water interfacial tension: Nanofluids stability considerations. *Petroleum*, 6(3), 293–303.

34. Anderson, D. M. W., & Grant, D. A. D. (1988). The chemical characterization of some Astragalus gum exudates. *Topics in Catalysis*, 2(5), 417–423.

35. Aghajafari, A. H., Shadizadeh, S. R., Shahbazi, K., & Tabandehjou, H. (2016). Kinetic modeling of cement slurry synthesized with henna extract in oil well acidizing treatments. *Petroleum*, 2(2), 196–207.

36. Emadi, S., Shadizadeh, S. R., Manshad, A. K., Rahimi, A. M., & Mohammadi, A. H. (2017). Effect of nano silica particles on interfacial tension (IFT) and mobility control of natural surfactant (Cedr extraction) solution in enhanced oil recovery process with nano-surfactant flooding. *Journal of Molecular Liquids*, 248, 163–167.

37. Uddin, A. J., Watanabe, A., Gotoh, Y., Saito, T., & Yumura, M. (2012). Green tea-aided dispersion of single-walled carbon nanotubes in non-water media: Application for extraordinary reinforcement of nanocomposite fibers. *Textile Research Journal*, 82(9), 911–919.

38. Fujigaya, T., & Nakashima, N. (2015). Non-covalent polymer wrapping of carbon nanotubes and the role of wrapped polymers as functional dispersants. *Science and Technology of Advanced Materials*, 16(2), 1–21.

39. Hendraningrat, L., & Torsæter, O. (2015). Metal oxide-based nanoparticles: Revealing their potential to enhance oil recovery in different wettability systems. *Applied Nanoscience (Switzerland)*, 5(2), 181–199.

40. Teodorescu, M., & Bercea, M. (2015). Poly(vinylpyrrolidone) – A versatile polymer for biomedical and beyond medical applications. *Polymer – Plastics Technology and Engineering*, 54(9), 923–943.

41. Lee, Y. L., Du, Z. C., Lin, W. X., & Yang, Y. M. (2006). Monolayer behavior of silica particles at air/water interface: A comparison between chemical and physical modifications of surface. *Journal of Colloid and Interface Science*, 296(1), 233–241.

42. Yousefvand, H., & Jafari, A. (2015). Enhanced oil recovery using polymer/nanosilica. *Procedia Materials Science*, 11(2010), 565–570.

43. Khan, M. (2018). Effects of nanoparticles on rheological behavior of polyacrylamide related to enhance oil recovery. *Academic Journal of Polymer Science*, 1(5), 0083–0093.

44. Mout, R., Moyano, D. F., Rana, S., & Rotello, V. M. (2012). Surface functionalization of nanoparticles for nanomedicine. *Chemical Society Reviews*, 41(7), 2539–2544.

45. Vidal-Vidal, J., Rivas, J., & López-Quintela, M. A. (2006). Synthesis of monodisperse maghemite nanoparticles by the microemulsion method. *Colloids and Surfaces A: Physicochemical and Engineering Aspects*, 288(1–3), 44–51.

46. Chen, Y., Renner, P., & Liang, H. (2019). Dispersion of nanoparticles in lubricating oil: A critical review. *Lubricants*, 7(1), 1–21.

47. Luo, T., Wei, X., Zhao, H., Cai, G., & Zheng, X. (2014). Tribology properties of Al_2O_3/TiO_2 nanocomposites as lubricant additives. *Ceramics International*, 40(7 PART A), 10103–10109.

48. Andryszewski, T., Iwan, M., Hołdyński, M., & Fiałkowski, M. (2016). Synthesis of a free-standing monolayer of covalently bonded gold nanoparticles. *Chemistry of Materials*, 28(15), 5304–5313.

49. Yazid, M. N. A. W. M., Sidik, N. A. C., & Yahya, W. J. (2017). Heat and mass transfer characteristics of carbon nanotube nanofluids: A review. *Renewable and Sustainable Energy Reviews*, 80(February), 914–941.

50. Kharisov, B. I., Kharissova, O. V., & Dimas, A. V. (2016). The dispersion, solubilization and stabilization in "solution" of single-walled carbon nanotubes. *RSC Advances*, 6(73), 68760–68787.

51. Mosnáčková, K., Kollár, J., Huang, Y. S., Huang, C. F., & Mosnáček, J. (2018). Synthesis routes of functionalized nanoparticles. In *Polymer Composites with Functionalized Nanoparticles*, Elsevier, 29.

52. Cao, J., Song, T., Zhu, Y., Wang, X., Wang, S., Yu, J., Ba, Y., & Zhang, J. (2018). Aqueous hybrids of amino-functionalized nanosilica and acrylamide-based polymer for enhanced oil recovery. *RSC Advances*, 8(66), 38056–38064.

53. Ma, X.-k., Lee, N. H., Oh, H. J., Kim, J. W., Rhee, C. K., Park, K. S., & Kim, S. J. (2010). Surface modification and characterization of highly dispersed silica nanoparticles by a cationic surfactant. *Colloids and Surfaces A: Physicochemical and Engineering Aspects*, 358(1–3), 172–176.

54. Chen, Y., Zhang, Y., Zhang, S., Yu, L., Zhang, P., & Zhang, Z. (2013). Preparation of nickel-based nanolubricants via a facile in situ one-step route and investigation of their tribological properties. *Tribology Letters*, 51(1), 73–83.

55. Li, Z., & Zhu, Y. (2003). Surface-modification of SiO_2 nanoparticles with oleic acid. *Applied Surface Science*, 211(1–4), 315–320

56. Li, C. C., & Chang, M. H. (2004). Colloidal stability of CuO nanoparticles in alkanes via oleate modifications. *Materials Letters*, 58(30), 3903–3907.

57. Chancellor, A. J., Seymour, B. T., & Zhao, B. (2019). Characterizing polymer-grafted nanoparticles: From basic defining parameters to behavior in solvents and self-assembled structures. *Analytical Chemistry*, 91, 6391–6402.

58. Lee, J. U., Huh, J., Kim, K. H., Park, C., & Jo, W. H. (2007). Aqueous suspension of carbon nanotubes via non-covalent functionalization with oligothiophene-terminated poly(ethylene glycol). *Carbon*, 45(5), 1051–1057

59. Niu, L., Luo, Y., & Li, Z. (2007). A highly selective chemical gas sensor based on functionalization of multi-walled carbon nanotubes with poly(ethylene glycol). *Sensors and Actuators, B: Chemical*, 126(2), 361–367.

60. Li, D., Sheng, X., Zhao, B., & Knox, V. (2005). Environmentally responsive "hairy nanoparticle: mixed homopolymer brushes on silica nanoparticles synthesized by living polymerization technology. *Journal of the American Chemical Society*, 7, 6248–6256.

61. Zhang, J., Grzybowski, B. A., & Granick, S. (2017). Janus particle synthesis, assembly, and application. *Langmuir*, 33(28), 6964–6977.

62. Gennes, P. D. (1992). Soft Matter (Nobel Lecture). 256, 5056, 495-497.

63. Cui, Z. G., Cui, Y. Z., Cui, C. F., Chen, Z., & Binks, B. P. (2010). Aqueous foams stabilized by in situ surface activation of $CaCO_3$ nanoparticles via adsorption of anionic surfactant. *Langmuir*, 26(15), 12567–12574.

64. Devendiran, D. K., & Amirtham, V. A. (2016). A review on preparation, characterization, properties and applications of nanofluids. *Renewable and Sustainable Energy Reviews*, 60, 21–40.

65. Javadian, S., Motaee, A., Sharifi, M., Aghdastinat, H., & Taghavi, F. (2017). Dispersion stability of multi-walled carbon nanotubes in catanionic surfactant mixtures. *Colloids and Surfaces A: Physicochemical and Engineering Aspects*, 531(August), 141–149.

66. Ohshima, H. DLVO theory of colloidal stability. In *Electrical Phenomena at Interfaces and Biointerfaces: Fundamentals and Applications in Nano-, Bio-, and Environmental Sciences*, First Edition, John Wiley and Sons, (27–34)..

67. Harikrishnan, A. R., Dhar, P., Gedupudi, S., & Das, S. K. (2017). *Effect of interaction of nano-particles and surfactants on the spreading dynamics of sessile droplets Langmuir*, 33, 12180-12192.

68. Sharma, B., Sharma, S. K., Gupta, S. M., & Kumar, A. (2018). Modified two-step method to prepare long-term stable CNT nanofluids for heat transfer applications. *Arabian Journal for Science and Engineering*, 43(11), 6155–6163.

69. Hwang, Y., Lee, J., Lee, J., Jeong, Y., Cheong, S., Ahn, Y., & Kim, S. H. (2008). Production and dispersion stability of nanoparticles in nanofluids. *Powder Technology*,186(2), 145–153.

70. Mourdikoudis, S., Pallares, R. M., & Thanh, N. T. K. (2018). Characterization techniques for nanoparticles: Comparison and complementarity upon studying nanoparticle properties. *Nanoscale*, 10(27), 12871–12934.

71. Dumonteil, S., Demortier, A., Detriche, S., Raes, C., Fonseca, A., Rühle, M., & Nagy, J. B. (2006). Dispersion of carbon nanotubes using organic solvents. *Journal of Nanoscience and Nanotechnology*, 6(5), 1315–1318.

72. Mochalin, V. N., Shenderova, O., Ho, D., & Gogotsi, Y. (2012). The properties and applications of nanodiamonds. *Nature Nanotechnology*, 7(1), 11–23

73. Ngyon, J. M., Lyuke, S. E., Neuse, W. E., & Yah, C. S. (2012). Covalent functionalization of multiwalled carbon nanotube – folic acid bound bioconjugate. *Journal of Applied Sciences*, 66, 37–39.

74. Bhattacharjee, S. (2016). DLS and zeta potential – What they are and what they are not? *Journal of Controlled Release*, 235, 337–351.

75. Zhang, L., He, R., & Gu, H. C. (2006). Oleic acid coating on the monodisperse magnetite nanoparticles. *Applied Surface Science*, 253(5), 2611–2617.

76. Kitabara, A., Amano, M., Kawasaki, S., & Kon-no, K. (1977). The concentration effect of surfactants on zeta-potential in non-aqueous dispersions. *Colloid and Polymer Science. Kolloid-Zeitschrift & Zeitschrift Für Polymere*, 255(11), 1118–1121.

10 Novel Nanosurfactants and their Industrial Applications

*Ashutosh Kumar[1], Priyanka Chhabra[2],
Vandana Yadav[3] and Rajni Srinivasan[4]*

[1]Department of Basic Sciences, School of Basic and Applied Sciences, Galgotias University, Greater Noida, Uttar Pradesh, India

[2]Department of Bio-Sciences, School of Basic and Applied Sciences, Galgotias University, Greater Noida, Uttar Pradesh, India

[3]The Bhopal School of Social Science, Bhopal, Madhya Pradesh, India

[4]Department of Chemistry, Geosciences and Environmental Science, College of Science and Technology, Tarleton State University, Stephenville, TX, United States

CONTENTS

DOI: 10.1201/9781003144878-10

10.1 INTRODUCTION

Nanotechnology is focused at atomic scale (0.1–100 nm) and examined and applied in all the scientific fields to explore the desired applications, such as materials engineering, health care, environment and electronics [1]. Burns reported that the global surfactant market has increased continuously by a growth rate of ~ 4%/year [2]. According to 'Acmite market intelligence', it is expected that market value of surfactant is reached up to approx. $40 billion in 2020 [3]. The main objective of surfactants is the preparation of nanoparticles, as a result of their high effect on dispersion. Nowadays nano-emulsions are directly used in consumer products, drugs, personal care, food industry for nutraceutical delivery, packaging, flavoring agents and agriculture field, after being industrialized for 25 years. They are very small droplets (50–100 nm) possessing a highly stable colloidal system, low viscosity and transparency or transparency with enhanced functional properties [4], which causes no deposition on storing, due to the lessening in gravity force and Brownian motion. The small droplet size blocks their unification. Afterwards, these droplets are non-deformable and therefore surface instability is avoided.

10.2 TYPES

On the basis of their origin, nanosurfactants may be broadly classified into following two groups.

10.2.1 NANOSURFACTANT TYPE 1(NANOPARTICLES IN SURFACTANT MOIETY)

Nanosurfactants can be defined as amphiphilic nanoparticles consisting of a hydrophilic region with a hydrophilic functional group bonded to the first portion of a surface of the nanoparticle along with a hydrophobic region of a surface of the nanoparticle. These nanosurfactants have abilities similar to surfactants in order to undergo molecular self-assembly and sensitivity towards surface properties such as interfacial tension (IFT), wettability, foaming and emulsion stability [5,6]. Nanosurfactants are covering the size range of 100–1000 nm [7].

10.2.2 NANOSURFACTANT TYPE 2 (FORMULATIONS THAT HAVE NANOPARTICLES IN SURFACTANT SOLUTIONS)

In some recent researches, the term nanosurfactant has also been used for the formulations that are made up of nanosized particles with natural conventional surfactants [8–11]. Addition of nanoparticles to surfactant formulations, typically stable metal oxides sized between 1 and 100 nm [12–15], has gained increased interest as an additive to EOR-surfactant formulations [16,17] as it displays different and better properties in comparison to the conventional bulk material counterparts. These formulations were found to have great chemical stability particularly in the formulations of suspensions or emulsions.

10.3 SYNTHESIS

10.3.1 Synthesis of Nanosurfactant Type 1

Very little literature is available on the nanosurfactants that have nanoparticles within the surfactants moiety. Dae-Yoon Kim et al. [6] synthesized C_{60} nanosurfactant (C_{60}NS). In this research, the author proposed an amphiphilic and asymmetric nanosurfactant that has both the LC (liquid crystals)-favoring organic functions as well as LC-repelling nanoparticles. LC-favoring group is directly tethered to the LC-repelling group. Modification of this nanosurfactant was performed through attaching alkyl chain and cyanobiphenyl mesogens to the nanobuilding block of C_{60}, resulted in the desired product (C_{60}NS), which were studied for their application in the fabrication of optoelectronic devices.

Lin et al. [18] designed a two-dimensional ultrathin charged zirconium phosphate nanoplatelet as nanosurfactant and studied its applications in order to disperse the reduced graphene oxide (RGO) agglomerates in liquid phase and assemble the exfoliated RGO in an ordered dispersed medium of chiral nematic LC.

10.3.2 Synthesis of Nanosurfactant Type 2

Till date, not much literature is available that describes the synthesis of such type of nanosurfactants using different nanoparticles.

Emadi et al. [19] synthesized nano-silica particle-based nanosurfactant formulations. In this work, they used different wt% of Cedr extraction (1, 2, 3, 4, 5, 6, 7, 8, 10) in 100,000 ppm of brine along with different concentrations of nano-silica (500, 1000, 1500, 2000, 2500 ppm). These surfactants were studied for their various surface-active properties such as CMC and IFT. These surfactants were also evaluated for their application in enhanced oil recovery (EOR).

Mohajeri and Hemmati [20] synthesized Zr nanoparticle-based nanosurfactants. In order to synthesize these surfactants, they first prepared ZrO_2 nanoparticles using sol gel method [20]. In this synthesis, zirconium oxychloride ($ZrOCl_2$) has been used as a source of Zr. Nanosurfactant was synthesized by mixing 2.1 g of citric acid, 1.2 g of succinic acid, 0.5 g of CTAB (cetyltrimethyl ammonium bromide) 10 ml of ethoxylated nonylphenol and 20 ml of ethylene oxide with 3.2 g of $ZrOCl_2$. This surfactant is evaluated on the basis of its surface-active properties like IFT, CMC and wettability. Comparison of their performance properties has also been made with commercially available cationic (CTAB) and anionic (SDS) surfactants in terms of their rheological properties along with their applications specifically in EOR.

Al Harbi et al. [11] prepared nanosurfactants formulations in seawater by following a proprietary method [10] and studied their EOR in carbonates using advanced NMR technique. In this procedure, sodium petroleum sulfonate and a co-surfactant (referred to as STRX) have been used. These nanosurfactants were evaluated for their IFT and EOR.

Thammachart et al., 2017 [21], prepared the nanosurfactant formulation by combining hydrophilic nano-silica and natural surfactant and studied the potential of their aqueous solutions in enhancing oil recovery scenarios in carbonate reservoirs.

10.4 APPLICATIONS OF NANOSURFACTANTS

10.4.1 In Pharmaceutical and Personal Care Field

Nano-emulsions are used in the vast variety of fields like cosmetics, pharmaceutics and gene delivery for topical and transdermal applications. They are used for transportation of active ingredients effectively through the skin and boosts penetration of constituents owing to large surface area and dimension. Nano-emulsions are advantageous as they cause low skin irritation, possess high drug loading capacity and potential for skin hydration. Penetration also improved as a result of less surface tension and IFT of the system. Nano-emulsions are also attracted for the application of hair care products because of their biophysical properties like hydrating. Nano-emulsions also can be used for the distribution of fragrant and alcohol-free perfumes in many personal care merchandises. These advantages of nano-emulsions have pulled attention in recent years in the pharmaceutical field [22]. Some difficulties are also present in their industrialization, for instance [23]:

1. Expensive equipment and high concentrations of emulsifiers is required for the cosmetic and personal care industry.
2. Lack of understanding in the mechanism of production of submicron droplets.
3. Less knowledge of interfacial chemistry for production of it.
4. Specific preparation techniques required for certain applications such as high-pressure homogenizers.

Different drug delivery mechanisms are plasticized for transportation of ingredients like transdermal, mucosal, oral and through injection. For all these systems, nano-emulsions are better than others ordinarily. They have several advantages, for example decent solubilization to insolvable drugs and high bioavailability [23]. Some applications are listed as follows.

10.4.1.1   Antimicrobial Nano-Emulsions
Antimicrobial nano-emulsions are emulsified droplets blended with oil, detergents and water ranging from 100 to 800 nm and show broad-spectrum antimicrobial activity against a variety of bacteria, virus and fungi. When nano-emulsions are applied on the microbial surface, they combine with the lipid bilayer cell membrane of the microbe and release the energy stored in the oil and detergent emulsion, which further destabilize the cell membrane of bacteria. A distinct property of nano-emulsions is discriminatory toxicity to microbes at nonirritating concentrations when applied to skin. Their safety is tested on various species and verified clinically. Nano-emulsion has enough energy to target microbes without harming healthy cells. These properties of nano-emulsion make it an ideal candid for wound management and surface decontamination

10.4.1.2   Inhibition of Murine Influenza A Virus Pneumonitis
Nano-emulsions with nonionic surfactants exhibit a wide range of antimicrobial properties. At an effective concentration, they are biocompatible and nontoxic with the skin and mucous membrane. In comparison to the liposome-based antimicrobial agent, nano-emulsions do not offer a vehicular delivery of antimicrobial agent but themselves act as active agents due to their physiochemical structures. They are chemically stable, heat resistant and can be modified for diverse environmental and clinical applications. Nano-emulsions are primarily nonphospholipids and so do show rapid oxidation of lipids. They are cost-effective and can be commercialized easily. Two nano-emulsions, 8N8 and 20N10 obtained from soybean oil, tributyl phosphate and Triton X-100, were tested to immunize murine influenza virus pneumonia in vivo. Animals pre-treated with 8N8 and 20N10 nano-emulsion showed significant decreased viral infection [23]

10.4.1.3   In controlling Particle Size of Polymeric Nanocapsules
Double emulsion solvent evaporation method is customized for the production of nanocapsules. In this process, dichloromethane acts as an organic solvent, whereas Tween and Span serve as surfactants. Nanocapsules filled with penicillin-G at low concentrations have a tendency to higher burst release. Dispersion of penicillin-G goes up to 60% and burst release falls below 40%, under favorable conditions with dimension of 130 nm of nanocapsules [24].

10.4.2   APPLICATION OF MICELLE IN DECREASING THE SIDE EFFECT OF ANTICANCER DRUG

Micelles and nano-emulsions have definite properties which allow them to participate as a carrier for anticancer drugs. They are the nanostructures possessing liquid cores and their submicron size accelerates the drug delivery of active substances at biological targets. The bioavailability of anti-cancer drugs is usually low in oral administration of drug due to decreased absorption. However, polymeric micelles-based drug delivery showed great advantages with increased bioavailability due to their small size over free drug delivery. Micelle carries anticancer drugs and reduces the consequences of cancer treatment within a short period of time. Drug carriers within the size of 40–70 nm improve the efficiency of encapsulation of drugs with drug delivery inside the tissues. Biomedical application of water-soluble nanosized drugs is affected by a poor encapsulation and fast release. It is reported that novel aerosol (AOT)-alginate nanoparticles boosted cellular delivery of water-soluble molecules [25]. Some basic molecules such as methylene blue, rhodamine, verapamil and doxorubicin are encapsulated efficiently in AOT-alginate nanoparticles [26]. It is observed that anionic drug molecules show poor drug delivery and efficiency in comparison to basic drugs [27]. Drug proclamation is decided by the calcium-sodium conversation among nanoparticle matrices and electrostatic interaction between drug and nanoparticles. Nowadays, some anticancer drugs have shown greater efficiencies loaded with nanoparticles, for example Cisplatin, Paclitaxel and Doxorubicin.

In double emulsion solvent evaporation method, dichloromethane acts as an organic solvent, whereas Tween and Span serve as surfactants. Nanocapsules filled with penicillin-G at low concentrations have a tendency to higher burst release. Dispersion of penicillin-G goes up to 60% and burst release falls below 40%, under favorable conditions with dimension of 130 nm of nanocapsules [26].

10.4.3   ROLE OF SURFACTANTS IN CARBON NANOTUBE TOXICITY

At the time of development, single-walled carbon nanotubes (SWCNTs) consist of several impurities, for example catalyst particles and amorphous carbon. So SWCNTs need to be purified for various biomedical and nanoelectronic device applications [28]. Several commercial surfactants are used for the impurity purification of SWCNTs such as sodium dodecyl sulfate (SDS), sodium cholate (SC) and sodium dodecylbenzene sulfonate (SDBS). SWCNTs immersed in SDS, SC and SDBS are toxic to 1321N1 human astrocytoma cells, biomolecules like DNA, siRNA and proteins due to the toxicity of SDS and SDBS on the nanotube surfaces, which, in turn, causes health complications for the worker, working with CNTs.

Toxicity of nanotube conjugates is due to the presence of specific functional groups and chemicals on the nanotube surfaces. However, there are few reports available on toxicity behavior of nanotube conjugates with the frequent use of surfactants such as SDS, SDBS and SC in industry. It is also difficult to understand the toxicity of these surfactants in laboratories. Dispersion of CNTs is a challenging task for their application at nanoscale. For this purpose, four favorable surfactants are as follows: Triton X-100, Tween 20, Tween 80 and SDS. Triton X-100 and SDS deliver maximum and minimum dispersions, whereas SC is an environmental-friendly reagent for purification

10.4.4   NANO-EMULSIONS IN AGRICULTURE INDUSTRY

Reduced crop yields due to increasing pests and plant diseases draw attention to develop eco-friendly, safe and green pesticide formulations. The agriculture industry is using various agrochemicals which possess a variety of broad-spectrum side effects towards the organism as well as the environment which causes major issues. To overcome this problem, researchers are now developing nanotechnology-based pesticide formulations. Development of nano-emulsions and nanodelivery systems aids in the formation of promising nano-pesticide formulations, further providing great opportunities in the agriculture industry. These remolding like nano-emulsion-based agrochemicals enhance the solubility of active ingredients and increase the bioavailability and thus increase stability and wettability properties. They result in increased efficacy of pest control and treatment.

Nano-emulsions are widely used in agrochemical commerce. Chemicals of superior specificity and lesser amount of persistence are needed to avoid side effects of extreme use of agrochemicals on the ecosystem. The comfort of handling, stability, high concentration of surfactants and less stinking solvents are benevolence of practice of nano-emulsions. Organic herbicides contain nano-emulsions that are more operative than other for plant growth. Small droplet size of nano-emulsion is the cause of higher absorptivity [10]. Nano-emulsions made of fatty-acid methyl esters (FAME) and alkyl polyglucosides (APG) are used as herbicide and glyphosate [28].

10.4.5 NANO-EMULSION IN LUBRICANTS

Nano-emulsions were extensively used in lubricants and cutting oils quite a few years ago. It is also applicable in corrosion inhibitors due to the presence of surfactant in nano-emulsion. Nano-emulsified resins overcome many shortfalls of the outdated water-based structures without affecting the well-being. Advantages of paint consisting of nano-emulsions are scrub resistance, color, stability and stain resistance owing to the small size [28]. Commercial surfactants are fulfilling two key functions in lubricants, one as reaction control additives limiting the growth of the inorganic particles and another as the essential stabilizer of the inorganic solid-hydrocarbon oil interface, for efficient and effective colloidal dispersion in the diluent oil. The purpose of these nano-detergents is mainly acid neutralization. As we know that fuel combustion produces numerous by-products, like acids originated by oxidation of sulfurous and nitrogenous contamination, these acids would cause corrosion during ignition of the engine. Commonly colloidal nanoparticles of calcium carbonate and calcium hydroxide (15–40 mass %) are used as nano-detergent stabilized by a surfactant layer. Since $CaCO_3/Ca(OH)_2$ nanoparticles are unresponsive to temperature, so they act as slow-release acid neutralizers [29,30]. The nano-detergents are effective oxidation inhibitors which provide protection against rust. It provides a clean engine, efficiency and trouble-free operation.

10.4.6 NANOPARTICLE PREPARATION

Today's nanoscience and bioscience rely on nanostructures with complex surface modifications. Nanoparticles of various chemical compositions, such as metals, polymers, chalcogenides and complex multi-shell, multi-ligand nanostructures containing elements from all over the periodic table, play a vital role in multidisciplinary areas like biotechnology, health care, energy sciences, catalysis, advanced materials, natural biosystems and ecosystems. Many traditional techniques for the grafting of ligands and their functionalization and initiation of cell targeting are available. Ionic, hydrophilic and hydrophobic and aromatic methods have been extensively used over a period of time to adjust the surface polarity of nanoparticles. However, the stability of controlled assembly of functionalized nanoparticles is achieved by the profusion of nanoparticle and surfactant interfaces. Surfactants play an important role in the interfacial process to stabilize the nanoparticle assemblies.

Platinum palladium and rhodium metal are the foremost colloidal solutions made [31,32] in the 1980s by the use of surfactants, micelles and reversed micelle microemulsions. From this groundbreaking work, a huge variety of nanoparticles are synthesized with these approaches in different liquid media. Noble metal nanoparticles have fascinated significant curiosity due to possible applications in the field of electronics, catalysis and medical applications for their bactericidal properties. Reverse micelles are commonly used for preparation of nanoparticles as the dimensions of the polar domains can be regulated over a range of 1–10 nm. Another approach for generating metallic nanoparticles involves preparation of a microemulsion which control exchange mechanism of materials and allowing nano-reactions to take place. There are many parameters that can influence the size of the nanoparticles, such as (i) type of solvent, (ii) surfactant, (iii) added electrolyte, (iv) composition/water content and (v) reagent concentration.

10.4.7   As a Charge Control Agent in Nonaqueous Media

The stabilization of ions in nonpolar solvents is more problematic in polar liquids like water. Nevertheless, charged species in nonpolar solvents illustrate a significant role in various applications [33], for example electronic inks and electrophoretic displays. The micron-sized colloidal particles are weakly charged. For stability, these charged species want to be isolated by large distances (approx. 40 times that in water). Therefore, charge stabilization in electronic inks is a significant challenge. Ionic surfactants find useful applications as inverse micelles can provide charge control additives. For example, aerosol OT, polyisobutylene succinimide, sorbitan oleate and zirconyl 2-ethylhexanoate are common surfactants.

10.4.8   Application for Cosmetics/Foods

Nanotechnology has provided many possible carrier systems to the field of cosmetics and personal care products, which is the world's fastest growing industry for the last two decades. Nano-emulsions possess massive potential and significant popularity among various cosmetic formulations due to their enhanced and controlled skin permeability, high stability and small size. Nano-emulsions properties are thoroughly familiarized with the use of cosmetics (moisturizers, lotions, creams etc.). Transparency and liquidity are significant properties respected by customers. Nano-emulsions permit a fast absorption of active ingredients through the skin due to decent hydrating power and their large surface area, as well as they can be sterilized by filtration [34,35]. Simonnet et al. granted a patent on nano-emulsions, based on alkyl ether citrates or fatty esters of sugars in cosmetic applications [36]. Likewise, nano-emulsions are also applied in the delivery of fragrant in care products and avoid use of alcohol. Nano-emulsions are also used in shampoos. Colloids, surfactants, proteins and block copolymers are extensively used in industrial application. They are prime candidates in the cosmetics and foods industry [20]. Biodegradability of wrapping material for food items can be boosted by application of surfactants which are useful in the alteration of layered silicate. Nano-emulsions also contribute significantly to the production of various topical therapies for the smooth progression of dermatologic disease healing. Nano-emulsions are a part of a soft-matter system which influences the skin dynamic in a positive manner due to its physico-chemical structure, which overcomes the limitations of traditional systems by representing its three-phase composition of oil surfactant and water. Many approaches have been made to augment the skin bioavailability of anti-acne substances. Nano-emulsion formulations have proved an ideal candidate for the anti-acne drug targeting.

10.4.9   Biotechnology Applications

The culture and expansion of circulating cells for *in vivo* and *ex vivo* studies play a vital role in precision medicine but it still represents a great challenge in translational research. So, researchers continue to find new approaches to enhance the cell growth and culture, which further assist in understanding the disease and cell dissemination. Lipids and fatty acids play an important role to facilitate cell growth and also esteem the cell culture and expansion. So, the use of nanostructure based on lipids and fatty acids supports the cell culture by augmenting the proliferation capacity. Nanosystems with oil-in-water nano-emulsion have the ability to mediate enhanced interface with cells and advance the intracellular delivery of encapsulated drugs. Nano-emulsion made from bioactive lipids and fatty acids by adjusting the composition of oil in water mimics the lipid droplets. They enhance the intracellular delivery and cell proliferation and have drawn noteworthy attention in cell culture technology. Nano-emulsions consist of a good combination of natural triacylglycerols: phospholipids, which are the prime factor of membranes [37]. Nano-emulsion can also be applied as antimicrobial agents in food products in oils. Nevertheless, there are some technological limitations such as low solubility in aqueous medium, reactivity and volatility. Essential oil with nano-emulsion shows stability, protection from active ingredients and antimicrobial activity. This bounded the effect on the organoleptic foods [38].

10.4.10 TEMPLATE TECHNOLOGY

Template technology has controlled the morphology and structure of nanomaterials during the synthesis [39]. Consequently, materials with explicit optical, magnetic, elastic or chemical characters can be easily manufactured. For drug delivery, versatile nanoparticle synthesis based on template technology methods is used. Nano-emulsion templates have readily adapted administration routes and target drugs by choosing the opposite construction method and surfactant [40]. This method is as well applicable for the synthesis of mesoporous materials used in electronic devices as catalysts. As an example, gel of N-lauroyl-L-glutamic acid, di-n-butylamine, is used as a template for the synthesis of small-pore mesoporous alumina [41].

10.4.11 WASTE WATER TREATMENT

Surfactants are frequently practicing in domestic and industrial purpose, such as detergent, shampoo, dishwashing liquids, cosmetics, pesticides, textiles, fibers, food, paints, polymers, pharmaceuticals, microelectronic, mining, oil recovery and pulp-paper industries. Due to vast applications, they may easily enter waste streams and aquatic environments. This may be harmful to rivers and other water resources [42]. Nowadays, one of the key challenges faced by waste water treatment is surfactant contamination. Consequently, the toxicity and environmental persistence of these surfactants are growing concerns. Currently, available treatment methods in the waste water industry have faced varieties of struggle due to a range of surfactants with different characteristics. The classification of surfactants is based on the electrical charges on the hydrophilic part of the surfactant molecule, i.e. anionic, cationic and nonionic. Surfactants are also zwitterionic with both negative and positive charges in the hydrophilic head [43]. Some of the extensively used surfactants are listed in **Table 10.1** [44].

Table 10.1 shows that the ideal surfactant treatment method is required for such a range of concentrations. This also highlights the issue of lack of treatment at a time when it deals with such a diverse class of surfactant during waste water treatment. From various treatment method, some of them are listed and discussed in **Figure 10.1**

TABLE 10.1
Occurrence of Surfactants and its Composition in Waste Water

Waste Water Type	Types of Surfactant	Compositions	Concentration of Surfactants (μg/L)
Industrial	Anionic	LAS-C10, C11, C12, C13	261, 456, 520, 499
	Cationic	BAC-C12, C14	0.48, 0.16
In-house	Anionic	LAS-C10, C11, C12, C13	203, 384, 386, 306
	Cationic	BAC-C12, C14	0.65, 0.22
Multiple	Anionic	LAS	4230
	Cationic	QACs	216
	Nonionic	Nonylphenols	4.54
		Ethoxylates	16.30
		Octylphenol	0.360
Medical	Only cationicz	BAC-C12, C14, DDAC-C10	2800, 1100, 210
Paper production and dairy	Cationic	BACs	2.58
		DDACs	0.33

Biological Treatment (For Domestic Wastewater)

Chemical Treatments

Combined Chemical and Biological Treatments

Some novel methods for wastewater treatment for surfactants ———▶ Carbon Nanotube (CNT) Treatment

Magnetic Removal Techniques

Micellar Enhanced Ultra-Filtration (MEUF)

By Using Zeolites

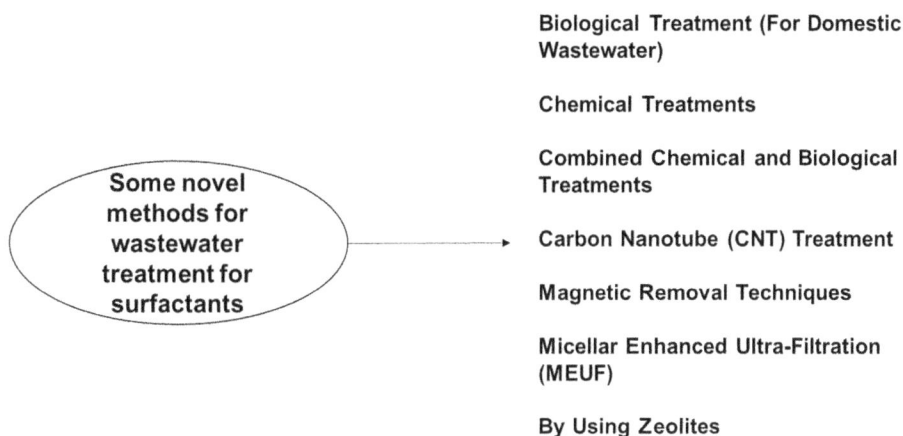

FIGURE 10.1   Novel methods of waste water treatment [47].

10.4.12   BIOLOGICAL TREATMENT (FOR DOMESTIC WASTE WATER)

Commonly, this method is useful for sewage waste water treatment. In this process, the stream is returned to the water source, whereas the remaining is used as a fertilizer. Biodegradation of surfactant happens in which microorganism uses surfactant as a nutrient or energy source [55]. There are various species of bacteria gifted in surfactant degradation, as an example strain *Citrobacter braakii* [56]. It is noted that activated tanks with trickling filters can show efficiency of 89.1–99.1% for LAS composition [57], whereas, alkylphenol ethoxylates have removal efficiency of 90% [58]. It is reported that these efficiencies can be reduced due to the occurrence of physiochemical processes [59]. Anaerobic degradation has a substantial role to remove LAS surfactant by using these mechanisms such as sulfur-limited conditions [60] and thermophilic upflow [40], using an acidogenic reactor injected with *Pseudomonas aeruginosa* [61], etc. Efficiency of LAS surfactant removal is between 40 and 85% by using anaerobic conditions [62]. Important limitations of the biodegradation method are the concentration level of surfactant, useful for some surfactants in waste water, and cost. Above 10 lakh µg/L, surfactant depolarizes the bacteria cell wall and destroys the function [63]. The concentration of surfactant at which this happens also depends on the composition of surfactants. There are also reports available for low and sufficient biodegradability of surfactants depending on feasibility issue [64,65].

10.4.13   CHEMICAL TREATMENTS

Coagulation-flocculation surfactant treatment method is used for surfactant removal, which is applied for erstwhile sedimentation. In the coagulation process, bulky mass forms to be settled or trapped during filtration. However, flocculation gentle agitation is done to boost the particles to form a large cluster which is filtered from the solution. A typical diagram of coagulation-flocculation is shown in **Figure 10.2** [66].

Ferrous chlorides showed a high level of efficiency for anionic surfactant removal [67]. Activated carbon or polyelectrolyte-based removal can also filter anionic surfactants more effectively [68]. Adsorption is a popular physico-chemical method for surfactant removal. Activated carbon showed a high affinity for surfactants at high concentrations as compared to anaerobic degradation or zeolite [69]. Polymeric resins [70], zeolite [71], silica [72] and alumina [73] are also used as adsorbents. Nevertheless, these traditional adsorbents cause environmental impact at a large scale [74]. Few reports are available for cationic surfactant removal as well as for nonionic and zwitterionic. More potential methods are required to deal with surfactants.

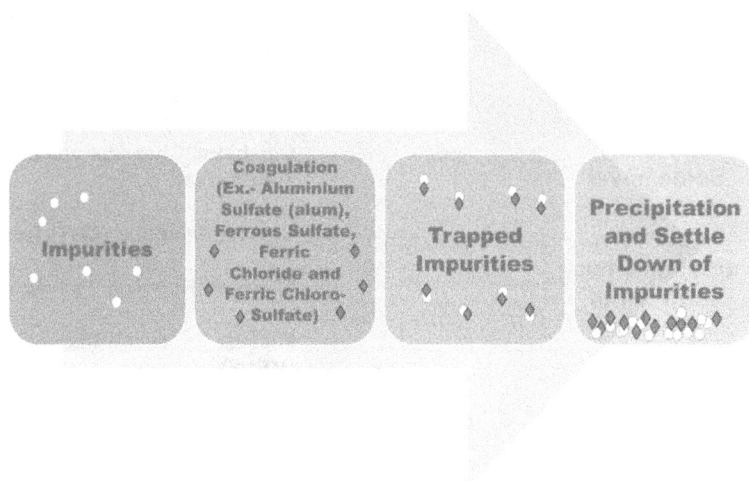

FIGURE 10.2 Typical diagram of Coagulation-Flocculation process.

10.4.14 COMBINED CHEMICAL AND BIOLOGICAL TREATMENTS

High concentrations of surfactants are used in the textile industry. This causes significant problems in waste water treatment. Due to the lack of efficiency at higher surfactant concentration in biological processes, it is not useful for industrial purposes. In the combined method, waste water is formerly treated chemically to convert them easily into biodegradable products but cannot be fully mineralized [75]. Advanced oxidation process (AOP) is the most common method used in this category. In this process, highly energetic oxidants generate OH$^-$ radical species, which degrade organic compounds, and as a result, ppm level of contamination reduces drastically [76]. The AOP process consists of a variable range of reactive species for different purposes; some of them are listed in **Table 10.2** [77].

10.4.15 CARBON NANOTUBE (CNT) TREATMENT

This method is based on the adsorption process by nanomaterial, in place of conventional AC, zeolite etc. [78]. It is applicable to both anionic and cationic surfactants. This method has drawn substantial attention due to large adsorption surface area and covers a wide range of contamination in the waste water industry [79,80]. CNT technology faces some challenges in waste water treatment such as efficiency inconsistency, aggregation [81], desorption [82] and environmental impact [83].

TABLE 10.2
Reactive Species in AOP Process for Different Purpose

AOP	Reactive Species	Area of Use
Fenton process: H_2O_2/Fe^{2+}	$^{\cdot}OH$, $HO_2^{\cdot}$	Mostly for acidic solutions, high and low concentrations for both LAS and ABS
Photo-Fenton process	$^{\cdot}OH$	Anionic surfactants
Ozone treatment: O_3	$^{\cdot}OH$, $HO_2^{\cdot}$, $HO_3^{\cdot}$, $O_2^{-\cdot}$, $O_3^{-\cdot}$	Laundry waste water [28]
UV/O_3, UV/H_2O_2 and $UV/O_3/H_2O_2$	$^{\cdot}OH$, $HO_2^{\cdot}/O_2^{-\cdot}$, $O_3^{-\cdot}$	Anionic/cationic surfactants [31–33]
Photocatalytic treatment UV/Vis light using TiO_2 and ZnO as catalysts	$^{\cdot}OH$, $HO_2^{\cdot}$, $HOOO^{\cdot}{}^{\cdot}$, $O_2^{-\cdot}$, e^-, h^+	Laundry waste water, nonionic surfactants [33]

10.4.16 MAGNETIC REMOVAL TECHNIQUES

As compared to CNT, this technology uses downstream processing with filters. It is more cost-effective. Magnetic technology has a variety of range to remove contaminants such as arsenic [84], red textile dye [85], radioactive pollutants [86] and metallic contaminants [87]. This method is also effectively applicable to cationic surfactants with 90% efficiency. It also shows removal inconsistency for anionic and nonionic surfactants [88].

10.4.17 MICELLAR ENHANCED ULTRA-FILTRATION (MEUF)

This process has attracted owing to the ultra-filtration of secondary solid waste pollutants to recover main chemicals from waste water [89]. This process is suitable for the textile and olive oil industry [90]. This method does not work on low molecular weight. With the addition of surfactants, previously mentioned critical micellar concentration of pollutants is easily removed due to large micelles' formation [91].

10.5 BIOSURFACTANTS

Biosurfactants are amphiphilic compounds which are formed by specific microorganisms [92]. Biosurfactants are from the part of the cell membrane of bacteria, yeasts and fungi. Biosurfactants are applicable in food, agriculture, cosmetic, pharmaceutics, oil industries, pollution control, petrochemicals and biotechnological processes [93]. Owing to several advantages over its chemical counterpart includes the following: ecological acceptability, high selectivity, environmentally friendliness, effectiveness at extreme environments, nontoxicity, extensive foaming activities, bioavailability, biocompatibility, biodegradability, and long lifetime [94]. Application of biosurfactants increased day-to-day in remediation of natural, inorganic contaminants and heavy metals from polluted water and soil [95,96]. They are classified on the basis of molecular weight as low molecular mass and high molecular mass biosurfactants [97]. Low molecular mass biosurfactant comprises glycolipids, phospholipids and lipopeptides, whereas high molecular mass is grouped in amphipathic polysaccharides, lipopolysaccharides, proteins, lipoproteins or composite combinations of these biopolymers. Some biosurfactants with their microorganism are listed in the table shown later [98–102].

10.5.1 MICROBIAL ENHANCED OIL RECOVERY

At present, crude oil is one of the vital sources of the world's energy production. While extracting the crude oil from the oil reservoir, a large proportion of treasured and nonrenewable oil is left behind the ground. So, there is a need to produce more crude oil to meet the present requirement of worldwide energy sources by making a progress in the EOR process. New and efficient methods which overcome the hindrances like high viscosity of crude oil, low permeability of some oil reservoirs and high oil to water surface tension may cause high capillary forces to keep the oil in the reservoir rock.

Microbial EOR (MEOR) process is customized to recover remaining oil from reservoirs after mechanical and physical recovery procedure [103]. It is observed that remaining oil is frequently trapped in pores by capillary pressure, in regions of reservoirs that are problematic to access [104]. Biosurfactants reduce capillary forces by decreasing interfacial pressure between oil/water and oil/rock. It prevents oil from moving through rock pores and binds tightly to the oil-water interface and forms emulsion, as shown in **Figure 10.3**. It is also observed that bacteria are active at high temperature and pressure. With this process, it is easy to remove oil along with injection water [106]. Efficiency of biosurfactants is measured by the amount of oil extracted from the column after mixing the aqueous solution of biosurfactants in it. Examples of such biosurfactants are as follows:

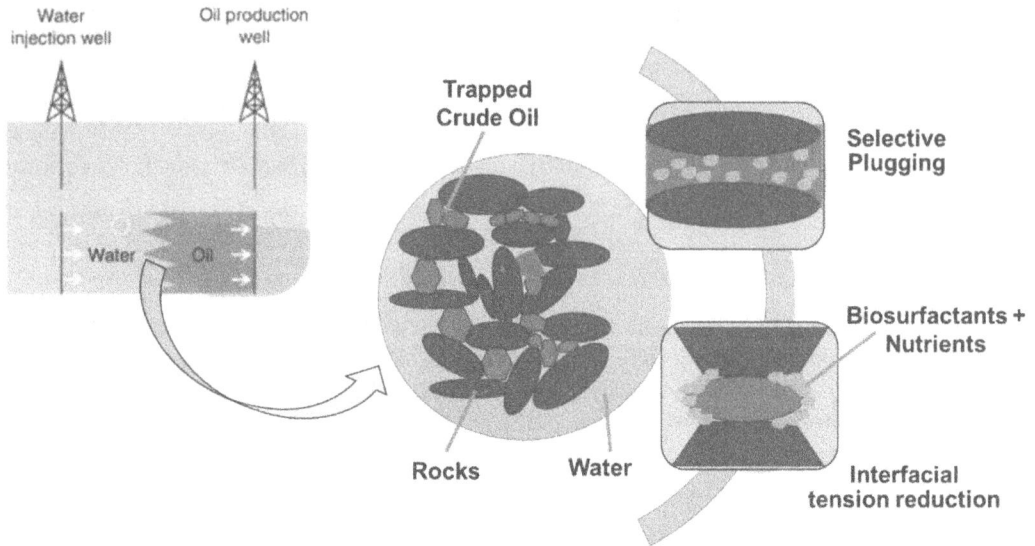

FIGURE 10.3 Process of MEOR by biosurfactants [105].

1. Mixture of *Arthrobacter* sp. (A02), *Pseudomonas* sp. (P15) and *Bacillus* sp. (B24) strain suspension at high temperature (73°C) [83].
2. *Bacillus subtilis* (PT2) and *P. aeruginosa* (SP4) with three synthetic surfactants:
   a. polyoxyethylene sorbitan monooleate (Tween 80), SDBS and sodium alkyl polypropylene oxidesulfate [107].
3. *P. aeruginosa* strains (MTCC7815, MTCC7814, MTCC7812 and MTCC8165) [85].

Biosurfactants are also extracted from hydrocarbon compounds as an option of petroleum energy from oil shales [108].

10.5.2 Metals' Remediation From Soil

Soil contamination with heavy metals is a serious concern for human beings and other living creatures in the ecosystem. They are very toxic in nature; even at low concentrations, it is hazardous [109]. These pollutants are not biodegradable; nevertheless with the help of microorganisms, they can change mobility and toxicity, from one chemical state to another. The properties of biosurfactants like biodegradability, low toxicity and effectiveness makes them potential candidates for the remediation from soil. Biosurfactant for remediation should be selected on the basis of characteristic properties of pollutants, treatment capacity, cost, time limitations and regulatory requirements. Furthermore, known mechanisms of interaction between biosurfactants and pollutants assist in the selection of suitable biosurfactants for remediation.

Microorganisms act on metals in numerous ways, such as by redox, alkylation or affecting pH value, or producing or releasing substances, which change mobility of the metals [110]. Remediation of metals from contaminated soil involved mainly two processes: soil washing and soil flushing. Biosurfactants are released on part of contaminated soil, where they can form complexes with metals by ionic bonds. These ionic bonds are stronger as compared to metal bonds with the soil. Now these metal-biosurfactant complexes are desorbed as a result of lower IFT, from the soil matrix to the soil solution. By the formation of biosurfactant micelles, toxic metal ions are easily removed from soil (**Figure 10.4**). Moreover, biosurfactants may be used to improve the sustainability of the remediation process by using recycled or waste substrates, in situ biosurfactant processing and

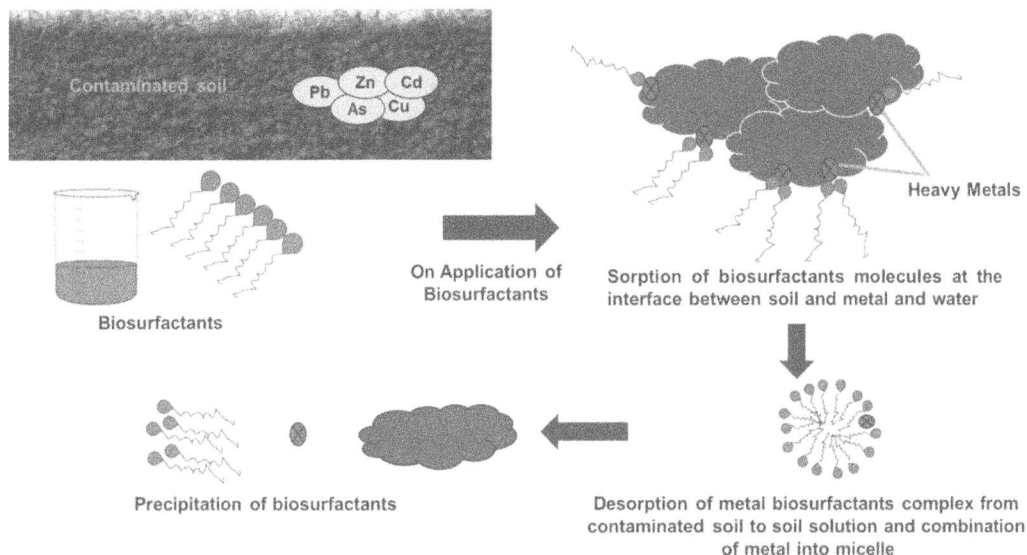

FIGURE 10.4   Process of metals' remediation from soil using biosurfactant.

greener biosurfactant production and recovery methods. Some examples are listed in **Table 10.3** [92,109]

10.5.3   FOOD SAFETY

Nowadays, due to increasing environmental concern, microbial-derived surfactants draw much attention as compared to chemical surfactants. Biosurfactants possess characteristic properties, like low toxicity, biocompatibility and biodegradability, which trigger their potential use in many sectors of the food industry. Biosurfactants have properties such as emulsion formation and stabilization, as well as antiadhesive and antimicrobial properties that could be used in food processing and formulation. They can control the accumulation of fat globules and stabilize the aerated systems. They also improve texture, modify viscosity and improve consistency and shelf life of starch content and fat-based products. Biosurfactants control the consistency of bakery and ice-cream formulations, retard staling and solubilize flavors. During the cooking of oils and fat, biosurfactants can act as fat stabilizer and anti-spattering agent.

Intake of heavy metals through food is vastly hazardous to human well-being. Biosurfactants can be utilized as washing agents to remove these contaminants effectively from food/vegetables [110]. It is observed that biosurfactants produced from Bacillus sp. MTCC 5877 washed toxic Cd

TABLE 10.3
Biosurfactants and Treated Metals

Biosurfactants	Treated Metals
Di-rhamnolipid	Cd, Cr, Pb, Cu and Ni
Marine bacterium	Chelate toxic, Pb and Cd
Rhamnolipid+1% NaOH	Cd and Ni
Rhamnolipid	Cd, Cu, La, Pb and Zn
Bacillus sp. MTCC 5514	Cr (III)

from vegetables at a percentage up to 73% [111]. For removal of zinc and iron contaminations from vegetables, rhamnolipid biosurfactants developed from *Pseudomonas putida* could play an important role. Similarly, *Bacillus licheniformis VS16* removes poisonous Cd from vegetables in the range of ~61% [112]. Rhamnolipid surfactants can increase the dough stability, texture, volume and conservation of bakery products [113].

10.5.4 NANOTECHNOLOGY

Setting up green technology for the synthesis of nanoparticles is expanding these days within the field of nanotechnology. The search for an ideal source with properties like stability, durability and ability to resist higher natural conditions with excellent characteristics is yet to be met. Therefore, it is necessary to develop an ecological approach to synthesize metal nanoparticles. Thought of 'green technology' could be fulfilled by the use of biosurfactants. Nanobiology is the next peer of biological method to fast synthesize nanoparticles. Characteristics of all green synthesis processes are cost-effective, environmentally friendly, zero-energy based, efficient and not using toxic chemicals. For nanoparticle synthesis, biosurfactants can be used as both stabilizer and reducing agent [109]. Biosurfactant *Brevibacterium casei MSA19* controls the formation of aggregates and increased the stability of nanoparticles for a longer time [110]. Microorganisms are also able to synthesize gold and silver nanoparticles [114]. Biosurfactant produced from *P. aeruginosa* stabilizes silver nanoparticles in microemulsions [111]. It has been analyzed that bacteria *B. subtilis* grown on agro-industrial wastes synthesized Ag nanoparticles, which can be used as nano agents against microbes [115].

10.6 CONCLUSIONS

It has been seen that nanosurfactants are extremely important in several facets of petroleum, cosmetic, food and pharmaceutical industries. Nanosurfactants possess key characteristics like very high surface area, cloud point and adsorption level, so having a widespread application. Nanosurfactants aid in the transportation of active ingredients effectively through the skin and augment diffusion of active constituents due to large surface area and dimensions. They also show a broad spectrum of antimicrobial properties in the form of nano-emulsion. Nowadays, nanosurfactants are used for the inhibition of murine A virus pneumonitis. In the pharmaceutical industry, nanosurfactants control particle size of nanocapsules and nanoparticles. The stability of controlled assembly of functionalized nanoparticles is also achieved by the profusion of nanoparticle and surfactant interfaces. Surfactants play an important role in the interfacial process to stabilize the nanoparticle assemblies. Nanosurfactants in the form of nano-emulsions also help in reducing the side effects of anticancer drugs by participating as carriers for drug delivery. In the agriculture industry, to increase the crop yield, nanotechnology-based pesticide formulations are being developed. So nanosurfactants have widespread applications in various domains and this chapter explores a range of such application areas.

REFERENCES

1. Philip, M. N. M., *J. Rep. Prog. Phys.*, 643, 297, 2001.
2. Burns, N. A., *Chem. Mag.*, 121, 32–35, 2016.
3. Acmite Market Intelligence, *Global Surfactant Market* (Vol. I, 4th ed.), Ratingen, Germany, 2016.
4. Muzaffar, F. A. I. Z. I., Singh, U. K., & Chauhan, L., Review on microemulsion as futuristic drug delivery. *Int. J. Pharm. Pharm. Sci.*, 5(3), 39–53, 2013.
5. Chakraborty, S., & Agrawal, G., U.S. Patent No. 9,428,383. Washington, DC: U.S. Patent and Trademark Office, 2016.
6. Kim, D. Y., Lee, S. A., Kim, S., Nah, C., Lee, S. H., & Jeong, K. U., Asymmetric fullerene nanosurfactant: interface engineering for automatic molecular alignments. *Small*, 14(1), 1702439, 2018.

7. Saeedi, L. H., Assadi, M. M., Heydarian, S. M., & Jahangiri, M., The production and evaluation of a nano-biosurfactant. *Petrol. Sci. Technol.*, 32(2), 125–132, 2014.
8. Mohajeri, M., Hemmati, M., & Shekarabi, A. S., An experimental study on using a nanosurfactant in an EOR process of heavy oil in a fractured micromodel. *J. Petrol. Sci. Eng.*, 126, 162–173, 2015.
9. Abdel-Fattah, A. I., Mashat, A., Alaskar, M., & Gizzatov, A., NanoSurfactant for EOR in Carbonate Reservoirs. In *SPE Kingdom of Saudi Arabia Annual Technical Symposium and Exhibition*. Society of Petroleum Engineers, 2017, June.
10. Cheraghian, G., & Nezhad, S. S. K., Improvement of heavy oil recovery and role of nanoparticles of clay in the surfactant flooding process. *Petrol. Sci. Technol.*, 34(15), 1397–1405, 2016.
11. Al Harbi, A. M., Gao, J., Kwak, H. T., & Abdel-Fattah, A. I., The Study of Nanosurfactant EOR in Carbonates by Advanced NMR Technique. In *SPE Abu Dhabi International Petroleum Exhibition & Conference*. Society of Petroleum Engineers, 2017, November.
12. Nwidee, L. N., Al-Anssari, S., Barifcani, A., Sarmadivaleh, M., Lebedev, M., & Iglauer, S., Nanoparticles influence on wetting behaviour of fractured limestone formation. *J. Pet. Sci. Eng.*, 149(2017), 782–788, 2016.
13. Das, S. K., Choi, S. U. S., & Patel, H. E., Heat transfer in nanofluids – a review, *Heat Transfer Eng.*, 27(10), 3–19, 2006.
14. Nwidee, L. N., Al-Anssari, S., Barifcani, A., Sarmadivaleh, M., & Iglauer, S., Nanofluids for enhanced oil recovery processes: wettability alteration using zirconium oxide. Offshore Technology Conference (OTC) Asia. One Petro, 2016.
15. Yu, W., & Xie, H., A review on nanofluids: preparation, stability mechanisms and applications. *J. Nanomater*, 2012, 1–17, 2012.
16. Sharma, T., Kumar, G. S., & Sangwai, J. S., Comparative effectiveness of production performance of Pickering emulsion stabilized by nanoparticle-surfactant-polymer over surfactant-polymer (SP) flooding for enhanced oil recovery for Brownfield reservoir. *J. Pet. Sci. Eng.*, 129, 221–232.32, 2015.
17. Sharma, T., Kumar, G. S., Chon, B. H., & Sangwai, J. S., Thermal stability of oil-in-water Pickering emulsion in the presence of nanoparticle, surfactant, and polymer. *J. Indust. Eng. Chem.*, 22, 324–334, 2015.
18. Lin, P., Yan, Q., Chen, Y., Li, X., & Cheng, Z., Dispersion and assembly of reduced graphene oxide in chiral nematic liquid crystals by charged two-dimensional nanosurfactants. *Chem. Eng. J.*, 334, 1023–1033, 2018.
19. Emadi, S., Shadizadeh, S. R., Manshad, A. K., Rahimi, A. M., & Mohammadi, A. H., Effect of nano silica particles on Interfacial Tension (IFT) and mobility control of natural surfactant (Cedr Extraction) solution in enhanced oil recovery process by nano-surfactant flooding. *J. Mol. Liquids*, 248, 163–167, 2017.
20. Mohajeri, M., Hemmati, M., & Shekarabi, A. S., An experimental study on using a nanosurfactant in an EOR process of heavy oil in a fractured micromodel. *J. Petrol. Sci. Eng.*, 126, 162–173, 2015.
21. Thammachart, M., Meeyoo, V., Risksomboon, T., & Osuwan, S., Catalytic activity of CeO_2–ZrO_2 mixed oxide catalysts prepared via sol–gel technique: CO oxidation. *Catal. Today*, 68(1), 53–61, 2001.
22. Bentra, S. & Levy, M. Y., *J. Pharm. Sci.*, 82, 1069, 1993.
23. Morsy, S. M. I., *Int. J. Curr. Microbiol. App. Sci.* 3(5), 237–260, 2014.
24. Khoee, S., & Yaghoobian, M., *Eur. J. Med. Chem.*, 45(12), 5541–6242, 2010.
25. Chavanpatil, M., Khdair, A., & Panyam, J., *Pharm. Res.*, 244, 803, 2007.
26. Guobin, S., Rao, Y. S., Rajeshwar, D. T., & Tian, C. Z., Nanomaterials for Environmental Burden Reduction, Waste Treatment, and Nonpoint Source Pollution Control: A Review, 33, September, 2009.
27. Piacenza, E., Presentato, A., & Turner, R. J. *Crit. Rev. Biotechnol.*, 38(8), 1137–1156, 2018.
28. Jiang, L. C., Basria, M., Omar, D., Abdul Rahman, M. B., Salleh, A. B., Zaliha Raja, R. N., & Abdul Rahman, A. S., *Pest. Biochem. Physiol.*, 102, 19–29, 2012.
29. Hudson, L. K., Eastoe, J., Dowding, P., Nanotechnology in action: overbased nanodetergents as lubricant oil additives. *Adv. Colloid Interface Sci.*, 123–126, 425–431, 2006.
30. Bearchell, C. A., Heyes, D. M., Moreton, D. M., & Taylor, S. E., Overbased detergent particles: experimental and molecular modelling studies. *Phys. Chem. Chem. Phys.*, 3, 4774–4783, 2001.
31. Eastoe, J., Hollamby, M. J., & Hudson, L. K., *Adv. Colloid Interface Sci.*, 128–130, 5–15, 2006.
32. Bumajdad, A., Eastoe, J., & Mathew, A., *Adv. Colloid Interface Sci.*, 147–148, 56–66, 2009.
33. Smith, G., & Eastoe, J., *Phys. Chem. Chem. Phys.*, 15, 424–439, 2013.
34. Chellapa, P., Ariffin, F. D., Eid, A. M., Almahgoubi, A. A., Mohamed, A. T., Issa, Y. S., & Elmarzugi, N. A., *Eur. J. Biomed. Pharm. Sci.*, 3, 08–11, 2016.

35. Sonneville-Aubrun, O., Simonnet, J.-T., & L'Alloret, F., *Adv. Colloid Interface Sci.*, 108–109, 145–149, 2004.
36. Simonnet, J.-T., Sonneville, O., & Legret, S., 'Nanoemulsion based on sugar fatty esters or on sugar fatty ethers and its uses in the cosmetics, dermatological and/or ophthalmological fields', US668371, L'Oréal, 2004.
37. Bidyut, K. P., & Satya, P. M., *Curr. Sci.*, 80, 8, 2001.
38. Zülli, F., Liechti, C., & Suter, F., *Int. J. Cosmetic Sci.*, 22, 265–270, 2000.
39. Donsì, F., Annunziata, M., Vincensi, M., & Ferrari, G., *J. Biotechnol.*, 159, 342–350, 2012.
40. Holmberg, K., Natural surfactants. *Curr. Opin. Colloid Interface Sci.*, 6, 148–159, 2001.
41. Anton, N., Benoit, J., & Saulnier, P., *J. Controll. Release*, 128, 185–199, 2008.
42. Cui, X., Tang, S., & Zhou, H., *Mater. Lett.*, 98, 116–119, 2013.
43. Scott, M. J., & Jones, M. N., *Biochim. Biophys. Acta (BBA)-Biomembranes*, 1508(1), 235–251, 2000.
44. Gerola, A. P., Costa, P. F. A., Quina, F. H., Fiedler, H. D., & Nome, F., *Curr. Opin. Colloid Interface Sci.*, 32, 39–47, 2017.
45. Ying, G. G., *Environ. Int.*, 32(3), 417–431, 2006.
46. Aloui, F., Kchaou, S., & Sayadi, S., *J. Hazard. Mater.*, 164(1), 353–359, 2009.
47. Palmer, M., & Hatley, H., *Water Res.* 147, 60–72, 2018.
48. Zhu, F.-J., Ma, W.-L., Xu, T.-F., Yi Ding, Zhao, X., Li, W.-L., Liu, L.-Y., Song, W.-W., Li, Y.-F., & Zhang, Z.-F., *Ecotoxicol. Environ. Safety*, 153, 84–90, 2018.
49. Clara, M., Scharf, S., Scheffknecht, C., & Gans, O., *Water Res.*, 41, (19), 4339–4348, 2007.
50. Kern, D. I., Schwaickhardt, R. D. O., Mohr, G., Lobo, E. A., Kist, L. T., & Machado, Ê. L. *Int. J. Environ. Analyt. Chem.*, 443, 566–572, 2013.
51. Aboulhassan, M. A., Souabi, S., Yaacoubi, A., & Baudu, M., *Int. J. Environ. Sci. Technol.*, 3(4), 327–332, 2006.
52. Kreuzinger, N., Fuerhacker, M., Scharf, S., Uhl, M., Gans, O., & Grillitsch, B., *Desalination*, 215, (1–3), 209–222, 2007.
53. Mungray, A. K., & Kumar, P., *Bioresour. Technol.*, 99 (8), 2919–2929, 2008.
54. González, S., Petrovic, M. & Barceló, D., *Chemosphere*, 67(2), 335–343, 2007.
55. Mantzavinos, D., Burrows, D. M., Willey, R., Biundo, G. L., Zhang, S. F., Livingston, A. G., & Metcalfe, I. S., *Water Res.*, 35(14), 3337–3344, 2001.
56. Denger, K., & Cook, A. M. J., *Arch. Microbiol.*, 86, 165–168, 1999.
57. Mogensen, A. S., & Ahring, B. K., *Biotechnol. Bioeng.*, 77(5), 483–488, 2002.
58. Almendariz, F. J., Meráz, M., Soberón, G., & Monroy, O., *Water Sci. Technol.*, 44(4), 183–188, 2001.
59. Giolando, S. T., Rapaport, R. A., Larson, R. J., Federle, T. W., Stalmans, M., & Masscheleyn, P., *Chemosphere*, 30(6), 1067–1083, 1995.
60. Charles, W., Ho, G., & Cord-Ruwisch, R., *Water Sci. Technol.*, 34(11 pt 7), 1–8, 1996.
61. Shang, X., Kim, H. C., Huang, J. H., & Dempsey, B. A., *Separat. Purif. Technol.*, 147, 44–50, 2015.
62. Aboulhassan, M A., Souabi, S., Yaacoubi, A., & Baudu, M., *Int. J. Environ. Sci. Technol.*, 3, 327–332, 2006.
63. Kaleta, J., & Elektorowicz, M., *Environ. Technol.*, 34(8), 999–1005, 2013.
64. Ochoa-Herrera, V., & Sierra-Alvarez, R., *Chemosphere*, 72(10), 1588–1593, 2008.
65. García-Delgado, R. A., Cotoruelo, L. M., & Rodríguez, J. J., *Sep. Sci. Technol.*, 27(8–9), 1065–1076, 1992.
66. Kawai, T., & Tsutsumi, K., *Colloid Polym. Sci.*, 273 (8), 787–792, 1995.
67. Goloub, T. P., Koopal, L. K., Bijsterbosch, B. H., & Sidorova, M. P., *Langmuir*, 12, 3188–3194, 1996.
68. Pham, T. D., Kobayashi, M., & Adachi, Y., *Colloid Polymer Sci.*, 293, 217–227, 2015.
69. Abd El-Lateef, H. M., Khalaf Ali, M. M., & Saleh, M. M., *J. Mol. Liquids*, 268, 497–505, 2018.
70. Adams, C. D., & Kuzhikannil, J. J., *Water Res.*, 34(2), 668–672, 2000.
71. Munter, R., *Proc. Estonian Acad. Sci. Chem.*, 50, 2, 59–80, 2001.
72. Krishnan, S., Chandran, K., Sinnathambi, C. M., *Int. J. Appl. Chem.*, 12, 4, 727–739, 2016.
73. Burakov, A. E., Galunin, E. V., Burakova, I. V., Kucherova, A. E., Agarwal, S., Tkachev, A. G., & Gupta, V. K., *Ecotoxicol. Environ. Safety*, 148, 702–712, 2018.
74. Gao, Q., Chen, W., Chen, Y., Werner, D., Cornelissen, G., Xing, B., Tao, S., & Wang, X., *Water Res.*, 106, 531–538, 2016.
75. Younis, A. M., Kolesnikov, A. V., & Desyatov, A. V., *Am. J. Anal. Chem.*, 5 (17), 1273–1284, 2014.
76. Koh, B., & Cheng, W., *Langmuir*, 30 (36), 10899–10909, 2014.
77. Shang, J. J., Yang, Q. S., Yan, X. H., He, X. Q., & Liew, K. M., *Nanomaterials*, 6(10), 177, 2016.
78. Liné, C., Larue, C., & Flahaut, E., *Carbon*, 123, 767–785, 2017.

79. Mayo, J. T., Yavuz, C., Yean, S., Cong, L., Shipley, H., Yu, W., Falkner, J., Kan, A., Tomson, M., & Colvin, V. L., *Sci. Technol. Adv. Mater.*, 8(1–2), 71, 2007.
80. Iram, M., Guo, C., Guan, Y., Ishfaq, A., & Liu, H., *J. Hazard. Mater.*, 181(1–3), 1039–1050, 2010.
81. Ambashta, R. D., & Sillanpää, M., *J. Hazard. Mater.*, 180(1–3), 38–49, 2010.
82. Oliveira, L. C. A., Petkowicz, D. I., Smaniotto, A., & Pergher, S. B. C., Magnetic zeolites, *Water Res.*, 38(17), 3699–3704, 2004.
83. Borghi, C. C., Fabbri, M., Fiorini, M., Mancini, M., & Ribani, P. L., *Sep. Purif. Technol.*, 83, 180–188, 2011.
84. Sandefur, H. N., Asgharpour, M., Mariott, J., Gottberg, E., Vaden, J., Matlock, M., & Hestekin, J., *Ecol. Eng.*, 94, 75–81, 2016.
85. Erkanlı, M., Yilmaz, L., Çulfaz-Emecen, P. Z., & Yetis, U., *J. Cleaner Prod.*, 165,1204–1214, 2017.
86. Víctor-Ortega, M. D., Martins, R. C., Gando-Ferreira, L. M., & Quinta-Ferreira, R. M., *Colloids Surf. A: Physicochem. Eng. Aspects*, 531, 18–24, 2017.
87. Parthipan, P., Preetham, E., Machuca, L.L., Rahman, P.K., Murugan, K., & Rajasekar, A., *Front. Microbiol.* 8, 193, 2017.
88. Mnif, I., Ellouze-Chaabouni, S., & Ghribi, D., *Probiotics Antimicrobial. Proteins* 5, 92–98, 2013.
89. Anjum, F., Gautam, G., Edgard, G., & Negi, S., *Bioresour. Technol.* 213, 262–269, 2016.
90. Araujo, H. W. C., Andrade, R. F. S., Montero-Rodriguez, D., Rubio-Ribeaux, D., Alves da Silva, C. A., & Campos-Takaki, G. M., *Microb. Cell Factories* 18, 2, 2019.
91. Sun, S., Wang, Y., Zang, T., Wei, J., Wu, H., Wei, C., Qiu, G., & Li, F., *Bioresour. Technol.* 281, 421–428, 2019.
92. Pacwa-Płociniczak, M., Płaza, G. A., Piotrowska-Seget, Z., & Singh Cameotra, S., *Int. J. Mol. Sci.*, 12, 633–654, 2011.
93. Sifour, M., Al-Jilawi, M. H., Aziz, G. M., *Pak. J. Biol. Sci.*, 10, 1331–1335, 2007.
94. Whang, L. M.; Liu, P. W. G., Ma, C. C., Cheng, S. S., *J. Hazard. Mater.*, 151, 155–163, 2008.
95. Franzetti, A., Gandolfi, I., Bestetti, G., Smyth, T. J., & Banat, I. M., *Eur. J. Lipid. Sci. Technol.*, 112, 617–627, 2010.
96. Toren, A., Navon-Venezia, S., Ron, E. Z., & Rosenberg, E., *Appl. Environ. Microbiol.*, 67, 1102–1106, 2001.
97. Banat, I. M., Makkar, R. S., & Cameotra, S. S., *Appl. Environ. Microbiol.*, 53, 495–508, 2000.
98. Sen, R., *Prog. Energ. Combust.*, 34, 714–724, 2008.
99. Suthar, H., Hingurao, K., Desai, A., & Nerurkar, A., *J. Microbiol. Methods*, 75, 225–230, 2008.
100. Jinfeng, L., Lijun, M., Bozhong, M., Rulin, L., Fangtian, N., & Jiaxi, Z., *J. Petrol. Sci. Eng.* 48, 265–271, 2005.
101. Pornsunthorntawee, O., Arttaweeporn, N., Paisanjit, S., Somboonthanate, P., Abe, M., Rujiravanit, R., & Chavadej, S., *Biochem. Eng. J.*, 42, 172–179, 2008.
102. Bordoloi, N. K., & Konwar, B. K., *Colloid Surf. B*, 63, 73–82, 2008.
103. Haddadin, M. S. Y., Arqoub, A. A. A., Reesh, I. A., & Haddadin, J., *Energ. Convers. Manage.*, 50, 983–990, 2009.
104. https://www.thechemicalengineer.com/features/meor-making-the-case/
105. Aşçı, Y., Nurbaş, M., & Açıkel, Y. S., *J. Environ. Manage.*, 91, 724–731, 2010.
106. Ledin, M., *Earth Sci.*, 51, 1–31, 2000.
107. Jimoh, A. A., Lin, J., *Ecotoxicol. Environ. Safety*, 184, 109607, 2019.
108. Meenakshisundaram, M., & Pramila, M., *Int. J. Curr. Microbiol. Appl. Sci.* 6, 402–411, 2017.
109. Giri, S. S., Sen, S. S., Jun, J. W., Sukumaran, V., & Park, S.C., *Front. Microbiol*, 8, 514, 2017.
110. Gomez-Grana, S., Perez-Ameneiro, M., Vecino, X., Pastoriza-Santos, I., Perez-Juste, J., Cruz, J. M., & Moldes, A. B., *Nanomaterials (Basel)*,7 (6), 139, 2017.
111. Rane, A. N., Baikar, V. V., Ravi Kumar, V., & Deopurkar, R. L., *Front. Microbiol.*, 8, 492, 2017.
112. Khademolhosseini, R., Jafari, A., & Shabani, M.H., *Procedia Mater. Sci*, 11, 171–175, 2015.
113. Van Haesendonck, I. P. H., & Vanzeveren, E. C. A., Rhamno lipids in bakery products. W.O. 2004/040984, *International application patent (PCT)*, 2004.
114. Kiran, G. S., Hema, T., Gandhimathi, R., Selvin, J., Thomas, T. A., Ravji, T. R., & Natarajaseenivasan, K., *Colloids Surfaces B Biointerfaces*, 73, 250–256, 2009.
115. Farias, C. B. B., Silva, A. F., Rufino, R. D., Luna, J. M., Souza, J. E. G., & Sarubbo, L. A., *Electron. J. Biotechnol*, 17, 122–125, 2014.

11 Pulmonary Surfactants
Development and Respiratory Physiology

Anjali Gupta[1], Divya Bajpai Tripathy[1], Anuradha Mishra[2] and Anujit Ghosal[3]

[1]School of Basic and Applied Sciences, Galgotias University, Greater Noida, India

[2]Department of Applied Chemistry, School of Vocational Studies and Applied Sciences, Gautam Buddha University, Greater Noida, India

[3]Department of Food & Human Nutritional Sciences, The University of Manitoba, Winnipeg, Canada

CONTENTS

11.1 INTRODUCTION OF PULMONARY (LUNG) SURFACTANTS

A composite of phospholipids and surfactant proteins, pulmonary surfactants are proved to be a great example of a life-saving material. They decrease both surface tension and required energy for breathing and prevent the alveolar collapse [1]. Pulmonary surfactants are of prime importance in the case of infant-related problems, meconium aspiration syndrome (MAS) and neonatal respiratory distress syndrome (NRDS) in which due to insufficient natural surfactant, artificial surfactants need to be introduced to save their lives [2]. NRDS is caused when the lungs collapse in case of premature babies, as lung surfactants are still not fully developed and oxygen levels fall to an extent that the baby dies. In MAS, pulmonary surfactants become deactivated in post-term babies due to meconium aspiration.

The current situation of a pandemic caused by SARS-CoV-2 has affected almost 160 million people worldwide. Pulmonary failure is a greater cause for post-COVID-19 mortality and we really have to look for supportive symptomatic therapies. Artificial surfactants can possibly help in improving lung disease so as to combat the deadly viruses [3].

The composition of the pulmonary surfactants was earlier thought to be only made up of lipids when, in the 1970s, little protein amount was found in natural surfactants [4,5]. Few years later, hydrophobic proteins (SP) B and C were revealed to be constituents of the lung tissue solvent extracts that were used to yield surfactant preparations [6]. Nowadays, lung surfactants are well known to be mainly composed of 80% phospholipids (w/w), along with 10% neutral lipids which is generally cholesterol and proteins (SP-A: 5%; SP-B: 2%; SP-C: 2% and SP-D: 1%). Phospholipids, amphiphilic in nature, have one non-polar and another polar part with glycerol linked to one esterified phosphate group (hydrophilic) and two fatty acid chains (hydrophobic). Surface Proteins A and D are hydrophilic in nature and help in innate lung immune response while B and C are hydrophobic in nature and decrease the alveoli surface tension [7]. This review provides an insight of using natural surfactants as an alternative candidate to treat different respiratory issues.

11.2 HISTORY

It was firstly formulated by Laplace and Young that the internal and external pressures in an airbag are directly proportional to surface tension and inversely proportional to the radius of the bag. This equation found its importance when van Neergaard, in 1929, described the respiratory mechanism as requiring lower surface tension [8]. Thereafter, Thannhauser et al. in 1946 reported the presence of a large amount of dipalmitoylphosphatidylcholine (DPPC) in lungs [9]; although this fact was unknown if there is any relation between pulmonary surfactants and DPPC. Gruenwald, in 1947, showed that the surfactants were found to decrease pressure required for entrance of the air inside the lungs [10].

It was in 1959, Mead and Avery found the root cause of infants' mortality due to deficiency of pulmonary surfactants, which started a new era of research in the field of lung surfactants [11]. They found higher surface tension of lungs in the case of newborn babies who died due to NRDS compared to others.

During a clinical trial, nebulized DPPC spray was used by Chu for the first time but this trial failed, which hampers the research in this direction for some time [12]. But, later in the 1970s, some researchers found the effectiveness of naturally formulated pulmonary surfactants in animal models of NRDS [13,14]. In 1973, structural components of pulmonary surfactants were discovered whereby surfactant proteins along with phospholipids were found to play key roles in the surface properties [15].

In 1980, Fujiwara et al. for the very first time isolated pulmonary surfactants from bovine lungs [16]. They added phospholipids to this artificial pulmonary surfactant to constitute Surfactant-TA and introduced it in ten new born babies with RDS. This was proved to be a milestone in the field of lung surfactants as the treatment was a great success resulting in improved oxygenation and prognosis.

These natural pulmonary surfactants formulations are generally isolated from animals like pigs or cows. Other natural pulmonary surfactants include surfacten, curosurf and survanta which are the derivatives of bovine-minced surfactants with added palmitic acid, tripalmitoylglycerol, and DPPC. Infasurf, Alveofact® and BLES are extracted from bovine pulmonary surfactant using $CHCl_3$–CH_3OH. The use of these artificial pulmonary surfactants significantly reduced the neonatal mortality rate due to RDS in the past 30 years [17,18].

11.3 COMPOSITION OF SURFACTANTS

Glycerophospholipids constitute almost 80% of the surfactants' part with sphingomyelin possessing minor contribution. Almost 80 wt.% of the phospholipids is found to be phosphatidylcholine while 10% is phosphatidylglycerol and phosphatidylethanolamine with minor contribution. Cholesterol is found to constitute about 10% of the mammalian surfactant. Surfactant proteins constitute 10% of the pulmonary surfactants: SP-A, SP-B, SP-C and SP-D. Among them, SP-A and SP-D are known to be calcium-dependent lectins in mammals only, found to be hydrophilic in nature which help in pulmonary defence mechanism, while SP-B and SP-C are found as hydrophobic in nature that enhance the adsorption of lipid to the alveoli surface [1, 19].

11.3.1   SP-A

SP-A, a glycoprotein, is being produced by the respiratory cells of the developing fetus. It has the ability for the phospholipids binding in a calcium-independent manner. It is also an antioxidant agent and, hence, prevents lungs from various inflammations and pollutants. It is also able to bind with ATII cells specifically and inhibit the lipid secretion and may potentially participate in the regulation of surfactant homeostasis. SP-A binds and aggregates the Gram-positive as well as Gram-negative bacteria and kills them through phagocytosis. *In vitro*, it was found to activate pulmonary macrophages and thereby increases the opsonization and killing of pathogenic virus, bacteria and fungus through phagocytosis and may directly play an important role in non-specific defence action in lungs [20,21]. It also works in cooperation with surface proteins B and C to lower the surface tension of surfactant specifically during pathological situations. The transcription of SP-A is controlled by thyroid transcription factor-1 and its expression increases in amniotic fluid during late gestation; hence, an important marker for the determination of fetal lung maturity [22,23]. On the contrary, its amount in amniotic fluid does not predict RDS.

11.3.2   SP-B

SP-B is a hydrophobic, made up of 79 amino acid proteins that lower down the phospholipids properties by increasing the surface tension [24]. A single gene, that is SFTPB, encodes SP-B, initially directs the preproprotein production which consists of 381 amino acids that endure proteolytic processing to produce mature peptide which is found in airspaces. Along with the presence of SP-A, calcium and phospholipids, SP-B is firmly connected with surfactant PLs [25]. SP-B is critical for unifying phospholipids in lamellar bodies to form tubular myelin as well as for distributing phospholipids to the pulmonary air-liquid interface. The deficiency of SP-B fallouts in abnormal processing of SP-C is also because of anomalous lamellar body creation. Electrostatic interactions of SP-B were found with polar head groups of phospholipids which resists surface tension and thereby enhances the lateral firmness of phospholipid monolayer [26] and based on Nuclear Magnetic Resonance spectroscopy data, their interaction with lipid head groups is as helical peptide conformation [27].

SP-B constitutes a useful component of artificial surfactants developed by organic solvent extraction of lung minces. Its complete deficiency can be highly fatal and lung transplantation becomes the only therapeutic option in this case.

11.3.3   SP-C

Another component of lung surfactant, SP-C, made up of 33–35 amino acid residues and highly hydrophobic in nature, not only reduces surface tension of lungs but also plays an important part in regulating various intracellular processes required for the surfactant production. The precursor (197 amino acid transmembrane peptide) for SP-C is proSP-C, having 21 kDa molecular mass which is encoded by a 2.4-kb gene known as SFTPC. The proSP-C precursor is cleaved proteolytically in multivesicular areas and N- and C-terminal peptides are removed, thereby generating a 35-amino acid peptide that is kept in lamellar bodies along with surfactant phospholipids till they are secreted into the pulmonary area where it increases the stability and dispersal of phospholipids [28]. The specific structure and extended hydrophobicity of SP-C decides the function of SP-C. SP-C destroys the lipid packing when introduced into phospholipid membranes, thereby enhancing the lipid movement.

Downregulation of SP-C expression level has been observed in the case of various animal models of lung infections. Different studies were carried out and it was shown that SP-C expression is also mediated by TNF-α which is a proinflammatory cytokine which inhibits SP-C expression. This downregulation results in chronic conditions of surfactant disproportion and respiratory distress. However, during lung injury, refurbishment of pulmonary function has been verified via administration of surfactant comprising either SP-B or SP-C. According to *in vitro* studies, SP-C portrays a great role in recycling and degradation of surfactants as they are found to increase the surfactant phospholipid reuptake in ATII cells. Hence, it has been conferred that deficiency of SP-C is being associated with RDS both in new born babies and adults. It has also been concluded that SP-C is expected to play a key parameter in alveolar immunology in acute as well as chronic pulmonary disease [29].

11.3.4   SP-D

SP-D is a hydrophilic molecule that comes under the collagen-containing superfamily of C-type lectins. Structurally, it is similar to SP-A and is formed in alveolar cells Type II. SP-D has a significant role in aggregating and enhancing the phagocytosis of microbes through binding to lipid and carbohydrate structures on the bacteria, fungi, viral particles and protozoan surface. SP-D are found in pulmonary and non-pulmonary tissues. Its expression is observed in endothelium of the cardiovascular system as well as muscle cells for the inhibition of inflammatory signalling. More clinical trials are yet to be done in future for the progress of SP-D-based therapy and the treatment of diverse problems including allergic asthma, pulmonary fibrosis and chronic obstructive lung disease [30].

11.4   AN OVERVIEW OF SURFACTANT METABOLISM

As a result of respiratory dynamics, compression and expansion cycles structurally reorganize the components of surfactant. Therefore, partially metabolized surfactant along with varied extracellular structures can be seen in alveolar space. This collected pulmonary surfactant from alveolar space can be fractionated into subtypes according to their densities and hydrodynamic properties. The heavier subtype called large aggregates can be extracted from tubular myelin, secreted lamellar bodies and large lipid vesicles. Furthermore, large aggregates are metabolized to small aggregates, which might happen with the help of plasma proteins and surface protein C and D [31]. According to surfactant fractions obtained in bronchoalveolar lavage from different species, the phospholipid percentage was found to be 15% small aggregates and 85% large aggregates. Moreover, SP-A remains intact in both the cases while SP-B and SP-C are only attached with large aggregates; hence, small aggregates lack hydrophobic proteins and so as surface activity.

For degradation of surfactant via alveolar macrophages, uptake of small aggregates occurs from alveolar space through various pathways. It has been assessed that T2 type pneumocytes are responsible for about 65% surfactant recycling and only 20% alveolar clearance occurs due to macrophages. The 15% which is remaining gets diffused through mucociliary escalator toward the upper airways. During lung injury or any kind of inflammation, seepage of some amount of the surfactant may occur in the blood stream [32]. Hence, the occurrence of surfactant proteins in blood indicates the possibility of lung injury and breathing pathology. Rapid recycling of the surfactant has been observed with an average of 10% recycling of alveolar pool per hour. Recycling of phosphatidylcholine depends upon species and age as its recycling is faster in older animals compared to neonates.

P63/CKAP4 as SP-A receptor mediates the surfactant uptake by T2 pneumocytes to internalize SP-A and phospholipids through clathrin-coated vesicles. Thereafter, resecretion of SP-A takes place via vesicles (recycling) toward plasma membrane from where lipids are transferred to lamellar bodies, where recycling and biosynthetic routes converge. SP-A is also found to modulate degradation of phospholipids by inhibiting peroxiredoxin 6 [33]. SP-C and SP-D are also found to be responsible for phospholipid recycling.

Macrophages also play a critical role in surfactant clearance so as to prevent intra-alveolar accretion of proteins and lipids, which are then inactivated at alveolus' highly oxidizing environment and require continuous replacement. Granulocyte-macrophage colony-stimulating factor (GM-CSF) and protein Bach2 have been found to be essential as the regulator of alveolar macrophages, and thus important for surfactant clearance.

11.5 APPLICATIONS OF PULMONARY SURFACTANTS

Scientists are working in the direction of using lung surfactants for the transportation of medicinal agents deeply into the distal airways. The hydrophobic portion of lung surfactants hosts a perfect environment for fairly soluble drugs and other therapeutic molecules for the treatment of respiratory pathologies and other treatments alongside the body as well.

11.5.1 Therapeutic Target in Asthma

It has been estimated that over 400 million people are suffering from asthma worldwide. The disease has deleteriously affected adults and children as well. Some patients (around 5%–10%) with severe conditions have to inhale high doses of corticosteroids and undergo other treatments every day, still the situation is not under control. Asthma treatment is still a challenge as further developments are required to improve individual's health.

In this regard, studies on lung surfactants suggest their role by supporting lung host defence in some conditions. Moreover, altered bronchoalveolar surfactant levels were found to be allied with multiple pulmonary diseases including asthma. It was clearly revealed that different allergens upon inhalation in the absence of lung surfactants cause damage of epithelium or enhance Type 2 cytokine production that may lead to airway inflammation and pulmonary dysfunction [34]. In one of the recent reports, it has been evident from animal models that decreased SP-A level modulates inflammation induced by IL-13 via facilitation of IL-6/STAT3 downstream signalling [35,36]. In another study, an ovine model was used to find the effects of allergic asthma in mothers on cell phenotypes probably with risks of neonatal lung disease and immune system as well as to study the development of allergy and asthma in progeny [37]. It was observed that Type 2 alveolar epithelial cells that produce surfactant are being reduced due to maternal asthma as well as delayed maturity of fetus lung surfactant system. Nevertheless, more advanced study has to be carried out in this direction due to some clinical limitations. First, the nebulizer or the tube used for the administration of lung surfactant may damage the patient's airways and cause lung injury. Second, there always remain concerns related with infections due to animal-derived proteins. Third the scaling up of natural surfactants extracted from porcine or bovine lungs. Still, pulmonary surfactants may arise as potential therapeutic in the case of asthmatic patients although further investigations are required to explore the probability of surfactants as a targeted therapy [34].

11.5.2 Probability of Surfactants against COVID-19

Washing hands with soap and water is what was recommended during the pandemic for the prevention from COVID-19. Technically, soaps are the surfactants which damage viral spike glycoproteins or wash out this virus by entrapping in micelles. Despite reported antiviral properties, surfactants are still not being explored for therapeutics against COVID-19. The lipid portion present in the lung surfactants is known to be a barrier to viral infections [38,39], for example, H1N1 influenza virus [40]. The phospholipids and protein part present in lung surfactants are known to inhibit the virus-mediated inflammations [41,42]. Hence, it could be clearly hypothesized that surfactant therapy would be promising against SARS-CoV-2 treatment [3, 43].

Moreover, several clinical trials have shown no deleterious effect of various antiviral drugs on COVID patients. So, we need to have supportive therapies to prevent lung failure which is the most common reason of COVID-19 mortality. SARS CoV-2 virus enters the body through the respiratory tract and destroys T2 alveolar cells [44,45] and thereby reduces production of lung surfactant and finally respiratory failure [46]. It has been observed that early administration of natural lung surfactants better control the situation compared to synthetic surfactants and significantly improve blood oxygenation and ventilation time [47]. Natural lung surfactant bovactant has also been planned to be administered in adult COVID patients with pneumonia and found to be successful in such cases [48].

11.5.3 LUNG SURFACTANTS AS THERAPEUTICS FOR RESPIRATORY DISTRESS SYNDROME (RDS)

Nowadays, surfactant replacement is a well-established, safest and effective treatment in neonatology for 26-week gestation having RDS. RDS has been the primary cause of perinatal mortality due to respiratory insufficiency. Earlier, it was observed that administration of more than one dose of natural surfactants decreases the pneumothorax risk and it decreases mortality rate also. Now, there are different animal-derived surfactants available commercially which vary in the concentrations of proteins and phospholipids. A survey was carried out comparing around 62 infants who were treated with three natural surfactants beractant, poractant alfa or calfactant reported a lower mortality for infants (birth weights 500–749 g) when treated with poractant alfa and beractant [49]. According to recent guidelines, it is suggested that only one dose of poractant alfa is sufficient (200 mg/kg) at an early stage and a second dose is administered only if there is persistent RDS issue. [50–55].

Furthermore, Beractant with doses higher than 100 mg/kg are difficult to use because of its comparatively lower concentration in the solution. On the contrary, 200-mg/kg dose of poroctant-alpha results in decreased complications and better survival like intraventricular hemorrhage and persistent ductus arteriosus as well as decreases the need of dose repetition. Surfactant treatment during less invasive surfactant administration and intubation-surfactant-extubation, may decrease the need for doses repetition [56].

11.6 CONCLUSIONS

Respiratory pathologies in preterm neonates as well as adults is undoubtedly a serious cause of concern to life in the universe and should be dealt with safe and sustainable ways. Out of the several methods and techniques reported, the latest one is the use of surfactant replacement therapy which is found to be very effective. Researchers have explored different animal-derived natural surfactants for lung treatment. The expectations with these emerging surfactants are very high as scientists are seeing them as the most apt materials to be used for various respiratory pathologies. In this review, we have given a description of properties and characteristics as well as applications of pulmonary surfactants.

REFERENCES

1. Bernhard, W. (2016). Lung surfactant: function and composition in the context of development and respiratory physiology. *Annals of Anatomy – Anatomischer Anzeiger*, 208, 146–150.
2. Kim, H. C., & Won, Y. Y. (2018). Clinical, technological, and economic issues associated with developing new lung surfactant therapeutics. *Biotechnology Advances*, 36(4), 1185–1193.
3. Pramod, K., Kotta, S., Jijith, U. S., Aravind, A., Tahir, M. A., Manju, C. S., & Gangadharappa, H. V. (2020). Surfactant-based prophylaxis and therapy against COVID-19: a possibility. *Medical Hypotheses*, 143, 110081.
4. King, R. J., & Clements, J. A. (1972). Surface active materials from dog lung. II. Composition and physiological correlations. *American Journal of Physiology-Legacy Content*, 223(3), 715–726.
5. King, R. J., Martin, H., Mitts, D., & Holmstrom, F. M. (1977). Metabolism of the apoproteins in pulmonary surfactant. *Journal of Applied Physiology*, 42(4), 483–491.
6. Berggren, P., Curstedt, T., Grossman, G., Nilsson, R., & Robertson, B. (1985). Physiological activity of pulmonary surfactant with low protein content: effect of enrichment with synthetic phospholipids. *Experimental Lung Research*, 8(1), 29–51.

7. Basabe Burgos, O. (2020). Development of novel synthetic lung surfactants for treatment of respiratory distress syndrome.
8. Neergaard, K. V. (1929). New conceptions about a basic concept of breathing mechanics. *Journal of All Experimental Medicine*, 66(1), 373–394.
9. Thannhauser, S. J., Benotti, J., & Boncoddo, N. F. (1946). Isolation and properties of hydrolecithin (dipalmityl lecithin) from lung; its occurrence in the sphingomyelin fraction of animal tissues. *Journal of Biological Chemistry*, 166(2), 669–675.
10. Gruenwald, P. (1947). Surface tension as a factor in the resistance of neonatal lungs to aeration. *American Journal of Obstetrics & Gynecology*, 53(6), 996–1007.
11. Avery, M. E., & Mead, J. (1959). Surface properties in relation to atelectasis and hyaline membrane disease. *AMA Journal of Diseases of Children*, 97(5_PART_I), 517–523.
12. Chu, J., Clements, J. A., Cotton, E. K., Klaus, M. H., Sweet, A. Y., & Tooley, W. H. (1967). Neonatal pulmonary ischemia: Part I: Clinical and physiological studies. *Pediatrics*, 40(4), 709–782.
13. Enhörning, G., & Robertson, B. (1972). Lung expansion in the premature rabbit fetus after tracheal deposition of surfactant. *Pediatrics*, 50(1), 58–66.
14. Adams, F. H., Towers, B., Osher, A. B., Ikegami, M., Fujiwara, T., & Nozaki, M. (1978). Effects of tracheal instillation of natural surfactant in premature lambs. I. Clinical and autopsy findings. *Pediatric Research*, 12(8), 841–848.
15. King, R. J., Klass, D. J., Gikas, E. G., & Clements, J. A. (1973). Isolation of apoproteins from canine surface active material. *American Journal of Physiology-Legacy Content*, 224(4), 788–795.
16. Fujiwara, T., Chida, S., Watabe, Y., Maeta, H., Morita, T., & Abe, T. (1980). Artificial surfactant therapy in hyaline-membrane disease. *The Lancet*, 315(8159), 55–59.
17. Singh, N., Halliday, H. L., Stevens, T. P., Suresh, G., Soll, R., & Rojas-Reyes, M. X. (2015). Comparison of animal-derived surfactants for the prevention and treatment of respiratory distress syndrome in preterm infants. *Cochrane Database of Systematic Reviews*, (12).
18. Bae, C. W., Kim, C. Y., Chung, S. H., & Choi, Y. S. (2019). History of pulmonary surfactant replacement therapy for neonatal respiratory distress syndrome in Korea. *Journal of Korean Medical Science*, 34(25), e175.
19. Parra, E., & Pérez-Gil, J. (2015). Composition, structure and mechanical properties define performance of pulmonary surfactant membranes and films. *Chemistry and Physics of Lipids*, 185, 153–175.
20. Stoller, J. K. (2015). Murray & Nadel's textbook of respiratory medicine. *Annals of the American Thoracic Society*, 12(8), 1257–1258.
21. Chi, H. M., Moore, M. L., & Peebles Jr, R. S. (2011). The intersection of respiratory syncytial virus infection, innate immunity and allergic lung disease. In *Allergens and respiratory pollutants* (pp. 229–243). Woodhead Publishing.
22. Chernick, V., & Kendig, E. L. (2019). *Kendig's disorders of the respiratory tract in children*. Ed. 9. Saunders/Elsevier.
23. Mason, R. J., & Dobbs, L. G. (2016). Alveolar epithelium and pulmonary surfactant. In *Murray and Nadel's textbook of respiratory medicine* (p. 134). Elsevier.
24. Kendig, E. L., Wilmott, R. W., & Chernick, V. (2012). *Kendig and Chernick's disorders of the respiratory tract in children*. Elsevier Health Sciences.
25. Nkadi, P. O., Merritt, T. A., & Pillers, D. A. M. (2009). An overview of pulmonary surfactant in the neonate: genetics, metabolism, and the role of surfactant in health and disease. *Molecular Genetics and Metabolism*, 97(2), 95–101.
26. Cochrane, C. G., & Revak, S. D. (1991). Pulmonary surfactant protein B (SP-B): structure-function relationships. *Science*, 254(5031), 566–568.
27. Antharam, V. C., Farver, R. S., Kuznetsova, A., Sippel, K. H., Mills, F. D., Elliott, D. W. & Long, J. R. (2008). Interactions of the C-terminus of lung surfactant protein B with lipid bilayers are modulated by acyl chain saturation. *Biochimica et Biophysica Acta (BBA) – Biomembranes*, 1778(11), 2544–2554.
28. Conkright, J. J., Bridges, J. P., Na, C. L., Voorhout, W. F., Trapnell, B., Glasser, S. W., & Weaver, T. E. (2001). Secretion of surfactant protein C, an integral membrane protein, requires the N-terminal propeptide. *Journal of Biological Chemistry*, 276(18), 14658–14664.
29. Mulugeta, S., & Beers, M. F. (2006). Surfactant protein C: its unique properties and emerging immunomodulatory role in the lung. *Microbes and Infection*, 8(8), 2317–2323.
30. Sorensen, G. L. (2018). Surfactant protein D in respiratory and non-respiratory diseases. *Frontiers in Medicine*, 5, 18.
31. Olmeda, B., Martínez-Calle, M., & Pérez-Gil, J. (2017). Pulmonary surfactant metabolism in the alveolar airspace: biogenesis, extracellular conversions, recycling. *Annals of Anatomy – Anatomischer Anzeiger*, 209, 78–92.

32. Cross, L. M., & Matthay, M. A. (2011). Biomarkers in acute lung injury: insights into the pathogenesis of acute lung injury. *Critical Care Clinics*, 27(2), 355–377.

33. Krishnaiah, S. Y., Dodia, C., Sorokina, E. M., Li, H., Feinstein, S. I., & Fisher, A. B. (2016). Binding sites for interaction of peroxiredoxin 6 with surfactant protein A. *Biochimica et Biophysica Acta (BBA) – Proteins and Proteomics*, 1864(4), 419–425.

34. Choi, Y., Jang, J., & Park, H. S. (2020). Pulmonary surfactants: a new therapeutic target in asthma. *Current Allergy and Asthma Reports*, 20(11), 1–8.

35. Francisco, D., Wang, Y., Conway, M., Hurbon, A. N., Dy, A. B., Addison, K. J., ... & Kraft, M. (2020). Surfactant protein – a protects against IL-13-induced inflammation in asthma. *The Journal of Immunology*, 204(10), 2829–2839.

36. Dy, A. B. C., Arif, M. Z., Addison, K. J., Que, L. G., Boitano, S., Kraft, M., & Ledford, J. G. (2019). Genetic variation in surfactant protein-A2 delays resolution of eosinophilia in asthma. *The Journal of Immunology*, 203(5), 1122–1130.

37. Wooldridge, A. L., Clifton, V. L., Moss, T. J., Lu, H., Jamali, M., Agostino, S., ... & Gatford, K. L. (2019). Maternal allergic asthma during pregnancy alters fetal lung and immune development in sheep: potential mechanisms for programming asthma and allergy. *The Journal of Physiology*, 597(16), 4251–4262.

38. Perino, J., Crouzier, D., Spehner, D., Debouzy, J. C., Garin, D., Crance, J. M., & Favier, A. L. (2011). Lung surfactant DPPG phospholipid inhibits vaccinia virus infection. *Antiviral Research*, 89(1), 89–97.

39. Glasser, J. R., & Mallampalli, R. K. (2012). Surfactant and its role in the pathobiology of pulmonary infection. *Microbes and Infection*, 14(1), 17–25.

40. Numata, M., Mitchell, J. R., Tipper, J. L., Brand, J. D., Trombley, J. E., Nagashima, Y., ... & Voelker, D. R. (2020). Pulmonary surfactant lipids inhibit infections with the pandemic H1N1 influenza virus in several animal models. *Journal of Biological Chemistry*, 295(6), 1704–1715.

41. Numata, M., Chu, H. W., Dakhama, A., & Voelker, D. R. (2010). Pulmonary surfactant phosphatidyl-glycerol inhibits respiratory syncytial virus-induced inflammation and infection. *Proceedings of the National Academy of Sciences*, 107(1), 320–325.

42. van Eijk, M., Hillaire, M. L., Rimmelzwaan, G. F., Rynkiewicz, M. J., White, M. R., Hartshorn, K. L., ... & Haagsman, H. P. (2019). Enhanced antiviral activity of human surfactant protein d by site-specific engineering of the carbohydrate recognition domain. *Frontiers in Immunology*, 10, 2476.

43. Zhou, P., Yang, X. L., Wang, X. G., Hu, B., Zhang, L., Zhang, W., ... & Shi, Z. L. (2020). A pneumonia outbreak associated with a new coronavirus of probable bat origin. *Nature*, 579(7798), 270–273.

44. Hoffmann, M., Kleine-Weber, H., Schroeder, S., Krüger, N., Herrler, T., Erichsen, S., ... & Pöhlmann, S. (2020). SARS-CoV-2 cell entry depends on ACE2 and TMPRSS2 and is blocked by a clinically proven protease inhibitor. *Cell*, 181(2), 271–280.

45. Fang, L., Karakiulakis, G., & Roth, M. (2020). Are patients with hypertension and diabetes mellitus at increased risk for COVID-19 infection?. *The Lancet*. Respiratory Medicine, 8(4), e21.

46. Mirastschijski, U., Dembinski, R., & Maedler, K. (2020). Lung surfactant for pulmonary barrier restoration in patients with COVID-19 pneumonia. *Frontiers in Medicine*, 7, 254.

47. Busani, S., Dall'Ara, L., Tonelli, R., Clini, E., Munari, E., Venturelli, S., ... & Girardis, M. (2020). Surfactant replacement might help recovery of low-compliance lung in severe COVID-19 pneumonia. *Therapeutic Advances in Respiratory Disease*, 14, 1753466620951043.

48. Dilli, D., Çakmakçı, E., Akduman, H., Oktem, A., Aydoğan, S., Çitli, R., & Zenciroğlu, A. (2021). Comparison of three natural surfactants according to lung ultrasonography scores in newborns with respiratory distress syndrome. *The Journal of Maternal-Fetal & Neonatal Medicine*, 34(10), 1634–1640.

49. Sweet, D. G., Carnielli, V., Greisen, G., Hallman, M., Ozek, E., Plavka, R., ... & Halliday, H. L. (2017). European consensus guidelines on the management of respiratory distress syndrome-2016 update. *Neonatology*, 111(2), 107–125.

50. Tridente, A., De Martino, L., & De Luca, D. (2019). Porcine vs bovine surfactant therapy for preterm neonates with RDS: systematic review with biological plausibility and pragmatic meta-analysis of respiratory outcomes. *Respiratory Research*, 20(1), 1–13.

51. Rodriguez-Fanjul, J., Jordan, I., Balaguer, M., Batista-Muñoz, A., Ramon, M., & Bobillo-Perez, S. (2020). Early surfactant replacement guided by lung ultrasound in preterm newborns with RDS: the ULTRASURF randomised controlled trial. *European Journal of Pediatrics*, 179(12), 1913–1920.

52. Dargaville, P. A., Aiyappan, A., De Paoli, A. G., Dalton, R. G., Kuschel, C. A., Kamlin, C. O., ... & Davis, P. G. (2013). Continuous positive airway pressure failure in preterm infants: incidence, predictors and consequences. *Neonatology*, 104(1), 8–14.
53. Niemarkt, H. J., Hütten, M. C., & Kramer, B. W. (2017). Surfactant for respiratory distress syndrome: new ideas on a familiar drug with innovative applications. *Neonatology*, 111(4), 408–414.
54. Hentschel, R., Bohlin, K., van Kaam, A., Fuchs, H., & Danhaive, O. (2020). Surfactant replacement therapy: from biological basis to current clinical practice. *Pediatric Research*, 88(2), 176–183.
55. Polin, R. A., & Carlo, W. A. (2014). Surfactant replacement therapy for preterm and term neonates with respiratory distress. *Pediatrics*, 133(1), 156–163.

12 Environmental Acceptability of Surfactants Based on Renewable Raw Materials
Their Biochemical and Biomedical Applications

Chandreyee Saha[1], Subhalaxmi Pradhan[1] and Shilpi Mishra[2]
[1]Division of Chemistry, School of Basic and Applied Sciences, Galgotias University, Greater Noida, Uttar Pradesh, India
[2]Biological and Chemical Science Department, Montgomery College, Rockville, MD, United States of America

CONTENTS

12.1 INTRODUCTION

Surfactants are organic compounds that contain hydrophilic groups (their "heads") as well as hydrophobic units (their "tails"), which make them soluble in both water and organic solvents. The water-repelling unit in any surfactant is made up of a hydrocarbon chain which can be aliphatic, aromatic

DOI: 10.1201/9781003144878-12

or a combination of both. Surfactants are classified into nonionic, cationic, anionic or zwitterionic categories depending on the presence or absence of a charged water-loving head unit. Surfactants have the capacity to lower the surface tension between two liquids, liquid and gas and liquid and solid. They show their effectiveness by breaking down the interface between water and oils or dirt. Oils and dirt are held in suspension by surfactants which aid in their removal. They are thus popular for their cleaning activities, which make them indispensable among different cleaning agents, including detergents [1]. While anionic surfactants have effective cleaning properties and high sudsing potential, nonionic surfactants have the potential to withstand water hardness making them the most preferred choices of surfactants for laundry detergents [2]. Detergents containing these surfactants gained popularity by providing better cleaning potential and more suds than normal soaps at affordable prices. Over the years, surfactants have found diverse applications in a variety of consumer products, such as cosmetics, and also in basic household cleaning products. This extensive usage of synthetic detergents led to water pollution as the waste was directly discharged into surface waters.

12.2 SURFACTANTS AND ENVIRONMENT

Leftover surfactants and their degraded products are discharged into sewage treatment plants or directly into surface waters after use. They are then dispersed into different environmental segments where they undergo various chemical and physical changes. Due to their wide range of usage and large consumption, surfactants have been reported in surface waters, sludge-treated soils and sediments at different concentrations [3]. The specific properties of these surfactant degradation compounds increase their mobility and cause free circulation in the environment [4].

12.2.1 GROUND AND WASTE WATER POLLUTION

Since surfactants are used extensively, large amounts of anionic surfactants are emitted into the environment increasing surface water pollution [5]. After the surfactants have been used, considerable amounts of their used products are disposed of in waste water treatment plants (WWTP). In rural areas, due to the absence of WWTP, waste waters containing different classes of surfactants are released directly to surface water bodies, eventually getting dispersed into different environmental segments [4]. If purification of waste water is incomplete, it can lead to groundwater contamination by anionic detergents. It is thus important that the efficiency of the waste water purification processes involving anionic surfactants be closely regulated and monitored [6]. In WWTP, compounds generated from surfactant usage are removed partially or completely by aerobic biodegradation and sorption [3, 7].

12.2.2 SURFACTANTS IN SEWAGE SLUDGE

A prominent amount of surfactant can enter the soil through sewage sludge, which in recent times is used as fertilizer. Surfactants are also added to agricultural chemicals to stabilize shelf life for agrochemical formulations, bind granules and disperse, solubilize and wet or emulsify active ingredients [2]. As soon as the surfactants enter the environment through sewage disposal into surface water bodies, pesticide application and sludge disposal on land, surfactants and their subparts spread into different parts of the environment [8, 9]. When surfactants get accumulated in the sewage sludge and its concentration increases, they inhibit sewage sludge microorganisms, thereby hindering the process of pollutant removal in the WWTP by decomposing sewage. Different types of surfactants have been reported in various concentrations in surface waters, WWTP effluents, sewage effluents, sludge-treated soils, dry sludge or sediment [2].

12.2.3 Impact of Surfactants on Environmental Flora and Fauna

Overusage of particular types of surfactants and their disposal in the environment, especially in different water bodies, posed a serious threat to the ecosystem. It is thus very important to monitor and regulate the amounts of different types of surfactants given out in sewage and water receivers. Organisms most affected by surfactant toxicity are aquatic fauna and flora and terrestrial plants [2]. Since cationic surfactants contain a positive charge, they are easily adsorbed onto solid surfaces containing negative charge, like soil, sediments, sludge, plastic and cell membranes, which lead to the phenomenon of bioaccumulation in living organisms. Cationic surfactants are made up of certain chemical agents and exhibit specific characteristic features which slow down the growth of microorganisms and eventually cause their death [4]. In the aquatic environment, excessive surfactant concentration in water disables the proper functionality of the gills and fins of fishes and other aquatic organisms, thereby impairing their respiration and swimming ability [3]. High concentration of certain surfactants also indicated alterations in fish behavior, including body torsion, muscle spasm and erratic behavior [2, 13]. Anionic surfactants, when present inside the system of organisms living in water, interact efficiently with their cell membranes, enzymes and other proteins. It disturbs different biological functions of aquatic organisms like damaging the cell plasma and breaking down of a cell, which causes death [4]. Also, by reducing the surface tension of anionic surfactants, the process of movement of toxic pollutants into living organisms is made easier [10, 11]. Nonionic surfactants like nonylphenol ethoxylates (NPEs) and octylphenol ethoxylates get converted to alkylphenol ethoxylates and other carboxylated equivalents after undergoing degradation in WWTP. These degradation products are more toxic than the starting surfactants and are likely to disrupt the normal functionality of an organism's endocrine system due to their estrogenic properties [12]. Nonionic surfactants create chronic toxicity in aquatic organisms due to bioaccumulation [4]. Anionic surfactants are also toxic towards algae. The presence of high concentration of anionic surfactants in water causes surfactant accumulation and inhibition in algal growth [13]. It was also found that enhancing concentration of anionic detergents caused an increase in the biomass of phytoplankton, density and primary production in the regions of polluted oceans [14, 15]. Certain nonionic surfactants also tend to affect terrestrial plants [16]. It was found that they are more harmful to plants when present in solution form rather than soil as adsorption from soil reduces its poisonous nature [2].

In the environment, surfactants are treated by microbial activity in raw sewage and sewage treatment plants, which leads to their degradation in the environment, thus reducing its impact on biota. This process of removal of surfactant is termed as biodegradation [2]. Surfactants are said to have undergone primary degradation when their chemical structure has changed significantly for the surfactant to lose its surfactant properties, whereas when they are rendered to CO_2, CH_4, water, mineral salts and biomass, ultimate degradation is said to have taken place. Biodegradation of surfactants in the environment depends on several chemical and environmental factors like chemical structure and physicochemical conditions of the environment [7]. Different classes of surfactants have different degradation behavior in the environment.

All surfactants are mostly degradable under aerobic conditions but are resistant to degradation under anaerobic conditions as exhibited in Table 12.1. Cationic surfactants are found to be toxic even at very low concentration and hence when applied to soil may have a severe impact on the soil ecosystem. Certain nonionic surfactants on the other hand produce biodegradation products mimicking estrogenic compounds which disrupt the normal functioning of the endocrine system in many living organisms. Due to the harmful impact of surfactants on the environment and the potential challenges encountered during the biodegradation of synthetic surfactants, the need for alternative eco-friendly surfactants surfaced [7].

TABLE 12.1
Fate of Surfactants in the Environment

Type of Surfactant	Oxygen-Rich Environment	Oxygenless Environment
SAS (anionic)	Degrades easily	Persists
LAS (anionic)	Degrades	Persists
Soap (anionic)	Degrades easily	Degrades easily
AS (anionic)	Degrades easily	Degrades
Fatty acid esters (FES)	Degrades easily	Persists
AES (anionic)	Degrades easily	Degrades
DTDMAC (cationic)	Degrades	Persists
AE (nonionic)	Degrades easily	Degrades
APE (nonionic)	Degrades	Degrades partly

12.3 ALTERNATIVE SYNTHESIS OF ENVIRONMENT-FRIENDLY SURFACTANTS

Petroleum-related raw materials are used for synthesizing surfactants which find application in industries. Some of these synthetic surfactants are nonbiodegradable and toxic to the environmental flora and fauna. Biosurfactants are a class of surfactants synthesized from renewable substances that are nontoxic, easily biodegradable and environment friendly, can be scaled up in production and can help in moderating environmental pollution [17, 18]. Biosurfactants score over synthetic surfactants due to their unique properties, mildness, selective functionality and ability to function efficiently at high temperature and pressure [19, 20].

12.3.1 AMINO ACID BASED ANIONIC SURFACTANT

For most surfactants, amino acids are the preferred starting compounds since they are easily bio-degradable and can lower the interfacial free energy [21]. They can be used for the preparation of many new surfactants which are mild, nonpoisonous, less irritating, easily biodegradable and friendly to the water ecosystem. Such amino acid based surfactants find widespread application in different fields.

Earlier cocoyl chloride made from phosphorus trichloride and fatty acid was used for the synthesis of glycinate amino acid based surfactant, but since fatty acid is known to be toxic to human beings [22], readily biodegradable and nontoxic coconut oil made from natural substances substituted the fatty acid as starting material for environment-friendly amino acid based biosurfactant synthesis. Anionic amino acid based surfactants, potassium cocoyl glycinate (CGK) and sodium cocoyl glycinate (CGN), were prepared starting from coconut oil by the synthetic route as depicted in Scheme 12.1(a) and (b).

The hydrocarbon chain R in the coconut oil as shown in Schemes 12.1 and 12.2 is made up of approximately 4.85% of caprylyl group, 15.68% of myristyl group, 5.15% of capryl group, 60.70% of lauryl group, 6.41% of stearyl group and 7.21% of palmitoyl group on the basis of their weight. Identification of their structure was carried out using standard spectroscopic techniques [23].

Static surface tension of surfactant solutions of CGK and CGN was measured to determine the critical micellar concentration (CMC). The CMC in mol/L of CGK and CGN surfactants were measured as 1.74×10^{-2} and 3.38×10^{-2}, respectively. The surface tensions of CGK and CGN surfactant systems under CMC condition were found to be 29.81 and 36.20 mN/m, respectively [23]. These surface tension values of CGN and CGK at their critical micellar concentration were similar to those of the synthetic surfactants used in detergents and hence were considered to have similar interfacial properties [24].

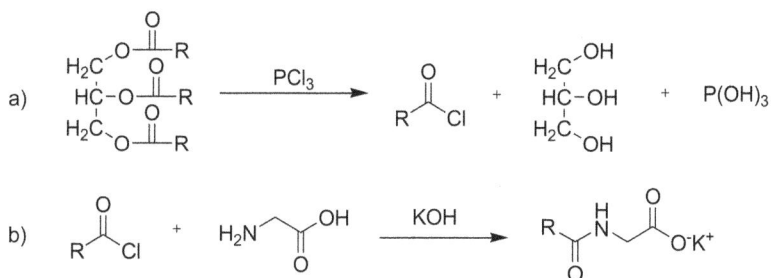

SCHEME 12.1 (a) Synthesis of cocoyl chloride and (b) synthesis of CGK.

Emulsification index (EI) was used to determine the emulsification activity of CGK and CGN surfactants, respectively. EI values for CGN and CGK were found to be $60.00 \pm 0.20\%$ and $61.71 \pm 0.14\%$, respectively [23]. To investigate the emulsification power of the CGN and CGK surfactants, NPE surfactants which are largely used in industrial applications was used as reference. The EI data reported for different NPE surfactants were NPE10 (59.98%), NPE9 (57.42%), NPE8 (55.45%), NPE7 (54.45%) and NPE6 (51.48%). The results indicated that both CGN and CGK have very good emulsification power. They were also very well comparable with the already established industrial emulsifiers [23].

A 1.0 wt.% surfactant solution was used for the calculation of foamability (average rate of formation of foam) at 25°C. Average foamability of CGN and CGK surfactant systems was calculated to be 66.69 ± 3.12 and 71.65 ± 2.78 s, respectively [23]. Stability of foam formation for the 1.0 wt.% surfactant solution was measured using a foam test apparatus [23].

The stability of foam formation of CGN and CGK surfactant systems was calculated to be 17.51% and 3.55%, respectively. The adsorption of the CGK surfactant molecules at the junction decreased the interfacial free energy at the junction of the air and water. This decreased the gas bubble size and enhanced foam stability. CGK surfactant was found to have superior foaming properties [23].

Biodegradability, acute skin and eye irritation, acute oral toxicity and tests for detergency were carried out using CGN and CGK surfactants to study their use as detergents and investigate their environmental compatibility. The reported test results have been summarized in Table 12.2 [23].

Since the biodegradability of both CGK and CGN was 99% (Table 12.2), they were considered to be easily biodegradable. It has been established that any surfactant which is more than 90% biodegradable can be safely used in detergent formulations [25]. Hence CGK and CGN could be potentially used as detergents. For toxicity measurement, LD_{50} measured for CGN and CGK were found greater than 2000 mg/kg. This result was suggestive of the fact that both CGK and CGN were very mild when compared to the other traditional surfactants used in detergents. Also, both CGN and CGK surfactants were found to be gentle and nonhazardous as skin and eye irritant (Table 12.2) [23].

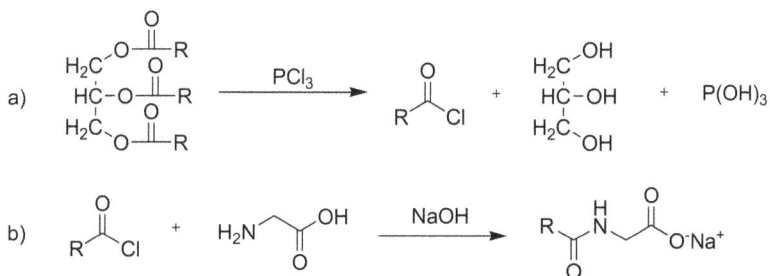

SCHEME 12.2 (a) Synthesis of cocoyl chloride and (b) synthesis of CGN.

TABLE 12.2
Characteristics of Surfactants Made from Amino Acid

Surfactant Type	Tendency to Biodegrade (%)	Detergency (%)	Toxicity (LD_{50}) (mg/kg)	Irritation to the Eye (MMTS)	Effect on Skin
CGN	99.0	86.5	>2000	0	Not harmful
CGK	99.0	85.9	>2000	0	Not harmful

12.3.2 SUGAR-BASED ANIONIC-NONIONIC SURFACTANT

As polyalcohols, sugar groups are largely used to synthesize different types of environment-friendly surfactants [26–29]. In the past, different anionic cum nonionic surfactants made from sugar have been taken up such as the synthesis of sodium methyl 2-acylamido-2-deoxy-6-O-sulfo-D-glucopyranoside. Adsorption properties and the tendency of such surfactants to form micelle were studied [30, 31]. The synthesis of dodecyl glucopyranoside carboxylate from dodecyl glucopyranoside, sodium chloroacetate and alkyl D-mannopyranosiduronate surfactants have also been reported in the literature [32]. Furthermore, the synthesis and properties of environment-friendly anionic cum nonionic surfactants made from sugar-like (DAGA-ES) have been studied in recent times where glucose was used as nonionic building blocks [33]. Glucose is nontoxic, easily biodegradable, and acts as a cheap raw material. Anhydrous glucose was used for making DAGA-ES along with dodecylamine and 2-ethyl chloride sulfonic acid sodium. The reaction was completed in two steps using ethanol, methanol and water as the preferred solvents.

Scheme 12.3 summarizes the synthesis of DAGA. Aldehyde group is a relatively active group in the glucose molecule and reacted with lauryl amine through hydrogenation reduction and generated glucose polyimide which eventually converted to glucose amine [33].

Scheme 12.4 summarizes the synthesis of DAGA-ES. DAGA synthesized as shown in Scheme 12.3 reacted with 2-chloride ethyl sulfonic acid sodium. The reaction proceeded through SN₂ substitution. Hydrogen chloride produced in the reaction was neutralized using sodium hydroxide to form DAGA-ES. The formed products were soluble in water and methanol. However, they were insoluble in ethanol, ether and acetone. Structure elucidation of the formed products was done by different spectroscopic techniques, respectively [33].

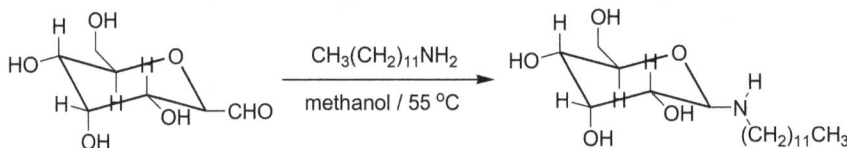

SCHEME 12.3 Synthesis of DAGA.

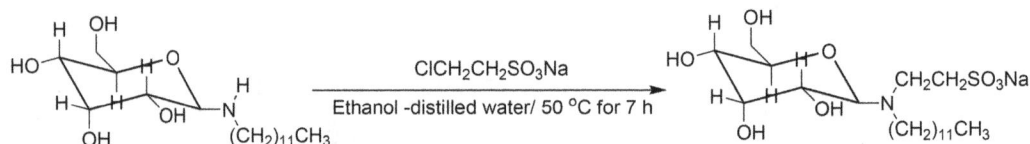

SCHEME 12.4 Synthesis of DAGA-ES.

Equilibrium surface tension measurement and critical micellar concentration of DAGA-ES were carried out at RT. Using brine solution, a gradual decrease in surface tension with concentration increase was observed [33]. Micelle forms when the concentration is greater than the critical micellar concentration; however, the surface tension remains constant. This observation is consistent with other normal surfactants [34, 35]. The surface tension at critical micellar concentration is termed γ_{CMC}. The critical micellar concentration of DAGA-ES in distilled water and solution of 3 mol/L sodium chloride was reported as 5.0 and 1.0 mmol/L, respectively. Their respective γ_{CMC} values were recorded as 24.9 and 22.3 mN/m. In NaCl solution, a marked decrease in critical micellar concentration was observed. However, the γ_{CMC} value changed only slightly. Such similar observation was reported for the traditional anionic surfactants as well [36, 37].

Hydrophile-lipophile balance (HLB) number measures the ratio of hydrophilic and lipophilic groups present in a surfactant molecule. HLB value of a surfactant is used to identify the surfactant affinity for water or oil.

The HLB value of DAGA-ES was determined using the CMC method [33]. The HLB value of DAGA-ES was calculated to be 11.72 [38]. The value of HLB for DAGA-ES thus obtained was between 10 and 13. It was concluded that the HLB value of DAGA-ES obtained by critical micellar concentration method was as per different solution properties [33, 38].

Surface excess (Γ_{max}) of any surfactant is the difference obtained between surface concentration and internal concentration. Γ_{max} at the air/water interface was calculated by the given equation [39, 40]:

$$\Gamma_{max} = -\left(\frac{1}{nRT}\right) \times \frac{dr}{d\ln c} \tag{12.1}$$

Here, Γ_{max} = maximum surface excess concentration (μmol/m^2),

R = universal gas constant (8.314 J/(mol K)),
T = absolute temperature (Kelvin) and
$dr/d\ln c$ = slope of the surface tension isotherm near critical micellar concentration.

The n value depends on solvent salinity and the type of surfactant used.

In distilled water, $n = 2$ because of a 1:1 ratio of ionic surfactants; however, in 3.0-mol/L brine solution, $n = 1$ [41–43]. It is observed that in distilled water, $dr/d\ln c = -2.45$ and calculated $\Gamma_{max} = 0.5 \pm 0.1$ μmol/m^2. However, in the 3.0-mol/L brine solution, $dr/d\ln c = -4.29$ and calculated $\Gamma_{max} = 1.7 \pm 0.1$ μmol/m^2. Since the brine solution has higher polarity than distilled water, the DAGA-ES adsorbs on air/water junctions due to their repulsive interactions with water [33].

Minimum area of any surfactant depicts the configuration of a molecule at the junction of air and water. If the value of A_{min} is small, surfactant density is large. A_{min} values for DAGA-ES in distilled water and 3.0-mol/L brine solution were calculated to be 3.3 ± 0.1 and 0.9 ± 0.1 nm^2, respectively [44]. The result suggested that in the presence of some electrolyte, a close arrangement of DAGA-ES molecules occurred at the junction of air and water. This happens because the electrolyte weakens the repulsive interactions between molecules. These results were consistent with other reported parameters in distilled water and 3.0-mol/L brine solution [33].

On account of the wettability property of DAGA-ES surfactant, it was observed that oil drops recoiled on hydrophilic surfaces, whereas they spread on hydrophobic surfaces without any surfactant. With increased surfactant concentration, the contact angles on the hydrophilic surfaces of DAGA-ES decreased from 79.9°C to 47.6°C. On hydrophobic surfaces, a decrease in contact angle went from 137.5°C to 80.9°C. The hydrophilicity on hydrophobic surface strengthened because of the presence of the long alkyl part of DAGA-ES which adsorbed on oil-wet surface through hydrogen bonds and hydrophobic interaction [45]. The hydrophilicity was also found to increase on hydrophilic surfaces of DAGA-ES [33].

The results deduced for all the earlier parameters suggest that DAGA-ES synthesized from anhydrous glucose could be used efficiently over any conventional anionic-nonionic surfactant.

12.3.3 Tannic Acid based Nonionic Surfactant

Most of the environmentally friendly nonionic surfactants are synthesized from natural and renewable sources like plant oils since they are compatible with the environment and are highly biodegradable in nature [46]. It has been established that tannic acid was used for the efficient production of nonionic surfactants. Tannin is a large polyphenolic moiety obtained from plants containing several hydroxyl groups along with other functional groups which can form strong complexes with various macromolecules. Tannic acid is a special form of tannin with weak acidity, found in the twigs of certain trees like the oak and chestnut trees. Tannic acid was reacted with polyethylene glycol fatty acids containing different numbers of ethylene glycol units as depicted in Scheme 12.5. The fatty acids used were octadecanoic, dodecanoic, hexadecanoic, and oleic acids respectively.

The structural analysis of the nonionic surfactant was carried out using standard spectroscopic techniques [47].

The surface activity of different nonionic surfactants synthesized from tannic acid was determined by surface tension measurements involving a vast range of surfactant concentration. On studying graphs illustrating the relationship between log C and surface tension of a range of surfactants at RT, two characteristic regions showing surface tension variation were highlighted [47]. Starting at lower concentration of surfactant, a continuous depression in the value of surface tension with an increase in surfactant concentration was observed. This depression in surface tension was because of the adsorption of surfactants at the solution interface, which decreased the communication between the polar phase and water-repelling chains [48]. However, at high surfactant concentration, no change in the value of surface tension was noticed with an increase in surfactant concentration. This is because when surfactant molecules saturate the surface completely, micellization takes place at CMC [49].

The hydrophobic chain length of the synthesized nonionic surfactants was found to have an effect on the values of CMC when the amount of ethylene glycol was fixed. It was observed that with a gradual increase in the length of the water-repelling chain of the surfactants, their corresponding CMC values decreased [47]. It was, however, noticed that γ_{CMC} values increased with an increase in their hydrophobic chain length. This was suggestive of the fact that an increase in alkyl chain length caused their HLB values to reduce. Their solubility also decreased causing a decrease in the adsorption of surfactant molecules at the surface. Geometry of the molecule was also responsible for the increase in the surfactant γ_{CMC} value. With the increase in hydrophobic chain length, coiling of the chains increased which decreased the actual size of the surfactant molecule due to which their tendency to cause surface tension depression was decreased [47].

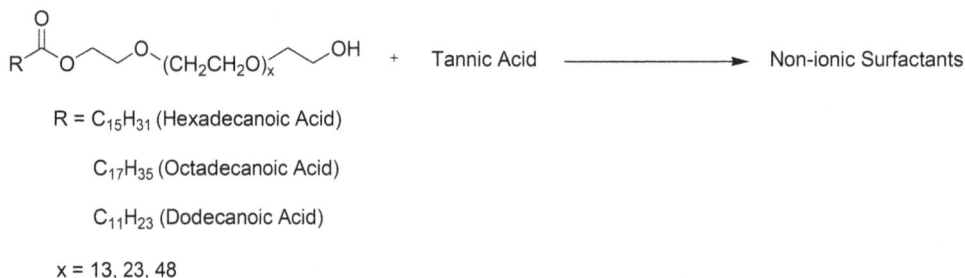

R = $C_{15}H_{31}$ (Hexadecanoic Acid)

$C_{17}H_{35}$ (Octadecanoic Acid)

$C_{11}H_{23}$ (Dodecanoic Acid)

x = 13, 23, 48

SCHEME 12.5 Tannic acid as a source of nonionic surfactant synthesis.

The surface area A_{min} possessed by different synthesized nonionic surfactants at the junction was calculated and it was observed that A_{min} values increased with the increase in the length of the water-repelling chain at constant polyethylene glycol content. However, when the ethylene glycol units were increased from 15 to 50 in the synthesized nonionic surfactant with constant hydrophobic chain length, A_{min} values were found to decrease. This was again due to the coiling of the surfactant molecule due to an increase in the nonionic chains [50].

It was thus established that the nonionic surfactants synthesized from tannic acid exhibited superior surface activities and low CMCs and were in accordance with the conventional nonionic surfactants synthesized synthetically.

12.4   APPLICATIONS OF RENEWABLE SURFACTANTS

Apart from being used as the usual cleaning and washing agents, surfactants synthesized from renewable sources find diverse applications in the fields of biochemistry and biomedicine.

12.4.1   PROTEIN CRYSTALLIZATION

Protein crystallization technique is used as a marker for the purity of chemical substances. It was developed as an important purity measurement tool in the later half of the nineteenth century. Crystallization of large biological molecules requires a supersaturated solution demonstrating conditions that do not change their natural state. Certain other additives which can alter the macromolecular structure like co-factors, inhibitors and different metal ions are also important factors to be considered during supersaturation. Surfactants are used for the crystallization of membrane proteins, which is important for their characterization. The main problem with finding out data about the protein molecules in membranes depends on how they are extracted from membranes containing fats. It causes denaturation and aggregation and also make them insoluble in aqueous solution. Surfactants can be efficiently used for protein crystallization since they can mimic the lipid bilayers of the membrane and can keep the structural and functional integrity of proteins intact [51]. The surfactant forms an aggregate surrounding the protein and forms a complex structure, thereby increasing protein stability. The complex structure of protein and surfactant can then crystallize depending on its uniformity and size [51]. Generally, surfactants with large values of critical micellar concentration (CMC) lead to denaturation of proteins in the membrane and make them unstable. Thus, low CMC surfactants are preferred for their ability to mimic the lipid bilayer due to which the protein remains in an active form. Surfactants with simple structures when used for protein crystallization can often lead to the membrane protein degradation [52]. Thus, over the last two decades, surfactants with complex structures like tripod amphiphiles or TPA, neopentyl glycol or NG and facial amphiphiles or FA have shown promising results in protein crystallization. Surfactants of the NG family were found to preserve the protein structure while producing high-quality crystals. FA surfactants exhibited "side polarity" containing nonpolar and polar parts on opposite surfaces of the molecule and were found to have excellent stabilization and solubilization ability over membrane proteins. TPA surfactants were constructed with a quaternary center in the middle. They are made up of one water-loving and three water-repelling groups. This arrangement restrained their moving abilities as compared to the traditional detergents with linear structure. Hence, they were used for effective protein crystallization [53].

12.4.2   GENE THERAPY AND TRANSFECTION VECTORS

Gene therapy involves the removal of a defective or missing gene into the nucleus of a cell. This method of introducing genetic material into cells is termed as transfection. In gene therapy, in order to replace a faulty gene, a normal gene can be introduced randomly, faulty gene can be exchanged for a good gene through homologous recombination or a faulty gene can be repaired. The major problem with gene therapy is the invasion of immune system. Since a new gene is a foreign body that is getting introduced into the body, thus the body's immune system gets activated and reduces the effectiveness of gene therapy. Also, diseases which are caused by combined variations in multiple genes are difficult to treat by conventional gene therapy. The new gene inside the cell produces proteins needed to rectify the disease. For successful gene therapy, effective gene delivery vectors are required. In recent times, specialized surfactants (gemini surfactants) have been used for gene delivery instead of viral vectors and liposomes [53]. Gemini surfactants are made up of two cationic head groups and two long hydrophobic tails joined together and are known to be nearly nontoxic and readily biodegradable [54]. Gemini surfactants based on carbohydrates, amino acids and organic acids [55–57] have been reported to be efficient nonviral vectors for their easy synthesis and high stability. They can easily attach and form compacted DNA assuring better surface properties as compared to the traditional surfactants having the same length of chain [53]. A large number of peptide-based gemini surfactants with different chain lengths were also shown to exhibit transfection activity, which increased with an increase in the hydrocarbon tail length [58]. Gene expression with such surfactants as vectors was found to be dependent on the bonds formed between the amide groups of the three molecules of lysine in their head unit. Transfection potential of new pH-sensitive gemini surfactants made from sugar was also explored [59]. It was found that such surfactants can effectively form complex structures with plasmid DNA and can perform transfection outside the system.

12.4.3 Drug Delivery

Drug delivery is a process of delivering pharmaceutical products into the body of living organisms for therapeutic purposes. Pulmonary and nasal passages are most commonly used for drug delivery purposes. Nanoemulsions are used for the purpose of drug delivery because their surface areas are increased and other factors like optical transparency, thermodynamic stability, large bioavailability and easy preparation are useful properties for drug delivery. As compared to the conventional drug delivery systems, nanoemulsions have reduced irritancy and toxicity, enhanced absorption and greater stability [60]. Using nanoemulsions as the drug delivery agent, cancer treatment can be improved. Since nanoemulsions have a very small size, they have size suitability to travel throughout the body and reach the specific target organs affected by cancer for treatment. In recent times, sucrose esters have been used as emulsifiers and stabilizers in nanoemulsions and nanosuspensions [61, 62]. Sucrose laureate and oleate have been reported to be used for the formation of octyl methoxycinnamate nanoemulsions. Sucrose laureate nanoemulsions exhibited the strongest penetration in the outermost layer of the skin as compared with other similar moieties [61]. Recent literature has also reported the use of green surfactants like glycolipids and fatty-acid esters for nanoemulsion formulation to be used in drug delivery [63, 64].

12.4.4 Template Technology

The technique of template technology can control form, structure and size of the materials synthesized on a nano-level. Surfactants can organize themselves into complex supramolecular units which eventually act as layouts used for nanoscale production of inorganic materials [65]. A significant number of nanoparticle preparation methods are based on nanoemulsion templates for the purpose of drug delivery. Mesoporous materials used in different electronic devices and metal catalysts are prepared by template technology technique using nanoemulsion. Traditionally hard templates like mesoporous silica are used; however, in recent times, the focus has shifted to the use of softer templates like surfactants and polymer for their good repeatability and controllability [66].

12.4.5 Antimicrobial Agents

Due to the fast growth of fungi and bacteria which are resistant to certain drugs, generation of new antimicrobial sources is necessary [67]. As compared to the antibiotics used currently, cationic surfactants derived from amino acids can be efficiently used as newer antibacterial and antifungal sources because of their ability to mimic natural antimicrobial peptides [67, 68]. The antimicrobial effect of cationic AAS largely depends on its size and structure. Other factors like the length of chain, amino acid residue, hydrophilic-lipophilic balance of the molecule and the cationic charge density also affect their activity [67]. Cationic antimicrobial AAS exhibits a close relationship between the water-repelling unit and the positive charge. This is essential for its function against the microorganisms [67]. Cationic AAS communicates with membranes of the cell, which causes depolarization and destruction leading to the death of the cell and hinders bacterial resistance [69, 70]. This is in contrast to the traditional antibiotics which target specific enzymes or DNA. Cationic surfactants prepared from arginine amino acid have remarkable biocidal and antiseptic properties because of the presence of a guanidine side chain [67, 71]. Cationic lysine-based surfactants have also been studied where the water-repelling unit is connected to the carboxylic group of lysine through an amide or ester bond [67, 72]. Cationic surfactants made from L-tyrosine and L-tryptophan act as very good gelators showing effective bactericidal effects [67]. Cationic surfactants made from tyrosine and phenylalanine are also known to have antimicrobial activity. Here, the alkyl chain is connected to the carboxylic group of the amino acid through an ester bond [73]. Generally, in most cases, it was observed that such cationic AASs were more effective against Gram-positive bacteria as compared to their corresponding negative strain.

12.4.6 NANOTECHNOLOGY

In the last few decades, nanoscience and nanotechnology have made significant advancements and indispensable contributions in diverse fields like electronic industries, medical and pharmaceutical sectors and energy divisions and also for environmental remodeling. Since surfactants have the tendency to self-assemble and can reduce the interfacial surface tension, they find application in the synthesis of nanoemulsions. Nanoemulsions are made up of small drops, the size of which is generally less than 300 nm and are thermodynamically unstable but kinetically stable [53]. Nanoemulsions are immune towards changes in concentration and temperature. The nature and ratio of surfactant and co-surfactant determine the size and stability of nanoemulsions. They are used in agrochemicals and personal care products. The only drawback of nanoemulsions is that over a long period of time, they are unstable [74].

12.4.6.1 Nanoemulsion in Agrochemicals

Agrochemical industries are always on the lookout for new kinds of pesticides which have high efficacy when used in agriculture, but at the same time, it is to ensure that the environment is not degraded due to their overuse. Once pesticides are added to the soil during crop production, large quantities of them are wasted due to factors like surface run-off and volatility, as a result of which all of the pesticides do not reach the plants. Nanoemulsions are used for effective delivery of agrochemicals to plants with low consumption of surfactant. A literature study was found discussing the use of alkylpolyglucosides and fatty-acid methyl esters to form a nanoemulsion which was used to deliver glyphosate – a herbicide [75]. Another study described the use of surfactants like polyoxyethylene 3-lauryl ether and alkylpolyglucoside and methyl laurate oil for the production of oil-in-water nanoemulsion. This arrangement was found to deliver promising results for the application of certain water-insoluble pesticide [76].

12.4.6.2 Nanoemulsion in Personal Care Products

Skincare products like moisturizers, creams and lotions make use of nanoemulsions since they have certain properties which make them a viable option for personal care products. Mostly, they have a gel-like nature and some are even transparent. Nanoemulsions are known to have superior hydrating power and they can penetrate deep into the skin because of their large surface area. Nanoemulsions made up of fatty esters of sugars, ethoxylated fatty ethers and alkyl ether citrates are predominantly found in cosmetics [77–79]. They also find use in hair care products like shampoos and conditioners. Nanoemulsions can also be used for enhancing the fragrance of certain personal care products without using alcohol in their formulations [53].

12.4.7 Biotechnology

Nanoemulsions have predominant application in the field of cell culture technology. They are used for the delivery of reproducible supplements which are oil soluble to cell cultures in human beings in a controlled manner. Phospholipids which are a class of natural triacylglycerols are largely used for making nanoemulsions. Since phospholipids are the primary components of cell membrane, thus nanoemulsions can be well imbibed by cells [80]. The use of certain surfactants as antimicrobial systems has already been discussed in Section 12.4.5. Nanoemulsions are also employed as potent antimicrobial agents. Essential oils have largely been used as antimicrobial agents in different food products. The problem with nanoemulsions as antimicrobial agents is their poor water solubility, volatility and reactivity. Inclusion of essential oils in nanoemulsions enhances their stability and also increases their antimicrobial properties in food products [81].

12.4.8 Miscellaneous Application

Interactions of surfactants, specifically amino acid surfactants with different biomolecules like proteins and phospholipids, make them indispensable in the biomedical industry. Surfactant-protein interaction helps in explaining the denaturation and solubilizing activities of surfactants [21]. The water-repellant part of the surfactant-protein complex activates protein unfolding by reacting with the amino acid remnants which are nonpolar in nature [82]. When interactions between hemoglobin and gemini surfactants made from glutamic acid were studied, it was observed that the denaturation property of the surfactant was reduced due to the bulky size. Also, interactions between certain globular protein BSA and gemini surfactant made from cystine were reported to be affected by temperature and pH. Interactions were also dependent on the stereochemistry of the gemini surfactant [21, 82]. Interactions between phospholipids and anionic gemini amino acid surfactants have also been studied. Interactions between gemini surfactants and a micelle-forming lipid, i.e. diheptanoyl phosphatidylcholine, and a vesicle-forming lipid, i.e. dimyristoyl phosphatidylcholine, were found to be harmonious [83]. These observations were because of reduced electrostatic repulsion between head groups of surfactants, which are anionic in nature because of the insertion of the zwitterionic phospholipid in the micelle [83]. Such interactions between surfactants and macromolecules generate results which can be used for various biomedical applications.

12.5 SUMMARY

Synthetic surfactants synthesized from petroleum feedstocks have been reported to have significant detrimental effects on the environmental flora and fauna. Also, all surfactants are not effectively biodegradable. After studying different surface properties, it was established that surfactants synthesized from renewable sources like amino acids, sugars and tannic acid are equally efficient as synthetic surfactants. They also have promising biochemical and biomedical applications. More research is thus underway to replace synthetic surfactants with the environment-friendly surfactants synthesized from renewable resources.

REFERENCES

1. Cowan-Ellsberry, C. et al.; *Critical Reviews in Environmental Science and Technology*; 2014; 44; 1893–1993.
2. Ivanković, T., & Hrenović, J.; *Archives of Industrial Hygiene and Toxicology*; 2010; 61; 95–110.
3. Ying, G.-G.; *Environment International*; 2006; 32; 417–431.

4. Olkowska, E., Ruman, M., & Polkowska, Z.; *Journal of Analytical Methods in Chemistry*; 2014; 2014; 1–15.
5. Odokuma, L. O., & Okpokwasili, G. C.; *Environmental Monitoring and Assessment*; 1997; 45; 43–57.
6. Zoller, U.; *Water Science and Technology*; 1993; 27; 187–195.
7. Scott, M. J., & Jones, M. N.; *Biochimica et Biophysica Acta*; 2000; 1508; 1–2; 235–251.
8. Gonzalez, S., Petrovic, M., & Barcelo, D.; *Trends in Analytical Chemistry*; 2007; 26; 2; 116–124.
9. Petrovic, M., & Barcelo, D.; *Emerging Organic Pollutants in Wastewaters and Sludges*, Barcelo, D., Ed., Springer, Heidelberg, Germany, 2004.
10. Nomura, Y., Ikebukuro, K., & Yokoyama, K. et al.; *Biosensors and Bioelectronics*; 1998; 13; 9; 1047–1053.
11. Jensen, J.; *Science of the Total Environment*; 1999; 226; 2–3; 93–111.
12. Loyo-Rosales, J. E., Rice, C. P., & Torrents, A.; *Chemosphere*; 2007; 68; 11; 2118–2127.
13. Cserha'ti, T., Forga'cs, E., & Oros, G.; *Environment International*; 2002; 28; 337–348.
14. Tkalin, A. V. et al.; *Marine Pollution Bulletin*; 1993; 26; 418–423.
15. Tkalin, A. V. et al.; *Marine Pollution Bulletin*; 1993; 26; 704–706.
16. Endo, R. M., Letey, J., Valoras, N., & Osborn, J. F.; *Agronomy Journal*; 1969; 61; 850–854.
17. Banat, I. M., Makkar, R. S., & Cameotra, S. S.; *Applied Microbiology and Biotechnology*; 2000; 53; 495–508.
18. Patel, Z. N., & Saraswathy, N.; *World Journal of Pharmaceutical Research*; 2014; 3; 1968–1977.
19. Vijayakumar, S., & Saravanan, V.; *Research Journal of Microbiology*; 2015; 10; 181–192.
20. Zhang, Q. Q., Cai, B. X., Xu, W. J., Gang, H. Z., Liu, J. F., Yang, S. Z., & Mu, B. Z.; *Colloids and Surfaces A: Physicochemical and Engineering Aspects*; 2015; 483; 87–95.
21. Perez, L., Pinazo, A., Pons, R., & Infante, M.; *Advances in Colloid and Interface Science*; 2014; 205; 134–155.
22. Zhang, G., Xu, B., Han, F., Zhou, Y., Liu, H., Li, Y., & Wang, N.; *American Journal of Analytical Chemistry*; 2013; 4; 445–450.
23. Yea, D. N., Lee, S. M., Jo, S. H., Yu, H. P., & Lim, J. C.; *Journal of Surfactant Detergent*; 2018; 21; 4.
24. Lim, J. C., Lee, M. C., Lim, T. K., & Kim, B. J.; *Colloids and Surfaces, A: Physicochemical and Engineering Aspects*; 2014; 446; 80–89.
25. Lee, S. M., Lee, J. Y., Yu, H. P., & Lim, J. C.; *Journal of Industrial and Engineering Chemistry*; 2016; 38; 157–166.
26. Salman, A. A., Tabandeh, M., Heidelberg, T., Hussen, R. S. D., & Ali, H. M.; *Carbohydrate Research*; 2015; 412; 28.
27. Ali, T. H., Tajuddin, H. A. B., Hussen, R. S. D., & Heidelberg, T.; *Journal of Surfactants and Detergents*; 2015; 18; 881.
28. Borges, M. R., & de Carvalho Balaban, R.; *Journal of Biotechnology*; 2014; 192; 42.
29. Zhu, D., Baryal, K. N., Adhikari, S., & Zhu, J.; *Journal of the American Chemical Society*; 2014; 136; 3172.
30. Bazito, R. C., & El Seoud, O. A.; *Carbohydrate Research*; 2001; 332; 95.
31. Zhang, L., & Somasundaran, P.; *Journal of Colloid and Interface Science*; 2006; 302; 25.
32. Behler, A., Hensen, H., & Seipel, W.; *Proceedings 6th World Surfactant Congress, Cesio*; 2004; 20–23.
33. Zhao, T., Gu, J., Pu, W., Dong, Z., & Liu, R.; *RSC Advances*; 2016; 6; 70165–70173.
34. Sagisaka, M., Narumi, T., Niwase, M., Narita, S., Ohata, A., James, C., Yoshizawa, A., Taffin de givenchy, E., Guittard, F., & Alexander, S.; *Langmuir*; 2014; 30; 6057.
35. Azira, H., & Tazerouti, A.; *Journal of Surfactants and Detergents*; 2007; 10; 185.
36. Gao, B., Yu, Y., & Jiang, L.; *Colloids and Surfaces A: Physicochemical and Engineering Aspects*; 2007; 293; 210.
37. Chauhan, S., Kaur, M., Kumar, K., & Chauhan, M. S.; *The Journal of Chemical Thermodynamics*; 2014; 78; 175.
38. Jiahua, Z., & Yingde, C.; *Speciality Petrochemicals*; 2001; 2; 4.
39. Ao, M., Huang, P., Xu, G., Yang, X., & Wang, Y.; *Colloid and Polymer Science*; 2009; 287; 395.
40. Shaban, S. M., Aiad, I., El-Sukkary, M. M., Soliman, E., & El-Awady, M. Y.; *Journal of Molecular Liquids*; 2015; 207; 256.
41. Fan, T., Chen, C., Fan, T., Liu, F., & Peng, Q.; *Journal of Hazardous Materials*; 2015; 297, 340.
42. Liu, X.-P., Feng, J., Zhang, L., Gong, Q.-T., Zhao, S., & Yu, J.-Y.; *Colloids and Surfaces A: Physicochemical and Engineering Aspects*; 2010; 362; 39.
43. Abe, M., Tsubone, K., Koike, T., Tsuchiya, K., Ohkubo, T., & Sakai, H.; *Langmuir*; 2006; 22; 8293.
44. Ao, M., & Kim, D.; *Journal of Chemical & Engineering Data*; 2013; 58; 1529.

45. Hou, B.-F., Wang, Y.-F., & Huang, Y.; *Applied Surface Science*; 2015; 330; 56.
46. Zhao, F., Clarens, A., Murphree, A., Hayes, K., Skerlos, S. J.; *Environmental Science Technology*; 2006; 40; 7930–7937.
47. Nabel, N. A., Ahmed, F. M. E., Mohammed, D. E., & Mohamad, H. N.; *Journal of Surfactants and Detergents*; 2012; 15; 4.
48. Yan, J., Wang, D., Bu, F., &Yang, F. F.; *Journal of Solution Chemistry*; 2010; 39; 1501–1508.
49. Negm, N. A., Aiad, I. A., & Tawfik, S. M.; *Journal of Surfactants and Detergents*; 2010; 13; 503–511.
50. Alsabagh, A. M.; *Polymer Advanced Technology*; 2000; 11; 48–56.
51. Blunk, D., Bierganns, P., Bongartz, N., Tessendorf, R., & Stubenrauch, C.; *New Journal of Chemistry*; 2006; 30; 1705–1717.
52. Parker, J. L., & Newstead, S.; *Protein Science*; 2012; 21; 1358–1365.
53. Guenic, S., Chaveriat, L., Lequart, V., Joly, N., & Martin, P.; *Journal of Surfactants and Detergents*; 2019; 22; 5–21.
54. Kirby, A. J., Camilleri, P., Engberts, J. B. F. N., Feiters, M. C., Nolte, R. J. M., Söderman, O., & van Eijk, M. C. P.; *Angewandte Chemie, International Edition*; 2003; 42; 1448–1457.
55. Fielden, M. L., Perrin, C., Kremer, A., Bergsma, M., Stuart, M. C., Camilleri, P., & Engberts, J. B. F. N.; *European Journal of Biochemistry*; 2001; 268; 1269–1279.
56. Camilleri, P., Kremer, A., Edwards, A. J., Jennings, K. H., Jenkins, O., Marshall, I., & Kirby, A. J.; *Chemical Communications*; 2000; 14; 1253–1254.
57. Buijnsters, P. J. J. A., García Rodríguez, C. L., Willighagen, E. L., Sommerdijk, N. A. J. M., Kremer, A., Camilleri, P., & Zwanenburg, B.; *European Journal of Organic Chemistry*; 2002; 2002; 1397–1406.
58. McGregor, C., Perrin, C., Monck, M., Camilleri, P., & Kirby, A. J.; *Journal of the American Chemical Society*; 2001; 123; 6215–6220.
59. Wasungu, L., Scarzello, M., van Dam, G., Molema, G., Wagenaar, A., Engberts, J. B. F. N., & Hoekstra, D.; *Journal of Molecular Medicine*; 2006; 84; 774–784.
60. Jaiswal, M., Dudhe, R., & Sharma, P. K.; *3 Biotech*; 2015; 5; 123–127.
61. Calderilla-Fajardo, S. B., Cázares-Delgadillo, J., Villalobos-García, R., Quintanar-Guerrero, D., Ganem-Quintanar, A., & Robles, R.; *Drug, Development and Industrial Pharmacy*; 2006; 32; 107–113.
62. Szuts, A., & Szabó-Révész, P.; *International Journal of Pharmaceutics*; 2012; 433; 1–9.
63. Ahmad, N., Ramsch, R., Llinàs, M., Solans, C., Hashim, R., & Tajuddin, H. A.; *Colloids and Surfaces B: Biointerfaces*; 2014; 115; 267–274.
64. Hadzir, N. M., Basri, M., Basyaruddin, M., Rahman, A., Salleh, A. B., Zaliha Raja, R. N., & Abdul Rahman, H. B.; *American Association of Pharmaceutical Scientists*; 2013; 14; 456–463.
65. Holmberg, K.; *Journal of Colloid and Interface Science*; 2004; 274; 355–364.
66. Xie, Y., Kocaefe, D., Chen, C., & Kocaefe, Y.; *Journal of Nanomaterials*; 2016; 1–10.
67. Pinazo, A., Manresa, M. A., Marques, A. M., Bustelo, M., Espuny, M. J., & Pérez, L.; *Advances in Colloid and Interface Science*; 2016; 228; 17–39.
68. Morán, M. C., Pinazo, A., Pérez, L., Pinazo, A., Clapés, P., Angelet, M., Garcia, M. T., & Vinardell, M. P.; *Green Chemistry*; 2004; 6; 233–240.
69. Faustino, C., Serafim, C., Ferreira, I., Pinheiro, L., & Calado, A.; *Colloids and Surfaces A: Physicochemical and Engineering Aspects*; 2015; 480; 426–432.
70. Serafim, C., Ferreira, I., Rijo, P., Pinheiro, L., Faustino, C., Calado, A., & Garcia-Rio, L.; *International Journal of Pharmaceutics*; 2016; 497; 23–35.
71. Castillo, J. Á., Pinazo, A., Carilla, J., Infante, M. R., Alsina, M. A., Haro, I., & Clapés, P.; *Langmuir*; 2004; 20; 3379–3387.
72. Colomer, A., Perez, L., Pons, R., Infante, M. R., Perez-Clos, D., Manresa, A., Espuny, M. J., & Pinazo, A.; *Langmuir*; 2013; 29; 7912–7921.
73. Joondan, N., Jhaumeer-Laulloo, S., & Camul, P.; *Microbiological Research*; 2014; 169; 675–685.
74. Gutiérrez, J. M., González, C., Maestro, A., Solè, I., Pey, C. M., & Nolla, J.; *Current Opinion in Colloid & Interface Science*; 2008; 13; 245–251.
75. Jiang, L. C., Basria, M., Omar, D., Rahman, M. B. A., Salleh, A. B., Zaliha Raja, R. N., & Abdul Rahman, A. S.; *Pesticide Biochemistry and Physiology*; 2012; 102; 19–29.
76. Du, Z., Wang, C., Tai, X., Wang, G., & Liu, X.; *ACS Sustainable Chemistry & Engineering*; 2016; 4; 983–991.
77. Simonnet, J.-T., Sonneville, O., & Legret, S.; Nanoemulsion based on alkyl ether citrates and its use in the cosmetics, dermatological, pharmacological and/or ophtalmological fields; US6413527; 2002; L'Oréal.

78. Simonnet, J.-T., Sonneville, O., & Legret, S.; Nanoemulsion based on ethoxylated fatty ethers or on ethoxylated fatty esters and its uses in the cosmetics, dermatological and/or ophthalmological fields; US6375960; 2002; L'Oréal.
79. Simonnet, J.-T., Sonneville, O., & Legret, S.; Nanoemulsion based on sugar fatty esters or on sugar fatty ethers and its uses in the cosmetics, dermatological and/or ophthalmological fields; US668371; 2004; L'Oréal.
80. Zülli, F., Liechti, C., & Suter, F.; *International Journal of Cosmetic Science*; 2000; 22; 265–270.
81. Donsì, F., Annunziata, M., Vincensi, M., & Ferrari, G.; *Journal of Biotechnology*; 2012; 159; 342–350.
82. Faustino, C. M. C., Calado, A. R. T., & Garcia-Rio, L.; *Biomacromolecules*; 2009; 10; 2508–2514.
83. Faustino, C. M. C., Calado, A. R. T., & Garcia-Rio, L.; *Journal of Colloid and Interface Science*; 2011; 359; 493–498.

13 Surfactants

Patent Landscape of the Most Versatile Class of Materials

Sonali Kesarwani[1], Divya Bajpai Tripathy[1],
Anuradha Mishra[2] and Anjali Gupta[1]
[1]School of Basic and Applied Sciences, Galgotias
University, Greater Noida, Uttar Pradesh, India
[2]Department of Applied Chemistry, SoVSAS, Gautam Budhha
University, Gautam Budh Nagar, Uttar Pradesh, India

CONTENTS

DOI: 10.1201/9781003144878-13

13.1 INTRODUCTION

Widespread use of surfactants from household and cosmetics to industrial processes and products has prompted researchers to synthesize new surface-active materials, tailor-made to specific industry [1–4]. The use of surfactants is a must/almost a must in formulations of detergents [5], cosmetics [6], agricultural [7] and petroleum products [8], lubricants [9], pharmaceuticals [10], drug delivery carrier [11], dispersants [12], corrosion inhibitors [13] and so on so forth. The widespread use of surfactants makes them and their markets constantly evolving, providing opportunities and challenges to the researchers. There are many types of surfactants available in the market such as cationic, anionic, zwitterionic, nonionic, gemini and polymeric [14–16]. Surfactants are being synthesized chemically [17], enzymatically [18] and chemoenzymatically [19]. Various routes have been implied in synthesis of surfactants, from conventional thermal to microwave assisted [20] and from multi-steps [21] to one-pot synthesis [22].

There is plethora of research papers and patents on surface-active materials. The present article reviews recent (during 2010–2020) patents issued on the synthesis and applications of surfactants. The manuscript brings light to the major outputs patented in the field of surfactants science and technology in the last 11 years. The bases taken for describing these outputs in the manuscript are the raw materials used for the synthesis and the routes followed in their preparation as well as the broad area of their applications.

13.2 METHODOLOGY

Patent lens [23] search tool was used to create the database. In addition to that, Google Patents and Escapenet like free tools have also been explored in this study. The broad areas taken were synthesis and applications of surfactants and the sub-areas considered in this study are:

   i. Duration: 01 January 2010 up to 31 December 2020
  ii. Number of patents published as per different surfactants category as well as on applications
 iii. Top jurisdictions such as:

- US: US Patent
- WO: WIPO
- AU: Australian Patent
- EP: European Patent
- CN: Chinese Patent
- KR: Korean Patent
- JP: Japanese Patent
- IN: Indian Patent
- MX: Mexico Patent
- FR: French Patent etc.

 iv. Number of patents under different classification systems such as Cooperative Patent Classification (CPC) [24], International Patent Classification Reform (IPCR) [25] and US Patent Classification (USPC) [26]. The codes used are described hereunder:

- A61: Medical or Veterinary Science Hygiene
- A61K: Preparations for medical, dental, or toilet purposes
- A61K 6/00: Preparations for dentistry
- A61K 8/00: Cosmetics or similar toilet preparations
- A61K 9/00: Medicinal preparations characterized by special physical form
- A61K 31/00: Medicinal preparations
- A61K 33/00: Medicinal preparations containing inorganic active ingredients

- A61K 35/00: Medicinal preparations containing materials or reaction products thereof with undetermined constitution
- A61K 36/00: Medicinal preparations of undetermined constitution containing material from algae, lichens, fungi or plants or derivatives thereof, e.g., traditional herbal medicines.
- A61K 41/00: Medicinal preparations obtained by treating materials with wave energy or particle radiation
- A61K 45/00: Medicinal preparations containing active ingredients not provided for in groups A61K 31/00–A61K 41/00
- A61K 47/00: Medicinal preparations characterized by the non-active ingredients used, e.g., carriers or inert additives; Targeting or modifying agents chemically bound to the active ingredient
- C07K 16/28: Against receptors, cell surface antigens or cell surface determinants
- C12Q 1/68: Involving nucleic acids
- A61P 29/00: Noncentral analgesic, antipyretic or anti-inflammatory agents
- A61P 35/00: Antineoplastic agents
- A61K 8/00: Cosmetics or similar toilet preparations
- A61K 45/00: Medicinal preparations containing active ingredients not provided for in groups
- A61K 38/00: Medicinal preparations containing peptides
- A61K 31/48: Containing macromolecular compounds having statistically distributed amino acid units
- A61K 31/74: Medicinal preparations containing antigens or antibodies
- A01N 2300/00: Combinations or mixtures of active ingredients
- A61Q 90/00: Cosmetics or similar toilet preparations
- A61Q 5/02: Preparations for cleaning the hair
- A61Q 19/10: Washing or bathing preparations
- B82Y 30/00: Nanotechnology for materials or surface science, e.g., nanocomposites
- 435/32: Testing for antimicrobial activity of a material
- 435/325: Process of propagating, maintaining or preserving an animal cell or composition thereof; process of isolating or separating an animal cell or composition thereof; process of preparing a composition containing an animal cell; culture media thereof
- 435/69.1: Recombinant DNA technique included in the method of making a protein or polypeptide.
- 530/350: Proteins
- 424/401: Cosmetic, antiperspirant, dentifrice
- 536/23.2: Encodes an enzyme
- 430/270.1-270.1: Radiation sensitive composition or product or process of making
- 514/12: Muscle contraction affecting
- 536/23.1: DNA or RNA fragments or modified forms thereof

v. Other criteria: The number of patents published by top inventors, applicants, owners etc. has also been discussed.

13.3 NUMBER OF PATENTS

In the last 11 years, the total number of patents published is 760,231. As given in Figure 13.1, it seems that 2019 was the most prominent year with 78,769 patents, whereas 2010 was the least prominent year with the number of patents, 55,283 only. With 13,115 patents, Procter and Gamble was the top applicant and the United States was the most prominent jurisdiction and W. N. Ulrike was the top inventor with 881 patents in the last 11 years.

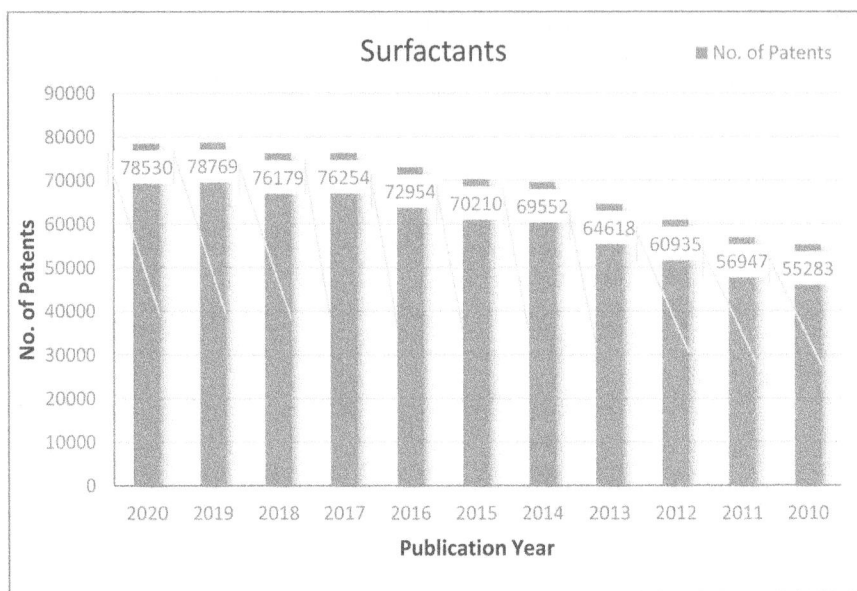

FIGURE 13.1   Number of patents on surfactant per year.

13.3.1   PATENTS ON THE SYNTHESIS OF SURFACTANTS

Within the last 11 years (2010–2020), a large number of patents (327,322) have been published on the synthesis of surfactants using different raw materials, types and methods of synthesis. Raw materials varied from fatty acids, alcohols, ethers, carbohydrates, proteins. Types of surfactants involve cationic, nonionic, anionic, amphoteric, gemini and polymeric surfactants. Methods of synthesis included conventional thermal methods to microwave and chemo to enzymatic synthesis. With results published in 11 different years, 2020 is most prominent with 33,731 patents (Figure 13.2).

13.3.1.1   Jurisdiction

Study on the data available since 2010 revealed that United States is on top with maximum number of patents (196,260) available on the synthesis of surfactants, whereas WIPO (1 of the 15 specialized agencies of the United Nations to promote the protection of intellectual property throughout the world) is the on second position with 57,081 patents, whereas China, Korea, Japan, Russia, France and Canada play the role of small fishes in this area (Figure 13.3).

13.3.1.1.1   Classification

In this study, three classification systems CPC, IPCR and US classification have been taken into account and results are summarized in Figures 13.4–13.6. Specification used in this classification has already been summarized in methods.

13.3.1.2   Applicants

Among the top 10 applicants, Fujifilm Corporation is on top with 4800 patents published, whereas Basf Se was the second top applicant with more than 4400 patents. List of 10 top applicants has been provided in Figure 13.7.

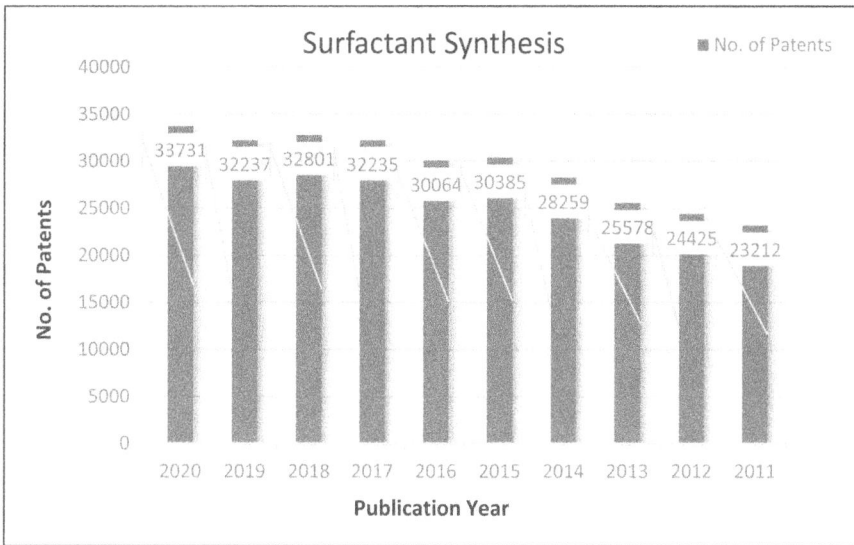

FIGURE 13.2 Total number of patents on surfactants synthesis.

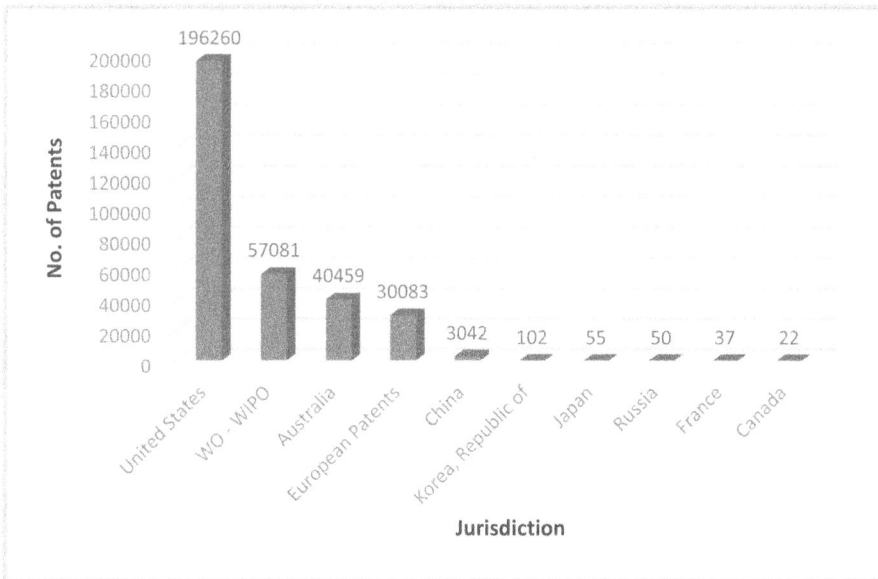

FIGURE 13.3 Top 10 jurisdictions on surfactants synthesis.

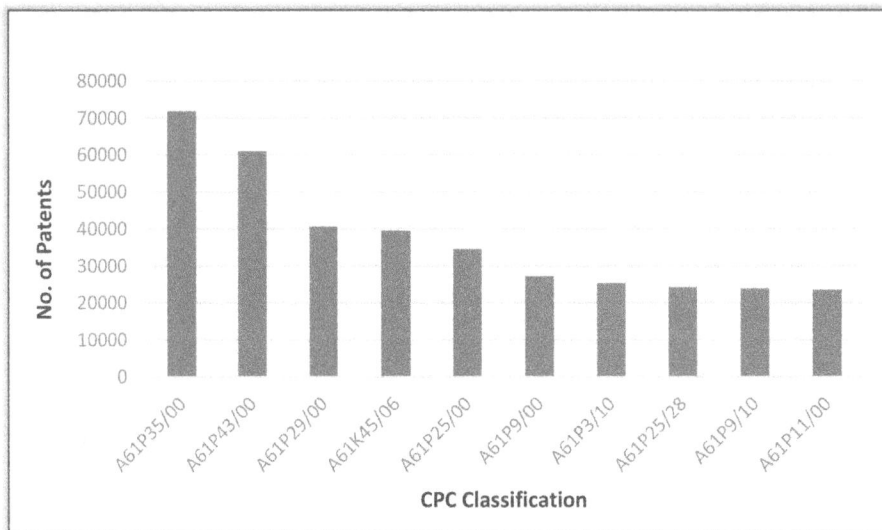

FIGURE 13.4 Top 10 categories under CPC classification.

13.3.1.3 Inventors

With 764 patents, W. N. Ulrike was on top in the list of inventors. Top 10 inventors in the field of surfactants synthesis are given in Figure 13.8.

13.3.1.4 Owners

With 3572 patents published, Fujifilm was at top in the list of owners. Novartis AG is in the second position with 2155 patents on the synthesis of surfactants since 2010. Rest top owners include Basf Se, Regents of University etc. (Figure 13.9).

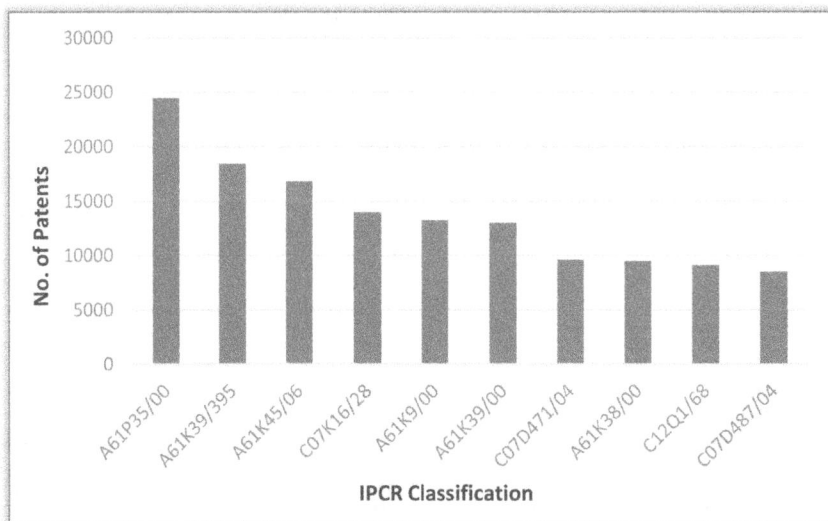

FIGURE 13.5 Top 10 categories under IPCR classification.

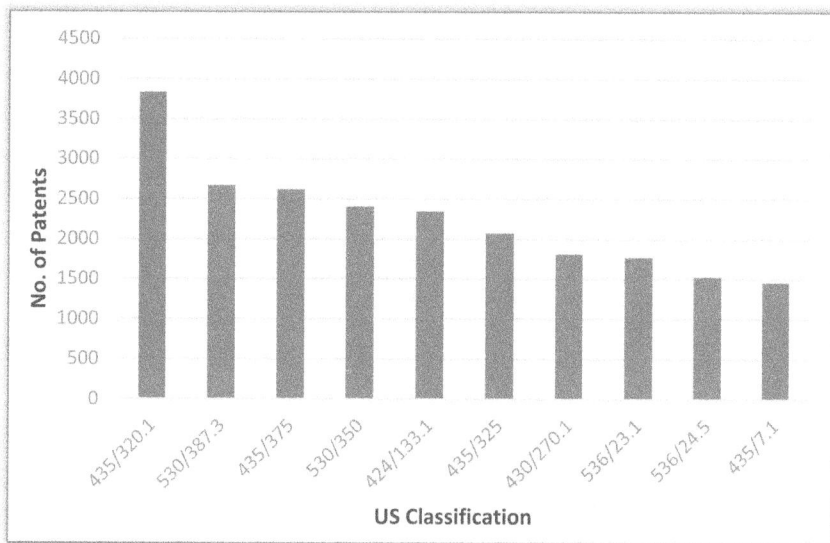

FIGURE 13.6 Top 10 categories under US classification.

13.4 CATIONIC SURFACTANTS

Study on the last 11 years data revealed that 2020 was the most prominent year with 18,612 patents on the synthesis of cationic surfactants, whereas in 2019, 18,104 patents were published on the cationic surfactant's synthesis (Figure 13.10). The total number of patents on the synthesis of cationic surfactants is 176,467.

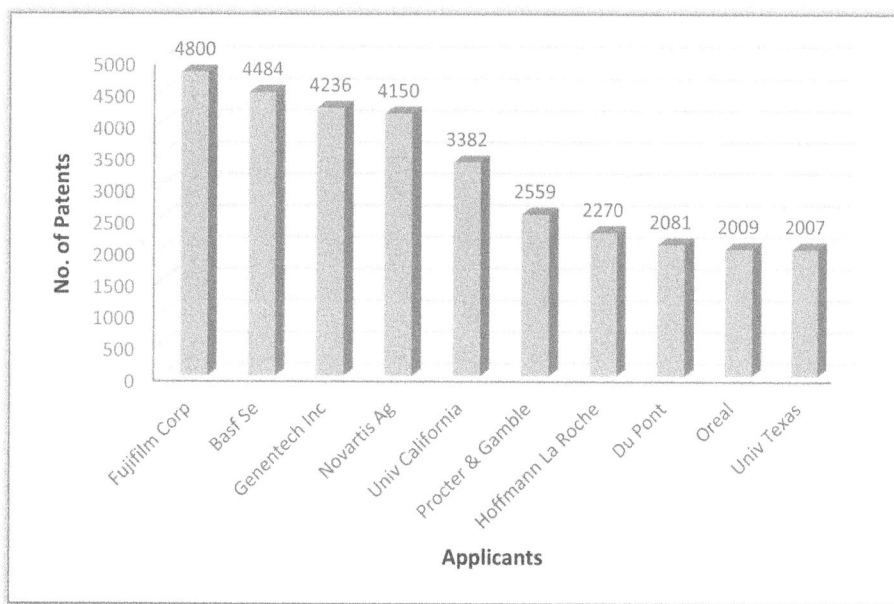

FIGURE 13.7 Top 10 applicants on surfactants synthesis.

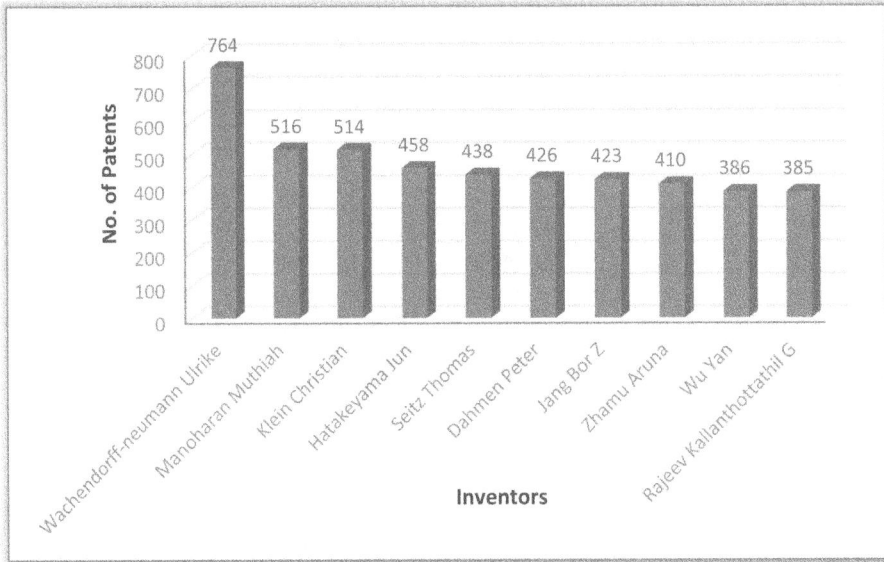

FIGURE 13.8 Top 10 inventors on surfactants synthesis.

13.4.1 JURISDICTION

The United States has published a maximum number of patents (106,244) on the synthesis of cationic surfactants, whereas WIPO was at second top with 31,581 patents. Australia published 22,583 and European patent office published 15,703 patents. Other top jurisdictions are China with 307 patents, Korea [10], France [8], Japan [7], European patent organization [5] and Russia with 3 patents (Figure 13.11).

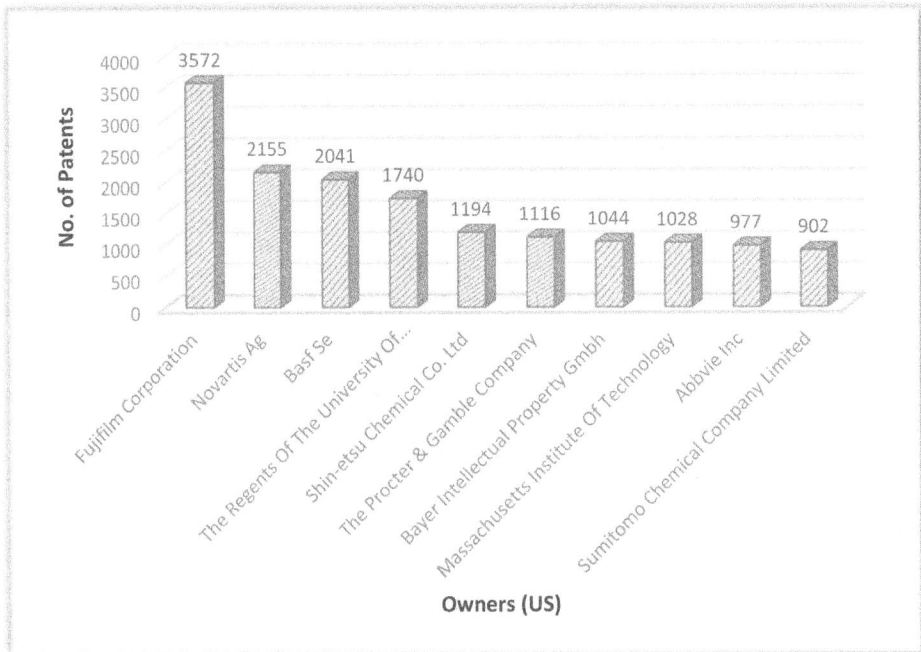

FIGURE 13.9 Top 10 owners on surfactants synthesis.

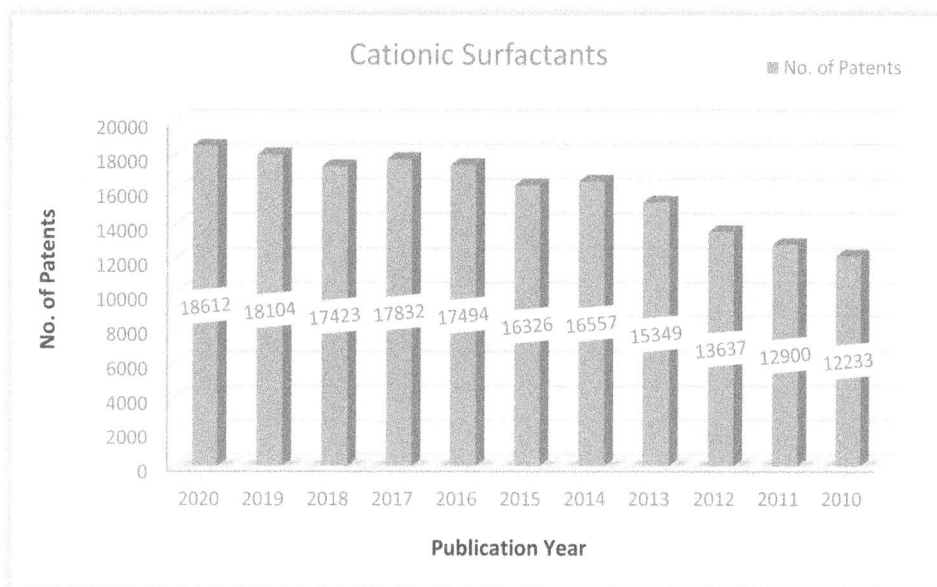

FIGURE 13.10 Number of patents on cationic surfactants synthesis.

13.4.2 Applicants, Inventors, Owners

As Fujifilm Corporation has published a maximum number of patents, top owners and applicants are from Fujifilm itself. Fujifilm Corp. is on top as applicant with 4045 patents as well as top owner with 2943 patents, whereas G. Wassilios is top inventor with 338 patents (Figures 13.12–13.14).

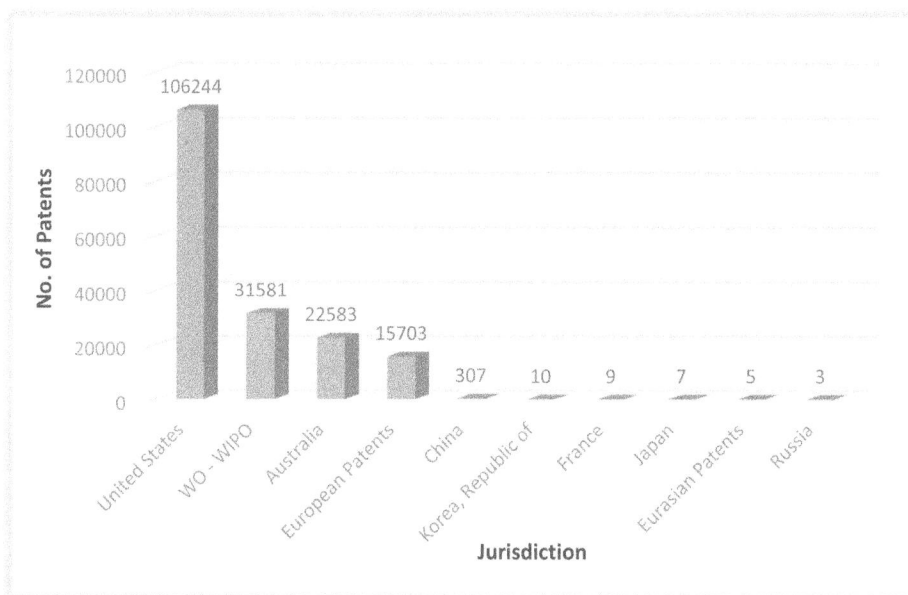

FIGURE 13.11 Top 10 jurisdiction on cationic surfactants synthesis.

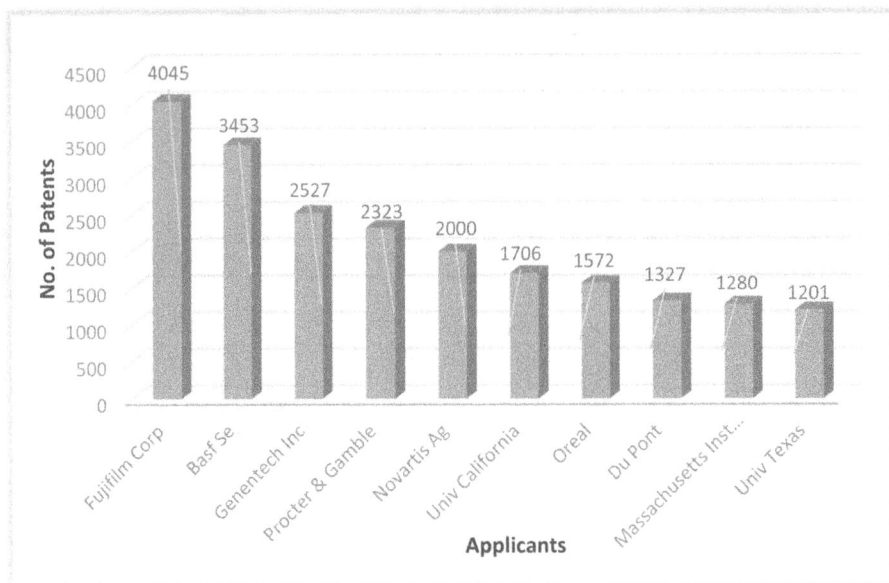

FIGURE 13.12   Top 10 applicants on the cationic surfactant synthesis.

13.5   ANIONIC SURFACTANTS

On the basis of the number of patents published on the synthesis of anionic surfactants, 2020 was the most prominent year with 15,467 patents. In 2010, this figure was about 10,000 (Figure 13.15).

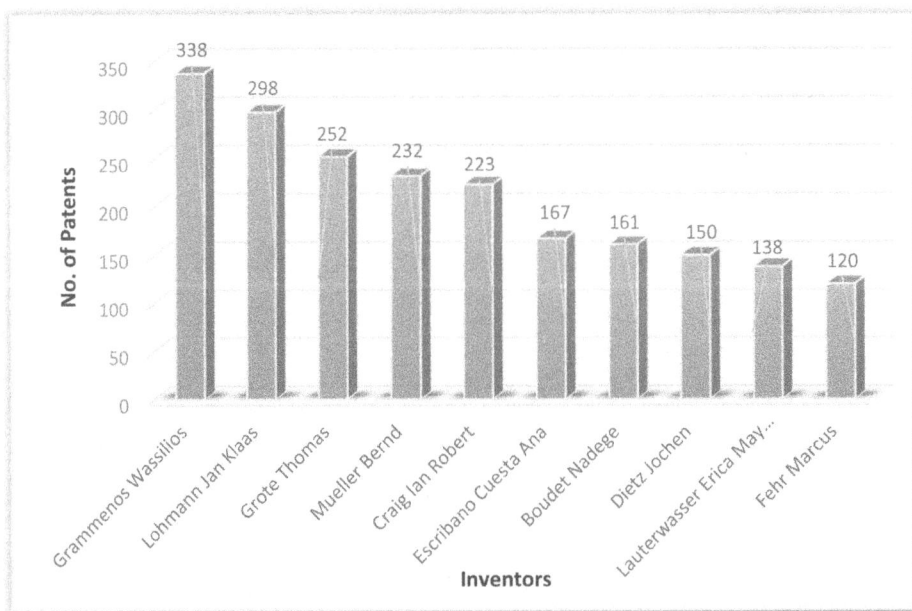

FIGURE 13.13   Top 10 inventors on the cationic surfactant synthesis.

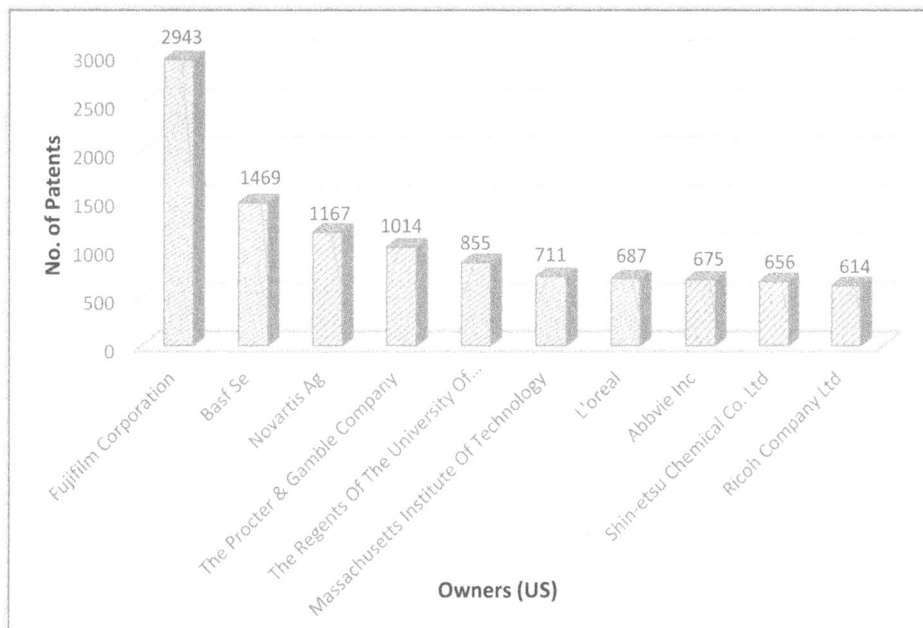

FIGURE 13.14   Top 10 owners on the cationic surfactant synthesis.

13.5.1   JURISDICTION

The United States has published the largest number of patents (92,276) on the synthesis of anionic surfactants, whereas WIPO was at second largest with 26,412 patents published. Australia published 18,612 and European patent office published 14,139 patents. Other top jurisdictions are China with 204 patents, France [8], Korea and Canada [6] and Germany with 5 patents. Russia has also published 3 patents (Figure 13.16).

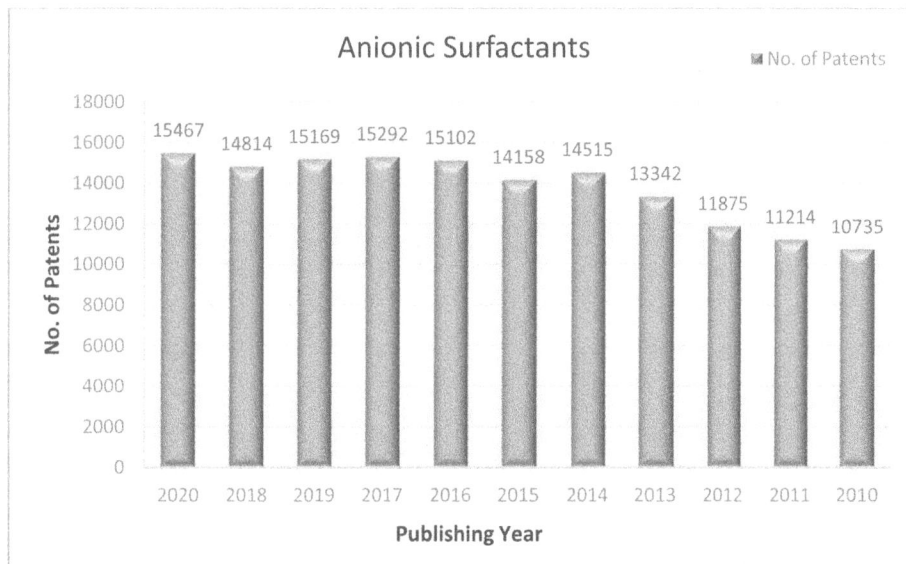

FIGURE 13.15   Number of patents on anionic surfactants synthesis.

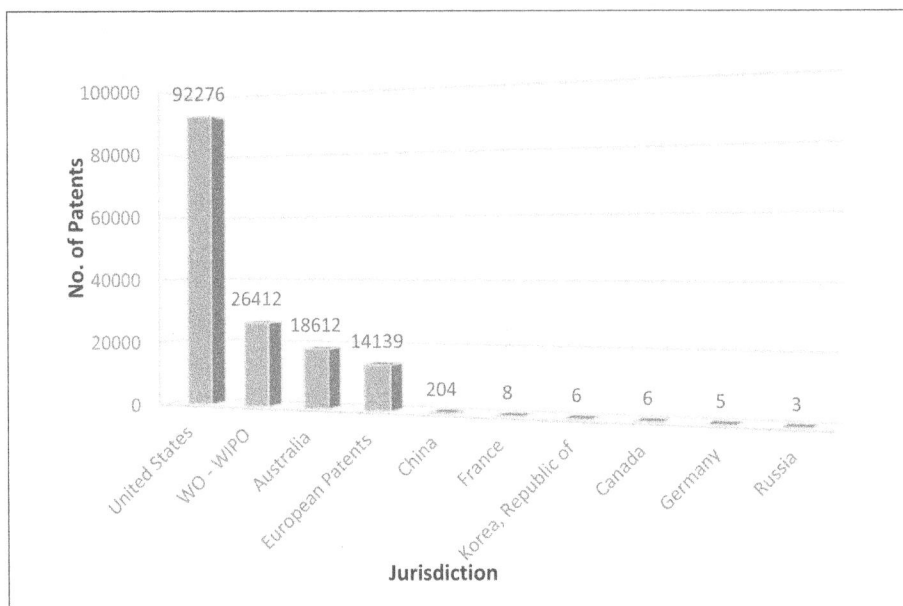

FIGURE 13.16   Top 10 jurisdictions on anionic surfactants synthesis.

13.5.2   APPLICANTS, INVENTORS, OWNERS

Fujifilm Corp. has top applicants and owners with Fujifilm Corp having 3920 patents published as top applicant as well as with 2844 patents as top owner. W. N. Ulrike is in the top 10 inventors with 557 patents (Figures 13.17–13.19).

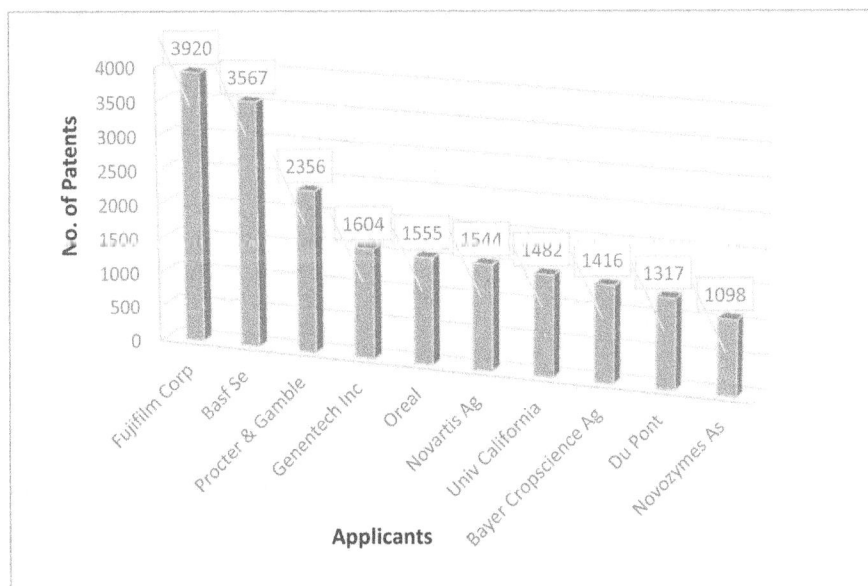

FIGURE 13.17   Top 10 applicants on anionic surfactants synthesis.

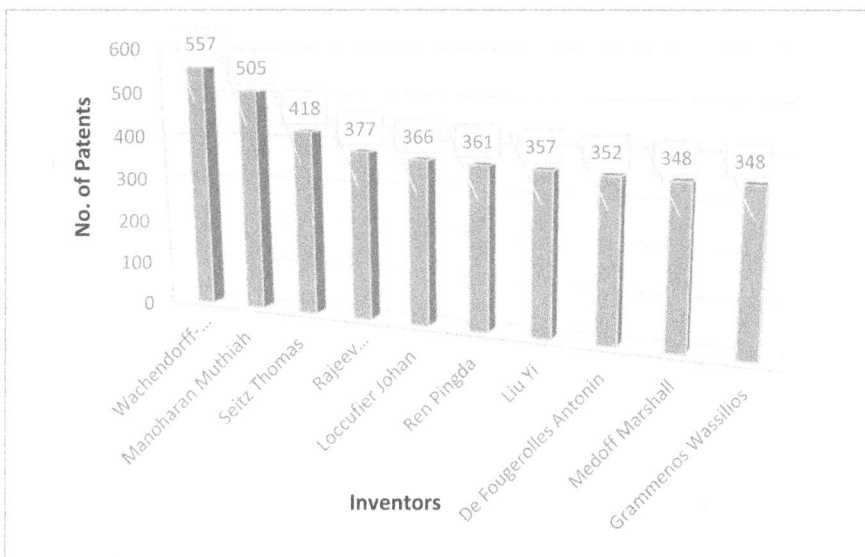

FIGURE 13.18   Top 10 inventors on anionic surfactants synthesis.

13.6   NONIONIC SURFACTANTS

On the basis of the number of patents published on the synthesis of nonionic surfactants, 2016 was the most prominent year with 7961 patents; however, this value is about half to the patent published on cationic/anionic surfactants synthesis. In 2010, this figure was less than 6000, whereas in 2020, more than 7800 patents will have been published (Figure 13.20).

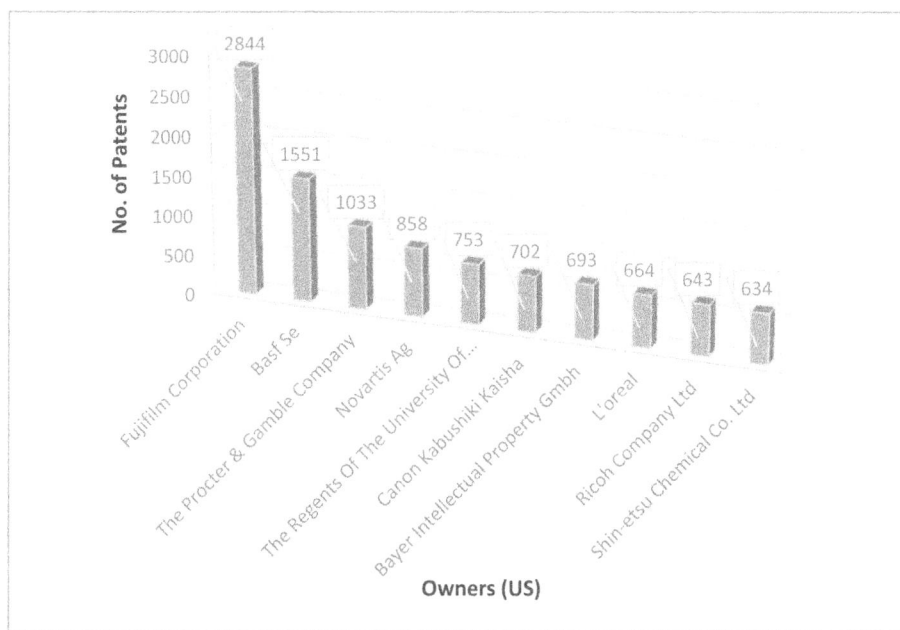

FIGURE 13.19   Top 10 owners on anionic surfactants synthesis.

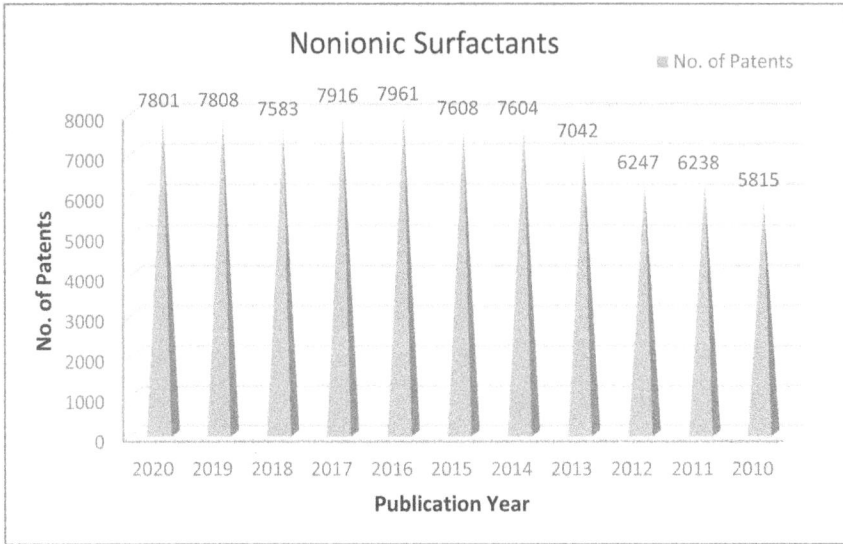

FIGURE 13.20 Number of patents on nonionic surfactants synthesis.

13.6.1 JURISDICTION

The United States has published the largest number of patents on the synthesis of nonionic surfactants and the number is 48,792. WIPO is the second largest with 12,853 patents published. Australia
published 9505 patents and European patent office published 8294 patents. Other top jurisdictions
are China with 161 patents, Germany and France with 4 patents, Canada with 3 patents. Russia and
Japan have also published 2 patents each on the synthesis of nonionic surfactants (Figure 13.21).

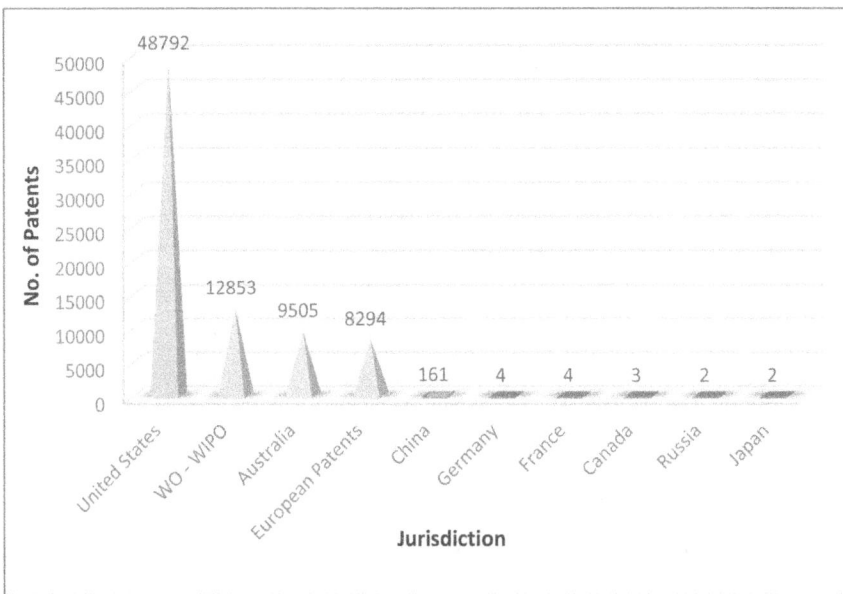

FIGURE 13.21 Top 10 jurisdictions on nonionic surfactants synthesis.

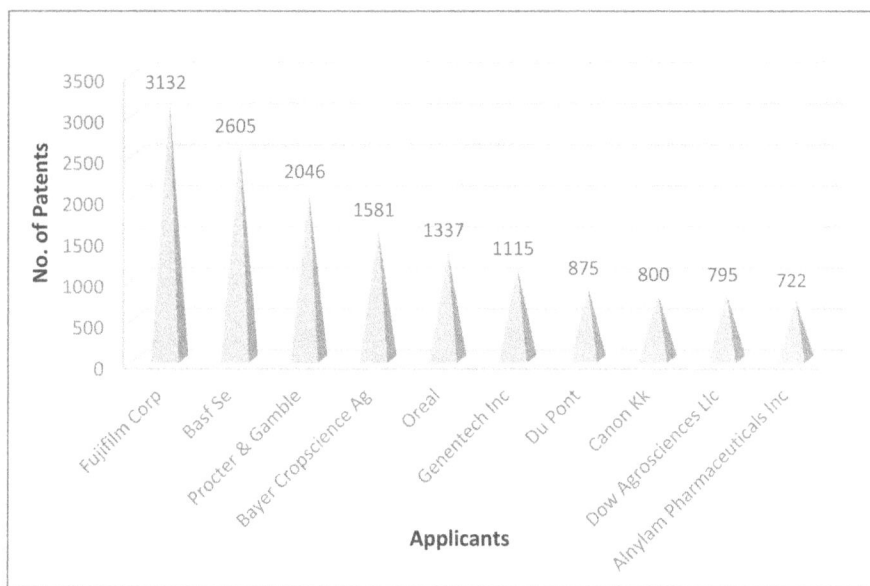

FIGURE 13.22 Top 10 applicants on nonionic surfactants synthesis.

13.6.2 APPLICANTS, INVENTORS, OWNERS

Fujifilm Corp. has top applicants and owners with 3132 patents published as top applicants as well as with 2185 patents as top owners. W. N. Ulrike is in top 10 inventors with 543 patents (Figures 13.22–13.24).

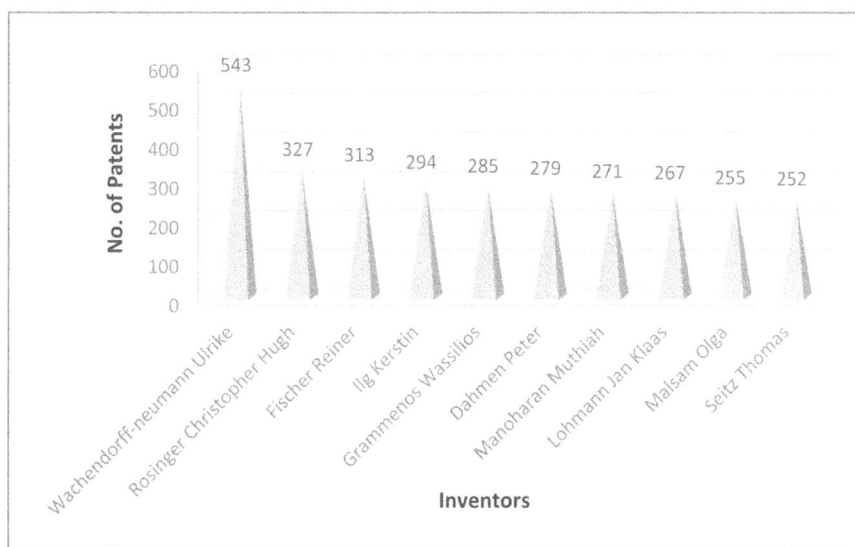

FIGURE 13.23 Top 10 inventors on nonionic surfactants synthesis.

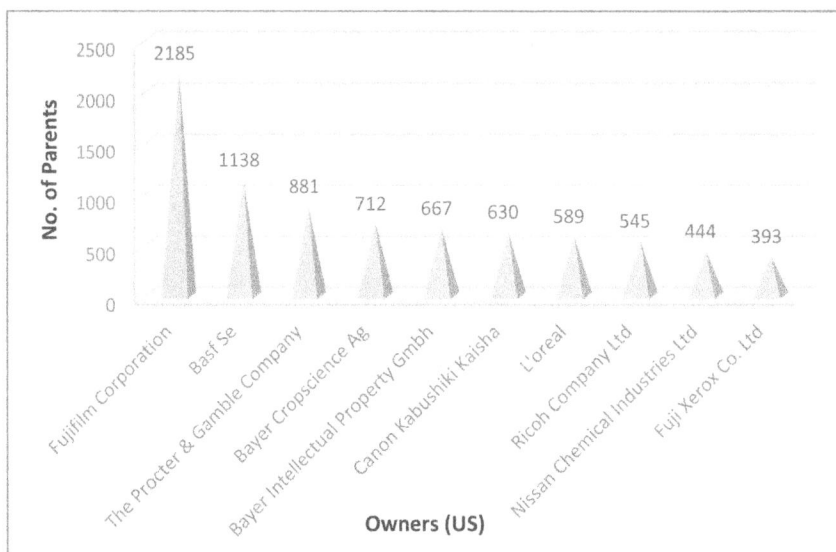

FIGURE 13.24   Top 10 owners on nonionic surfactants synthesis.

13.7   AMPHOTERIC

On the basis of the number of patents published on the synthesis of amphoteric surfactants, the year 2020 was the most prominent year with 3556 patents. This value is least among patents published on cationic/anionic/nonionic surfactants synthesis. In 2010, there were about 2309 patents published in this area, whereas in the year 2019, a total number of 3384 patents have been published (Figure 13.25).

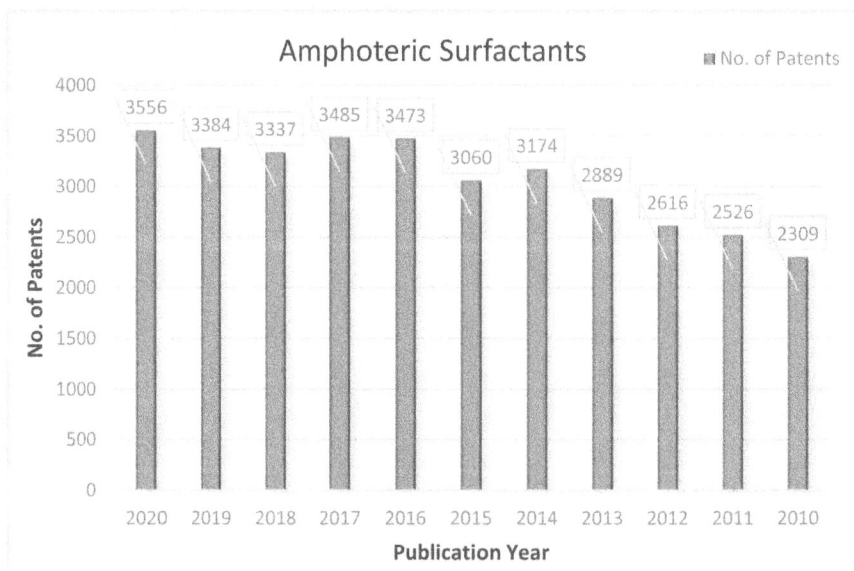

FIGURE 13.25   Number of patents on amphoteric surfactants synthesis.

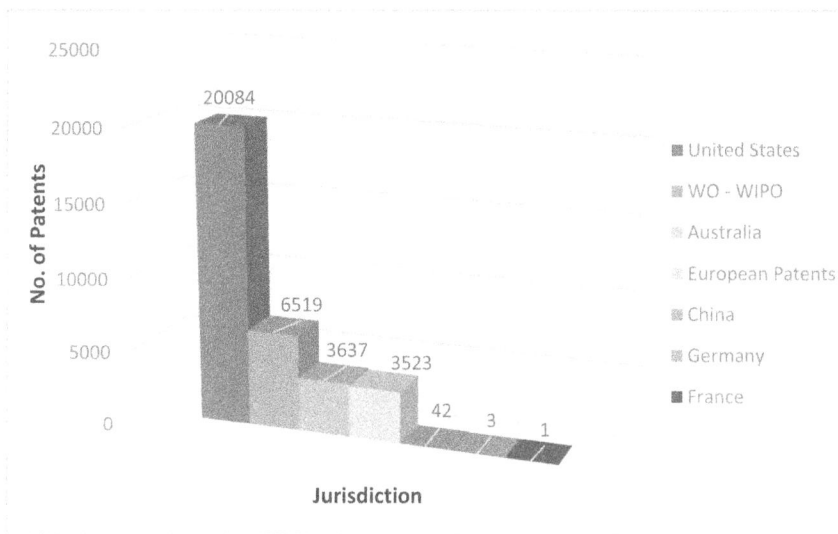

FIGURE 13.26 Top jurisdiction on amphoteric surfactants synthesis.

13.7.1 JURISDICTION

The United States has published the largest number of patents on the synthesis of amphoteric surfactants and the number is 20,084. WIPO is the second largest with 6519 patents published. Australia published 3637 and European patent office published 3523 patents. Other top jurisdictions are China with 42 patents, Germany and France with 3 and 1 patents, respectively (Figure 13.26).

13.7.1.1 Applicants, Inventors, Owners

Basf Se has top applicants with 1955 patents published. Top inventor in this area is G. Wassilios with 274 patents, whereas Fujifilm Corp. is topmost among the top 10 owners with 1002 patents (Figures 13.27–13.29).

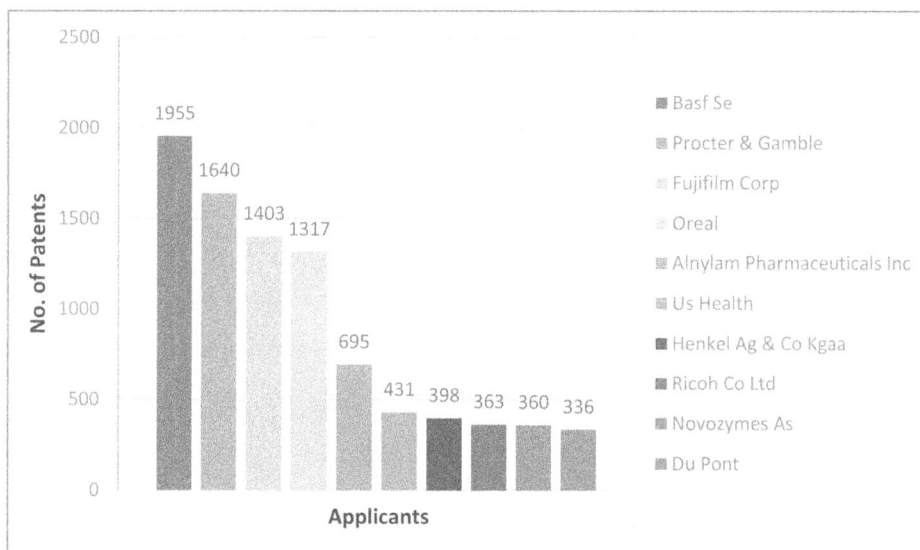

FIGURE 13.27 Top 10 applicants on amphoteric surfactants synthesis.

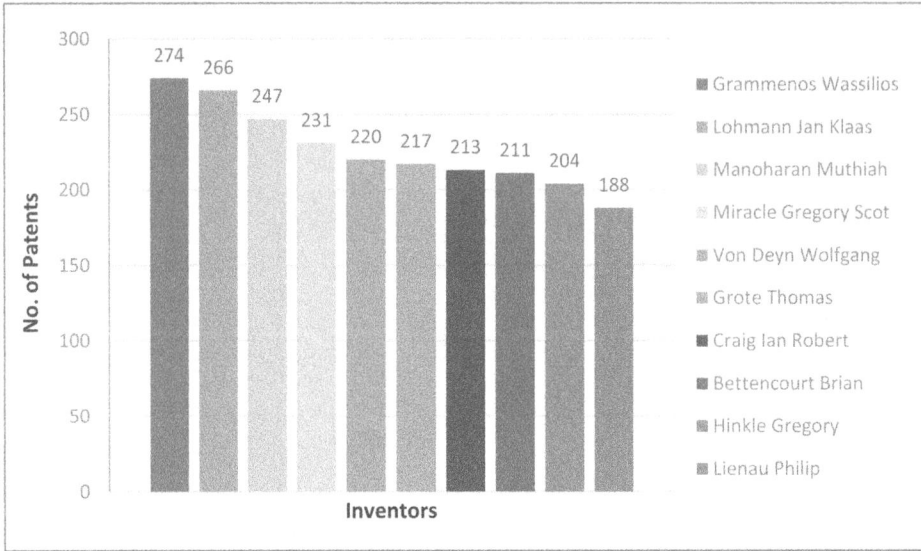

FIGURE 13.28 Top 10 inventors on amphoteric surfactants synthesis.

13.8 GEMINI SURFACTANTS

For gemini surfactant synthesis, there are 11,567 patents published and 2020 is the most prominent year with 1233 patents, whereas 2010 is having the least number of patents i.e., 673. Figure 13.19 shows that the number of patents increases consistently from 2009 to 2017. In 2018, the number of patents published was 1186 patents in this area (Figure 13.30).

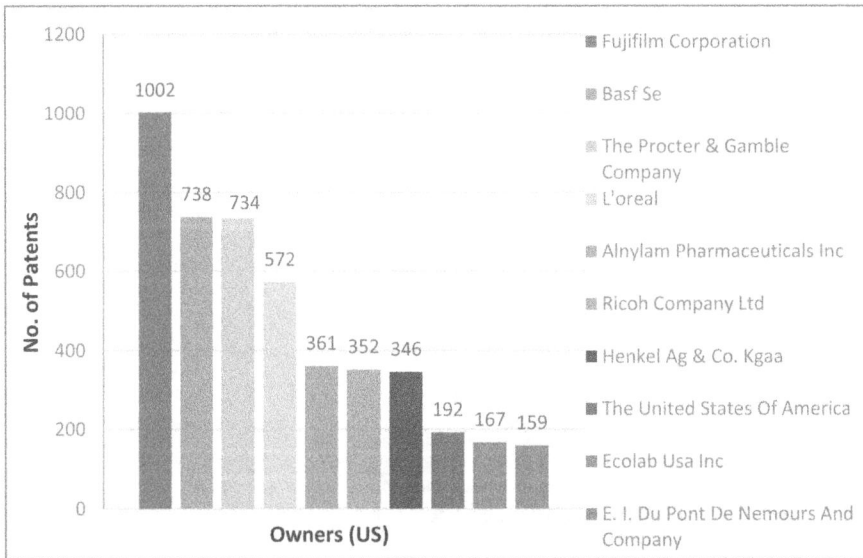

FIGURE 13.29 Top 10 owners on amphoteric surfactants synthesis.

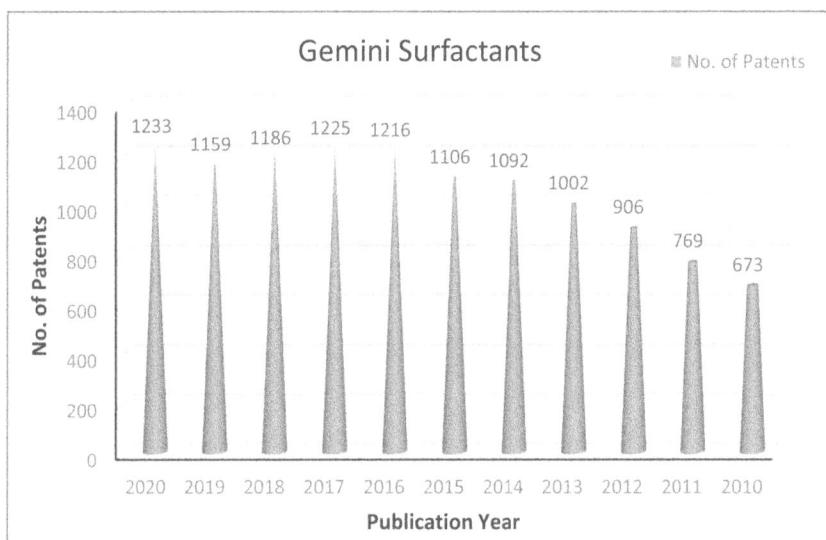

FIGURE 13.30 Number of patents on gemini surfactants synthesis.

13.8.1 JURISDICTION

The United States has published the largest number of patents on the synthesis of gemini surfactants and the number is 6617. WIPO is the second largest with 2160 patents published. Australia published 1566 and European patent office published 1029 patents. Other top jurisdictions are China with 190 patents and Japan with 2 patents. Russia, United Kingdom and France with 1 patent each (Figure 13.31).

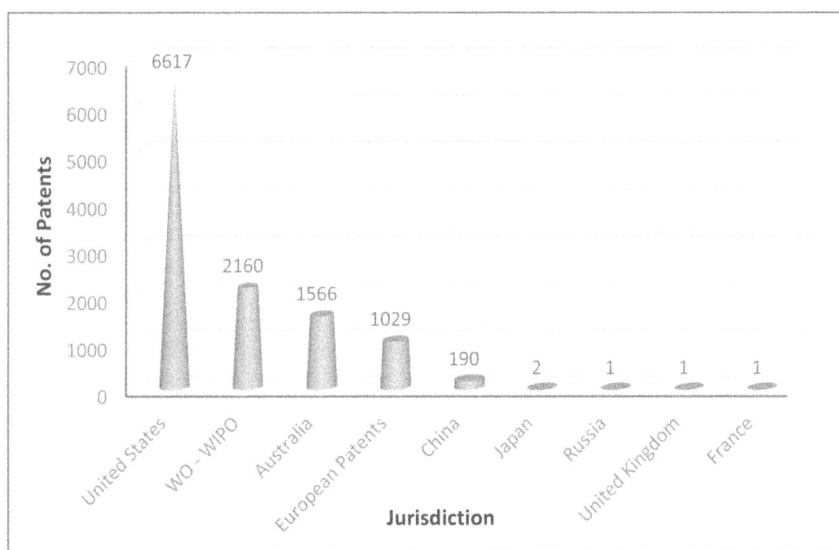

FIGURE 13.31 Top jurisdiction on gemini surfactants synthesis.

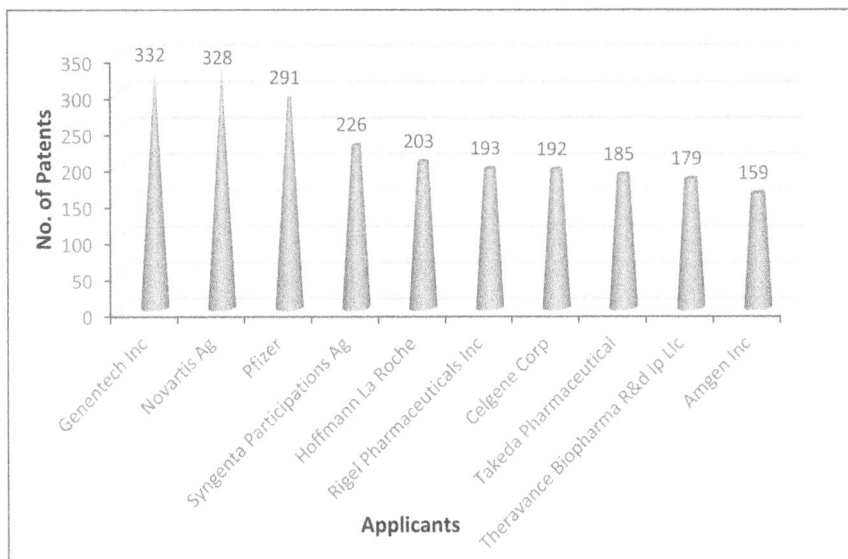

FIGURE 13.32   Top 10 applicants on gemini surfactants synthesis.

13.8.2   Applicants, Inventors and Owners

Genentech Inc. is the top applicant for gemini surfactant synthesis with 332 patents. Fleury Melissa is the topmost inventor with 192 patents. Novartis AG and T. Biopharma are the utmost top owners with 142 and 134 patents, respectively (Figures 13.32–13.34).

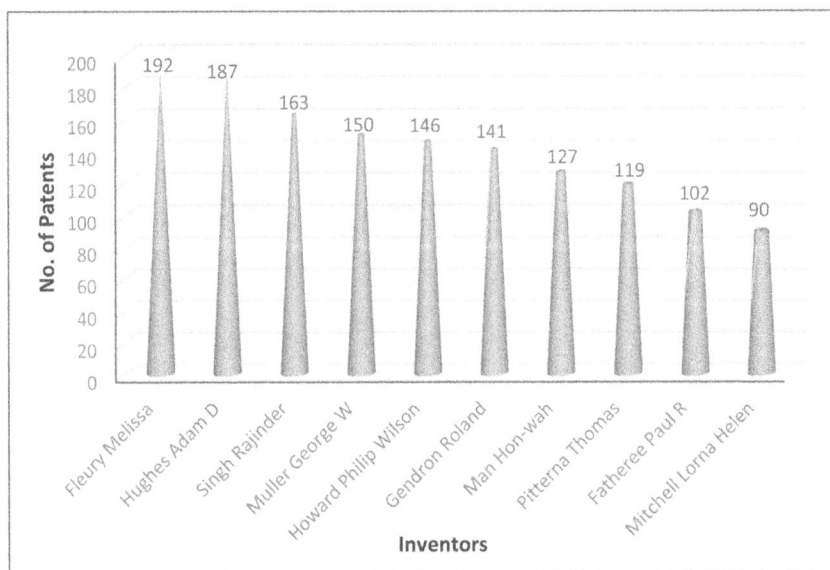

FIGURE 13.33   Top 10 inventors on gemini surfactants synthesis.

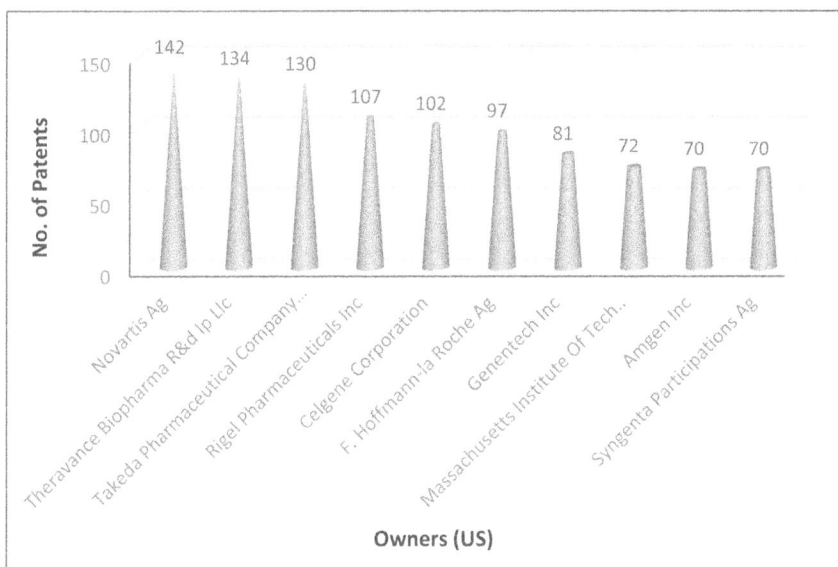

FIGURE 13.34 Top 10 owners on gemini surfactants synthesis.

13.9 POLYMERIC SURFACTANTS

Data available on the synthesis of polymeric surfactants for the last 11 different years revealed that 2020 and 2019 are at top with 19,925 patents and 19,405 patents, respectively. The total number of patents published in the last 11 years is 190,214 (Figure 13.35).

13.9.1 JURISDICTION

On the synthesis of polymeric surfactants, the United States has the largest number of patents, i.e., 117,182. WIPO is the second largest with 33,403 patents. Australia has 22,276, European patent

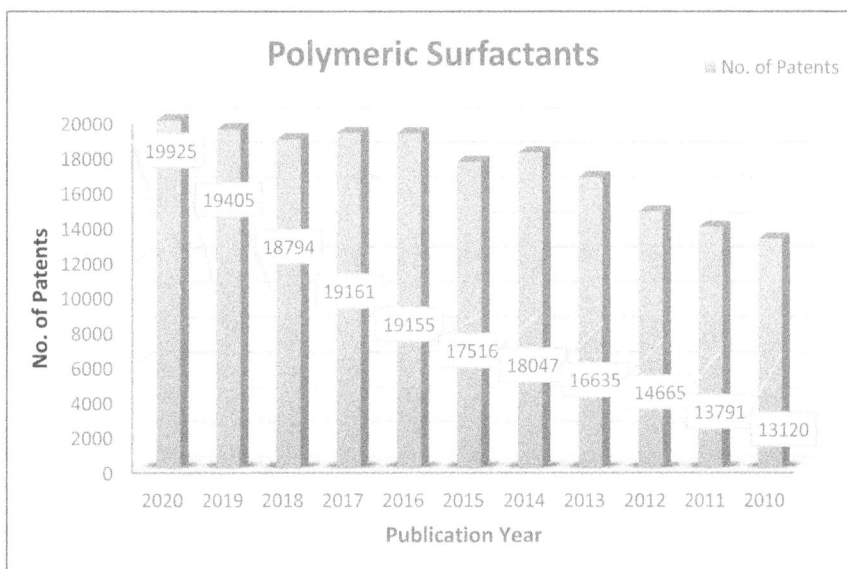

FIGURE 13.35 Number of patents on polymeric surfactants synthesis.

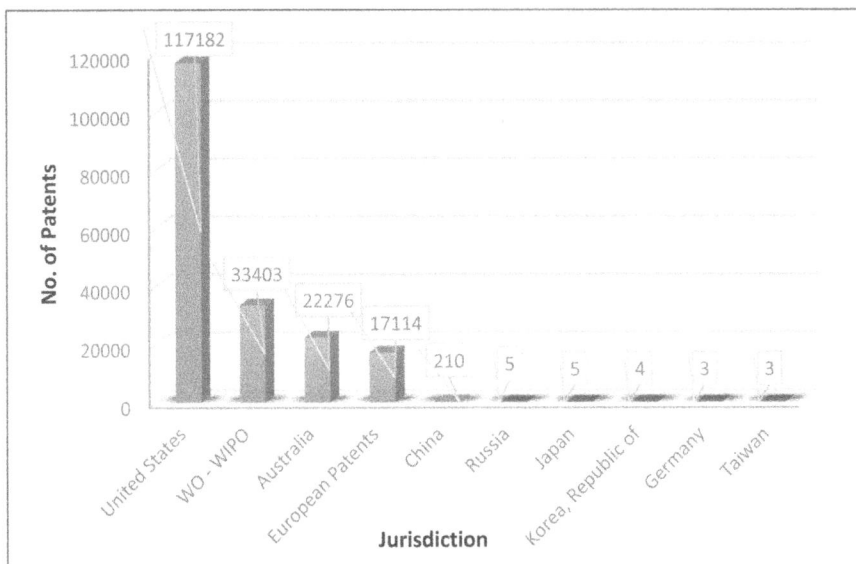

FIGURE 13.36   Top 10 jurisdiction on polymeric surfactants synthesis.

office has 17,114 patents and China has 210 patents on synthesis of polymeric surfactants. Small players in this field are Russia and Japan with 5 patents, Korea with 4 patents, and Germany and Taiwan with 3 patents (Figure 13.36).

13.9.2   APPLICANTS, INVENTORS AND OWNERS

Top two applicants for polymeric surfactants synthesis are Fujifilm Corporation and Genentech Inc. with 4290 and 3401 patents, respectively. W. N. Ulrike is the topmost inventor with 486 patents, whereas the topmost owner for polymeric surfactant is Fujifilm Corporation with 3217 patents (Figures 13.37–13.39).

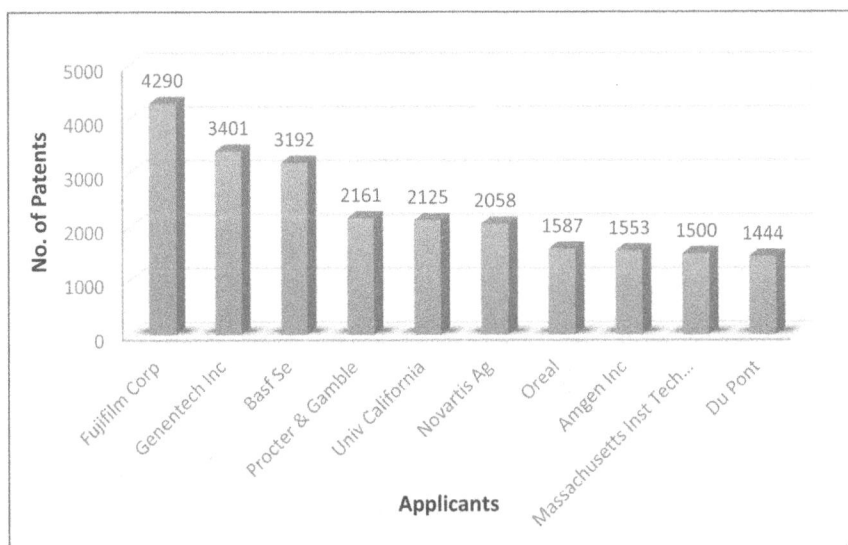

FIGURE 13.37   Top 10 applicants on polymeric surfactants synthesis.

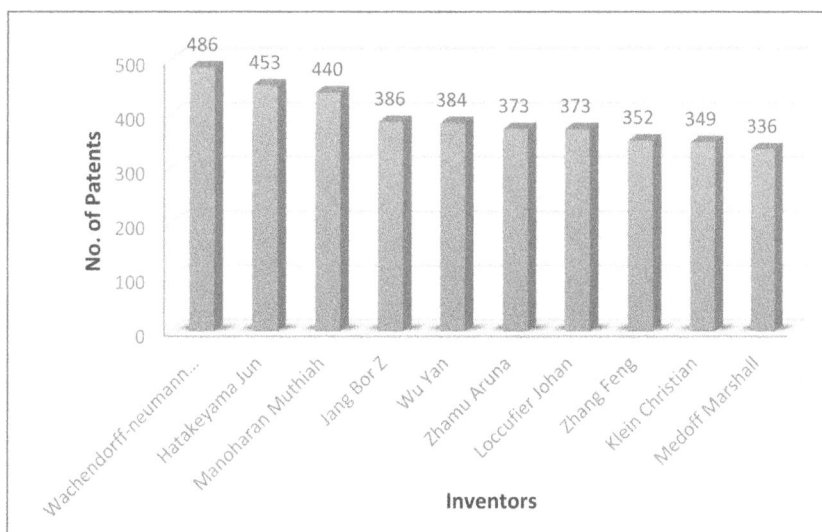

FIGURE 13.38 Top 10 inventors on polymeric surfactants synthesis.

13.10 RECENT PATENTS ON THE APPLICATIONS OF SURFACTANTS

In the last 11 years (2010–2020), a large number of patents have been filed and granted in the application area of surfactants (Figure 13.40) that include agricultural spray, in pharmaceuticals as drugs and in drug delivery systems, in laundry as detergents, in cosmetics, in petroleum industry as corrosion inhibitions and in oil recovery. There are approximately 513,865 patents published in the different areas of applications on surfactants. The year 2020 claimed 68,135, the largest number of patents and 2019 ranked second with 67,037 patents. The year 2010 only has 47,587 patents on surfactants applications (Figure 13.41).

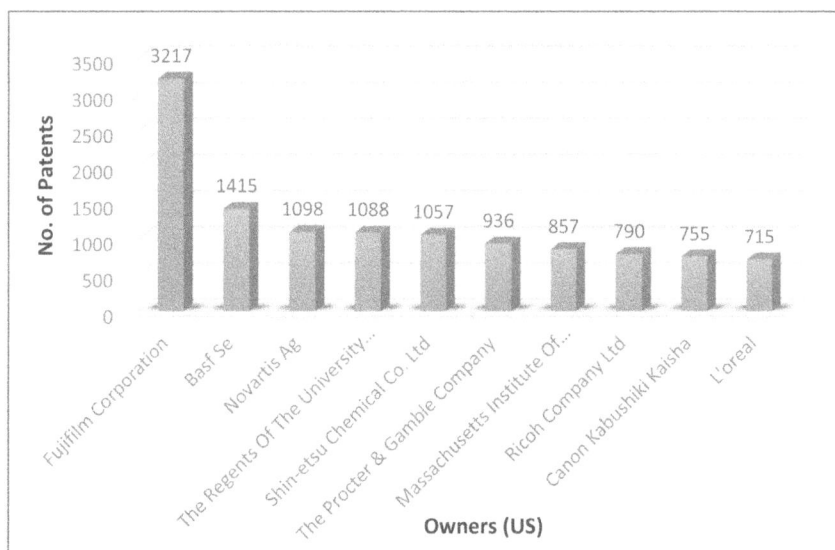

FIGURE 13.39 Top 10 owners on polymeric surfactants synthesis.

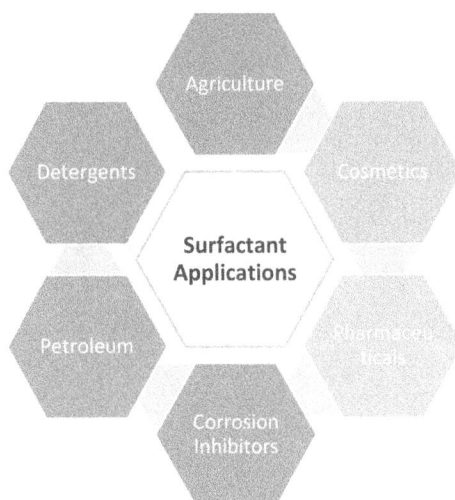

FIGURE 13.40   Areas of surfactants applications.

13.10.1   Jurisdiction

Results from jurisdictions show that the United States is most prominent with 391,265 patents. Rest top jurisdictions are WIPO and Australia with 109,605 and 75,348 patents, respectively. European patent has 63,126 and China has 11,096 patents, whereas China, Japan, Canada, Russia, Mexico and Korea have less than 500 patents (Figure 13.42).

13.10.2   Classifications

Top 10 CPC, IPCR and US classifications on the surfactant's applications in the last 11 years are given in Figures 13.43–13.45, respectively.

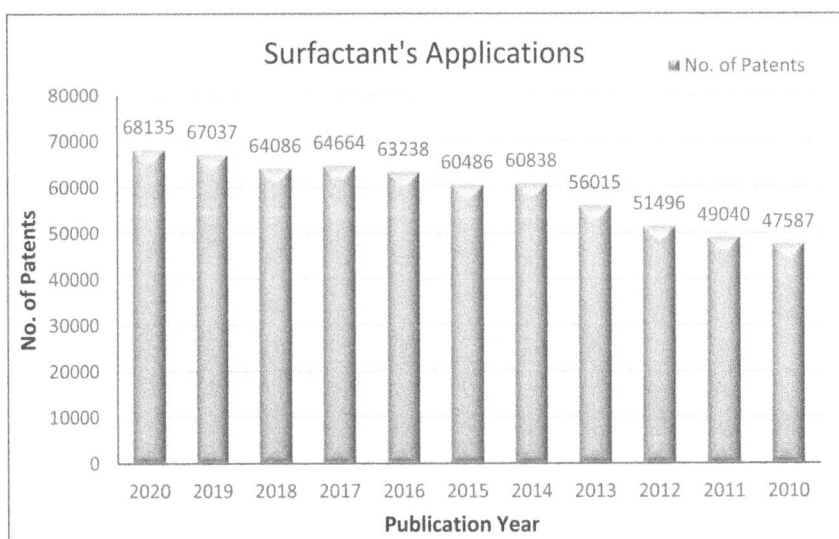

FIGURE 13.41   Number of patents on surfactants applications.

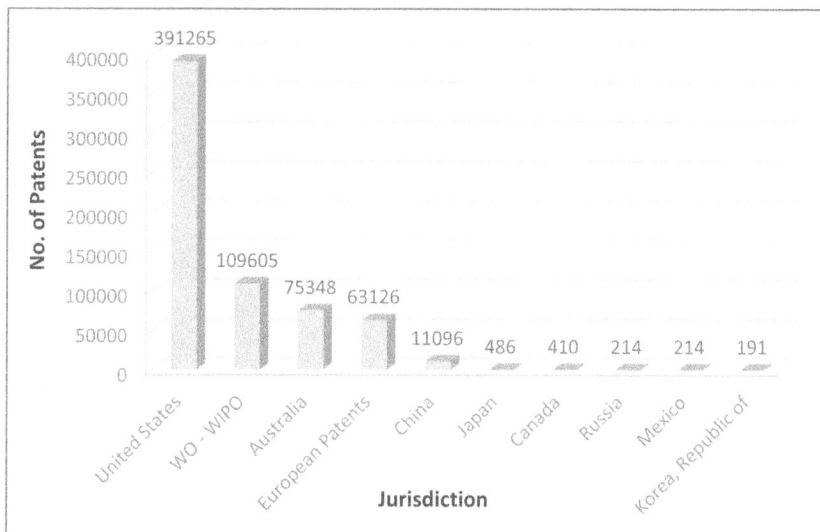

FIGURE 13.42 Top jurisdiction on surfactants applications.

13.10.3 APPLICANTS, INVENTORS AND OWNERS

Procter and Gamble is at top among the top 10 applicants with 10,715 patents to its credit, whereas Basf Se has the second-highest number of patents applications 8240 in the last 11 years. Rest top applicants are given in Figure 13.46. The top inventor in this field is S. Michael with 864 patents (Figure 13.47) and Fujifilm Corp. is at top among the top 10 owners having 5404 number of patents (Figure 13.48).

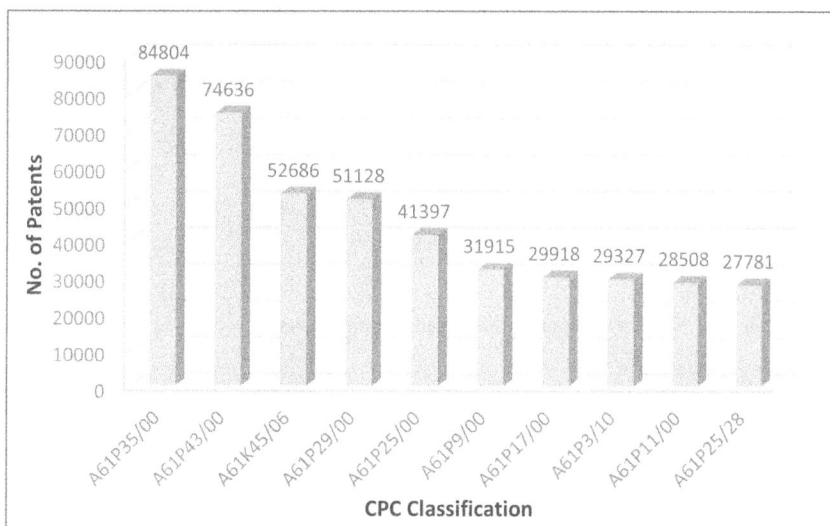

FIGURE 13.43 Top 10 CPC classifications on surfactants applications.

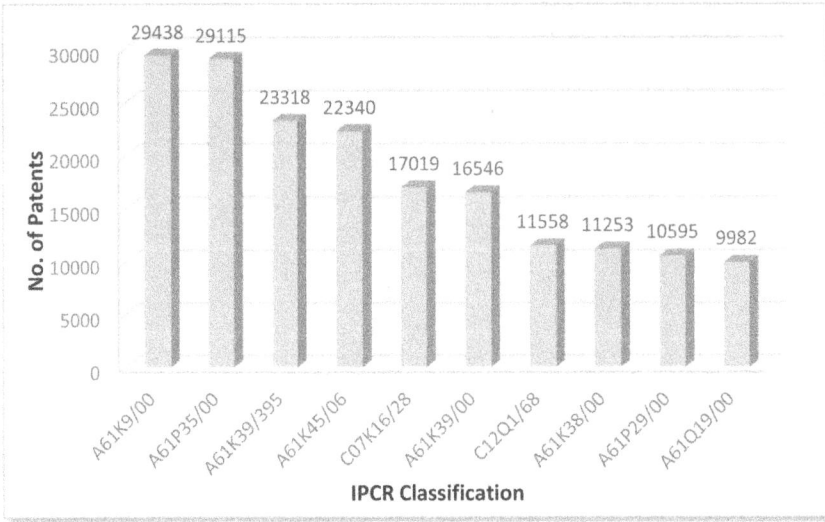

FIGURE 13.44 Top 10 IPCR classifications on surfactants applications.

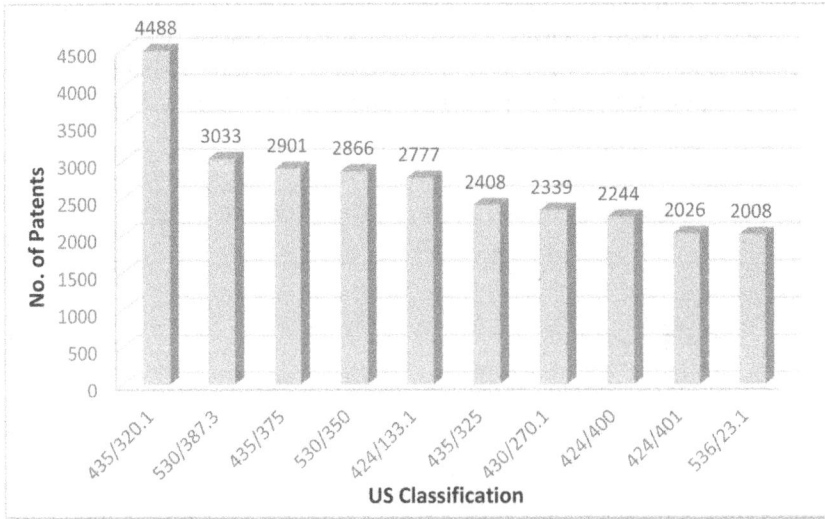

FIGURE 13.45 Top 10 US classifications on surfactants applications.

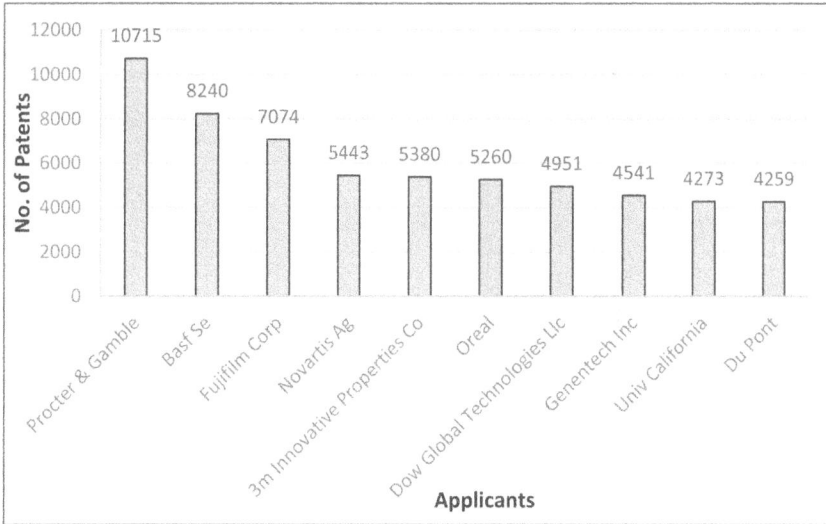

FIGURE 13.46 Top 10 applicants on surfactants applications.

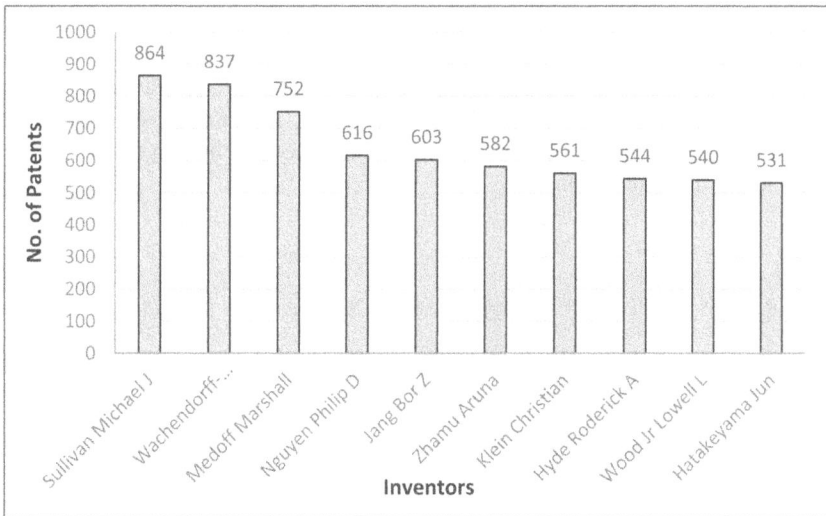

FIGURE 13.47 Top 10 inventors on surfactants applications.

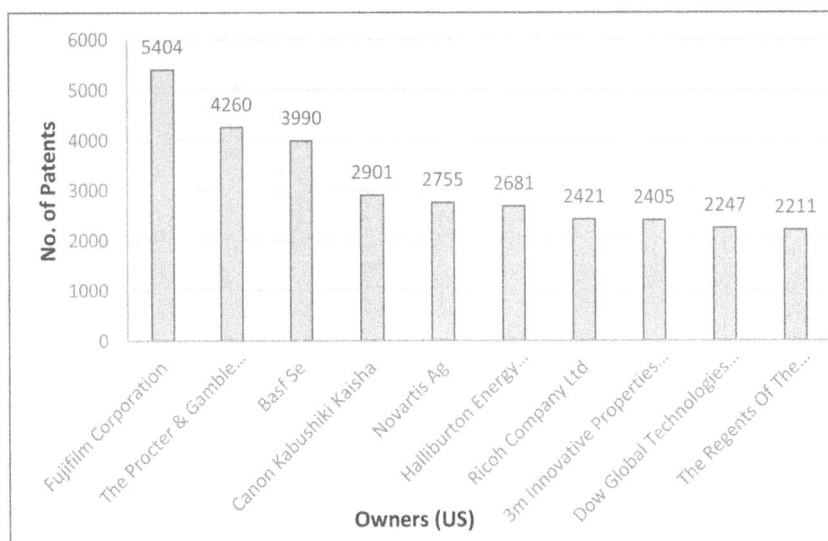

FIGURE 13.48 Top 10 owners (United States) on surfactants applications.

13.11 CONCLUSIONS

The study on patents on the synthesis and applications of surfactants for the last 11 years revealed that due to the specific amphiphilic nature of molecules, surfactants have enormous scope for scientists to make them think of their innovative use in several Industries. Current review is an effort to combine data available on the number of patents published year wise, top jurisdictions, applicants, owners and inventors. This review would definitely help the researchers working on surfactants to explore further to produce many more new surfactant molecules having desired properties and tailor-made for a specific application.

REFERENCES

1. Karsa, D. R. (Ed.). (1999). *Industrial Applications of Surfactants IV (No. 230)*. Elsevier. Cambridge, UK.
2. Ziaee, F., Ziaee, M., & Taseidifar, M. (2021). Synthesis and application of a green surfactant for the treatment of water containing PFAS/hazardous metal ions. *Journal of Hazardous Materials*, *407*(6) 124800.
3. Fait, M. E., Bakas, L., Garrote, G. L. et.al. (2019). Cationic surfactants as antifungal agents. *Applied Microbiology and Biotechnology*, *103*, 97–112
4. Lebeuf, R., Liu, C., Pierlot, C., & Rataj, V. N. (2018). Synthesis and surfactant properties of nonionic biosourced alkylglucuronamides. *ACS Sustainable Chemistry & Engineering*, *6*(2), 2758–2766. DOI: 10.1021/acssuschemeng.7b04456
5. Asselah, A., Pinazo, A., Mezei, A., Pérez, L., & Tazerouti, A. (2017). Self-aggregation and emulsifying properties of methyl ester sulfonate surfactants. *Journal of Surfactants and Detergents*, *20*(6), 1453–1465.
6. Eldridge, J. M. (2017). The analysis of surfactants in cosmetics. In *Surfactants in Cosmetics, Second Edition* (pp. 103–124). Routledge, California.
7. Székács, A. (2017). Mechanism-related teratogenic, hormone modulant and other toxicological effects of veterinary and agricultural surfactants. *Insights in Veterinary Science*, *1*, 24–31.
8. Permadi, P., Fitria, R., & Hambali, E. (2017, May). Palm oil-based surfactant products for petroleum industry. In *Organic Materials as Smart Nanocarriers for Drug Delivery* (Vol. 65, No. 1, p. 012034). IOP Publishing, Bristol, UK.

9. Aditya, A., Rene, M., & John, C. (2017). Use of boundary lubricants for the reduction of shear thickening and jamming in abrasive particle slurries. *Colloids and Surfaces A: Physicochemical and Engineering Aspects*, 537, 13–19.

10. Zakharova, L. Y., Pashirova, T. N., Fernandes, A. R., Doktorovova, S., Martins-Gomes, C., Silva, A. M., & Souto, E. B. (2018). Self-assembled quaternary ammonium surfactants for pharmaceuticals and biotechnology. In *Organic Materials as Smart Nanocarriers for Drug Delivery* (pp. 601–618). William Andrew Publishing.

11. França, M. T., Nicolay, R. P., Riekes, M. K., Pinto, J. M. O., & Stulzer, H. K. (2018). Investigation of novel supersaturating drug delivery systems of chlorthalidone: The use of polymer-surfactant complex as an effective carrier in solid dispersions. *European Journal of Pharmaceutical Sciences*, 111, 142–152.

12. Yan, X., Zhai, Z., Song, Z., Shang, S., & Rao, X. (2017). Synthesis of comb-like polymeric surfactants with a tricyclic rigid core and their use as dispersants in pymetrozine water suspension concentrates. *RSC Advances*, 7(88), 55741–55747.

13. Zhu, Y., Free, M. L., Woollam, R., & Durnie, W. (2017). A review of surfactants as corrosion inhibitors and associated modeling. *Progress in Materials Science*, 90, 159–223.

14. Amaral, M. H., das Neves, J., Oliveira, Â. Z., & Bahia, M. F. (2008). Foamability of detergent solutions prepared with different types of surfactants and waters. *Journal of Surfactants and Detergents*, 11(4), 275.

15. Moghadam, T. F., Azizian, S., & Wettig, S. (2017). Effect of spacer length on the interfacial behavior of *N,N'*-bis (dimethylalkyl)-α, ω-alkanediammonium dibromide gemini surfactants in the absence and presence of ZnO nanoparticles. *Journal of Colloid and Interface Science*, 486, 204–210.

16. Shehzad, F., Hussein, I. A., Kamal, M. S., Ahmad, W., Sultan, A. S., & Nasser, M. S. (2018). Polymeric surfactants and emerging alternatives used in the demulsification of produced water: A review. *Polymer Reviews*, 58(1), 63–101.

17. Negm, N. A., El Hashash, M. A., Abd-Elaal, A., Tawfik, S. M., & Gharieb, A. (2018). Amide type nonionic surfactants: Synthesis and corrosion inhibition evaluation against carbon steel corrosion in acidic medium. *Journal of Molecular Liquids*, 256, 574–580.

18. Watanabe, Y., & Adachi, S. (2017). Amphiphilic acyl ascorbates: Their enzymatic synthesis and applications to food. In *Food Biosynthesis* (pp. 381–408). Academic Press.

19. Tripathy, D. B., Mishra, A., Clark, J., & Farmer, T. (2018). Synthesis, chemistry, physicochemical properties and industrial applications of amino acid surfactants: A review. *Comptes Rendus Chimie*, 21(2), 112–130.

20. Jia, W., Rao, X., Song, Z., & Shang, S. (2009). Microwave-assisted synthesis and properties of a novel cationic gemini surfactant with the hydrophenanthrene structure. *Journal of Surfactants and Detergents*, 12(3), 261–267.

21. Brunsveld, L., Schill, J., van Dun, S., Pouderoijen, M., Janssen, H., Milroy, L. G., & Schenning, A. (2018). Synthesis and self-assembly of bay-substituted perylene diimide gemini-type surfactants as off-on fluorescent probes for lipid bilayers. *Chemistry–A European Journal*, doi.org/10.1002/chem.201801022

22. Scharfenberg, M., Wald, S., Wurm, F. R., & Frey, H. (2017). Acid-labile surfactants based on poly(ethylene glycol), carbon dioxide and propylene oxide: Miniemulsion polymerization and degradation studies. *Polymers*, 9(9), 422.

23. https://www.lens.org/lens/search/patent/structured

24. https://www.uspto.gov/web/patents/classification/cpc/html/cpc-A61Q.html#A61Q5/02

25. http://www.wipo.int/ipc/itos4ipc/ITSupport_and_download_area/20180101/pdf/scheme/full_ipc/en/

26. https://www.uspto.gov/web/patents/classification/uspc536/us536toipc8.htm#C536S029100

Index

Note: Locators in *italics* represent figures and **bold** indicate tables in the text.

For Product Safety Concerns and Information please contact our EU
representative GPSR@taylorandfrancis.com
Taylor & Francis Verlag GmbH, Kaufingerstraße 24, 80331 München, Germany

www.ingramcontent.com/pod-product-compliance
Lightning Source LLC
Chambersburg PA
CBHW061354210326
41598CB00035B/5982

9 780367 702052